NewSpace Systems Engineering

Ignacio Chechile

NewSpace Systems Engineering

Springer

Ignacio Chechile
Helsinki, Finland

ISBN 978-3-030-66897-6 ISBN 978-3-030-66898-3 (eBook)
https://doi.org/10.1007/978-3-030-66898-3

This Springer imprint is published by the registered company Springer Nature Switzerland AG
The registered company address is: Gewerbestrasse 11, 6330 Cham, Switzerland

To Leo, the only system I have contributed to that works as intended

Preface

I have mixed feelings at the thought of adding yet another book of Systems Engineering to the bookshelf of Systems Engineering. I feel there are plenty, perhaps too many. It soothes my mind that this book is not actually about Systems Engineering in the same way most of the books you will find around, but about *NewSpace* System Engineering. And, ironically, by saying this I join them: they all state to be different than the rest.

If I could define NewSpace in one line, I would say: NewSpace is the art of cutting corners. Engineering small spacecraft is a good problem to have, but not an easy one. Designing and developing space systems is a problem and knowledge-intensive activity. In reality, when you do NewSpace you do not have the luxury of dealing with one or just a few problems, but with a *network* of interconnected problems. Russell Ackoff coined the term "messes" (Ackoff, 1981), which are complex systems of interacting problems. NewSpace is about dealing with *messes*.

There is a lot of "analysis of the obvious" in this book. I take some time to think about some things we never really stop to analyze a lot, for example life cycles. This book is a sort of *eulogy* of early stage organizations; for they are truly something. They are like bomb squads, nervously trying to cut the right cable while looking and the clock ticking, of course without really knowing which cable to cut. The latent chance of total destruction is always around the corner. Hence, NewSpace early stage projects require careful consideration of those somewhat obvious things, since they get little attention at the beginning and they slowly fester until they become too cumbersome to be tackled. NewSpace projects must streamline their engineering practice in ruthlessly minimalistic ways: runways are short; i.e. the bomb ticks. NewSpace organizations cannot spend their energy on things that add no value. If they do so, they will find themselves out of the game very soon. An early stage startup is the epitome of uncertainty: there are so many things to define, to learn and to understand. From business plan to the technology, to office daily activities. No wonder most of them never make it beyond infancy. Hopefully, this book will help some stay in the game. Or, for new actors entering the game, to be aware of the challenges they will surely find ahead of the road. Pertinent is to say that most of the book is targeted to and inspired by very early stage small NewSpace companies at equally early stages of their design process, developing spacecraft of different kinds. If the occasional reader belongs to a NewSpace organization making components or bits and parts of spacecraft, you (luckily) deal with a lower complexity compared with what I cover in the next chapters. You may still find some parts useful; just tailor the context accordingly. If the reader is more advanced in the organizational and design process, for example he or she is part of a scale-up or a bigger space organization, the myriad of more classic Systems Engineering books can be a valuable insight.

But, how do you get yourself in the trouble of joining a NewSpace project? Let me guess: you are a founder, you have a great idea which has, in your view, great market potential. It is about Earth observation, or data collection, or connectivity. You strongly believe it will make a difference, but you have little clue what to do next, because you have not done space before. Or, you are an engineer who is about to quit your current job in a big space company or similar and you are planning to join a NewSpace company, being in total fear about the step you are about to take. Maybe you are an investor considering the idea of investing your hard-earned capital in some NewSpace related project. Or you are an outsider, curious how NewSpace works. Hopefully, this book will provide all of you with some answers. Not all the answers for sure. Perhaps in the end you will have more questions than before. If that is the case, I leave a reasonably decent collection of references for you to dive deeper as needed.

I proposed myself to write the book I would have loved to have in my hands when I started working in NewSpace 5 years ago. I also wanted to write a book that would not keep the reader busy for too long, or get her or him bored too quickly; I wanted to keep it concise, but I might have failed; it crept quickly out of boundaries. Hopefully, chapters are reasonably self-contained in a way the reader can skip them according to the schedule. I wanted to provoke some thoughts: I might have accomplished my goal, or not, it's going to be you who will decide. This is neither a doctoral thesis nor a scientific article, so please have a critical eye and rely on the references for more details on topics I have only briefly covered.

So, what type of book is this? Is it a technical book? A management book? A design book? A compilation of lessons learned? I guess it is a combination of all that together. It is a bit of a hybrid book where you will find different things. I try to revisit some preestablished practices we sometimes perform a bit *robotically* without thinking or questioning. I visit topics such as engineering habits, modularity, conceptual design, testing, and at the end of the book I talk about the bullshit that sometimes engulfs the NewSpace world. Take this book, ultimately, as a collection of thoughts. I have, for sure, put my best effort in writing it; hundreds of liters of coffee were consumed in the process in many places around Helsinki and at home, due to the COVID-19 pandemic. The outcome may still be mediocre at best, it will be the readers' role to say. Any feedback to lessen its mediocrity (if possible) will be more than welcome.

As a final thought for this preface: very early stage NewSpace organizations are like tightrope walkers; one wrong step and you surely go down. Their bank accounts are leaking by the minute, they burn money and struggle, swimming in a sea of uncertainty and facing disbelief. At the beginning, they lack the most basic structure such as HR department, or basically any department outside engineering. You must be a bit out of your mind to join these types of places, but at the same time you must be out of your mind not trying at least once in your life.

Helsinki, Finland
October 2020

Ignacio Chechile

Acknowledgements

I shall acknowledge first all the borderline bankrupt Small and Medium Enterprises I worked for during my younger years back in Buenos Aires, for they have taught me some invaluable lessons I apply every day. First, they taught me to survive with what I had at hand instead of waiting for panaceas that will never come. And second, they taught me, perhaps the hard way, that true engineering is about making things work in the field and not about making promises they will work, and that the moral responsibility of the engineer spans all the way from ideation to retirement of the systems we create. Despite the not-so-great salaries and the huge stress of studying and working at the same time back in those days, I do not regret a single day of having worked there. I am proudly formatted by the idiosyncrasies of one of the backbones of every economy in the world: The Small and Medium Enterprise.

I shall acknowledge the Argentinian free education system for teaching me (absolutely for free) all I know, all the way from kindergarten to University: I am a full product of free education; hence this book also is. Often criticized, the free education system remains one of my country's biggest assets. Education shall stay free for everyone, forever.

I shall acknowledge my parents Carlos and Alicia and my siblings Ramiro and Sol for supporting me throughout my career with all its twists and turns which, long story, have taken me quite far away from them. It is not easy to be this far, but their support makes it easier.

Last, but definitely not least, I shall acknowledge my wife Kaisa for accompanying every crazy idea I (frequently) come up with, including this book. Who knows what is coming next.

Contents

About the Author

Ignacio Chechile is a 40-year-old engineer from Quilmes, a city at the southern part of Greater Buenos Aires, by the Rio de la Plata, in Argentina. He studied Electrical Engineering at Universidad Tecnólogica Nacional, where he graduated, not without sweat and tears, with largely unimpressive grades in 2007. From 2002 to 2009, he zigzagged through different jobs in Buenos Aires, mostly in industrial embedded systems, as an all-rounder, where he made first contacts with the Systems Engineering discipline. In 2009, he went for a job in the Guidance and Control group at INVAP SE, a space and defense company located in Bariloche, in the Patagonia Argentina, where the contributed, for the good or the worse, to governmental LEO and GEO missions, notably Aquarius-SACD, SAOCOM, ARSAT1&2, and also spending some valuable time at the defense radar division. In 2015, he joined a NewSpace startup located in Finland, where he led the software efforts until late 2017, then to move to the position of Vice President of Spacecraft Engineering. During his tenure there, Ignacio helped design and develop the company's proof-of-concept spacecraft which would turn out to be the world's first Synthetic-Aperture Radar (SAR) microsatellite, launched in early 2018. He contributed for several subsequent missions.

In mid-2020, he decided he needed new challenges, so he joined a pre-seed startup called ReOrbit where he is currently holding the position of Chief Technology Officer (CTO). At ReOrbit, Ignacio is focusing on designing space architectures for a sustainable use of space, exploring concepts as modularity, reusability, segmented architectures, and in-orbit assembly.

Ignacio is a member of the International Council of Systems Engineering (INCOSE) and a Certified Systems Engineering Professional (CSEP).

Chapter 1
The Hitchhiker's Guide to NewSpace

La gloria o Devoto.
— Carlos Salvador Bilardo
(translates to: *all or nothing*).

Abstract If there is something NewSpace has redefined is the way risks are taken when it comes to designing and building space systems. NewSpace has created its own flavor of Systems Engineering, which has almost nothing to do with the "historical" discipline of Systems Engineering school of thought. Early stage NewSpace organizations do not only have the challenge to design and build a mission-critical complex artifact such as a spacecraft but also to design the whole organization around it, which makes them very vulnerable and prone to failure. It all indicates our dependency on satellites will only increase, which means orbits are getting crowded. NewSpace has the responsibility of making an ethical use of space to ensure sustainable access for the generations to come.

Keywords NewSpace systems engineering · Systems thinking · Clean space · Clean tech · Sustainable space · Cybernetics

It was a gray and wet November of 2015 when I visited Helsinki for the first time. Perhaps not the greatest time in the year to visit this beautiful city; and it was a puzzling experience. I spent 3 days visiting the company and the people who were willing to hire me, in a building which was a shared office space with other startups, right at the border between Helsinki and Espoo (on Espoo side). Three intense days, where I presented myself, what I knew, and what I thought I could contribute to the project, and so on. I must admit, I had never considered myself an adventurous person, but there I was, "selling" myself and thinking about quitting my (good) job back in a beautiful city in Patagonia, in a well-respected company. Taking this offer would mean getting rid of all my stuff and moving to the other side of the world, in more or less 2 months. During those 3 days in Finland, I corroborated there was a totally different way of doing space systems; I had heard and researched about NewSpace for some time, I was aware of what it was at least from a theoretical perspective. But this looked different: like a university project, but with a strong

I. Chechile, *NewSpace Systems Engineering*,
https://doi.org/10.1007/978-3-030-66898-3_1

and unusual (for me) commercial orientation university projects never have. It felt like a mixture of corner-cutting, naivety and creativity; a combo which looked very appealing for me. Having worked in "classic" space for quite some years, where time tends to go very slow and things are tested multiple times and documents are the main *produce* of projects, this opportunity looked like a chance to bring back the joy to do engineering. And here probably lies the key: time pace. I can say without a single doubt I am an anxious person; I continuously question why things have to take so long, whatever it is. I consider time a precious resource: not because my time is particularly precious, but because life is just too short. I never liked the idea of having to wait five years to see a project being launched into space. I did not want to wait that long anymore. NewSpace was the answer for my engineering anxiety, it moved at speeds too interesting not to be part of.

One of the most distinctive things about NewSpace was how risk, a (buzz)word you hear a lot in classic space, meant just nothing. Risk-taking was so extensive, that the term was seldom mentioned anywhere. It was taken for granted there were plenty of risks taken all over the place. It was a bit of a "shame" to talk about risk, people would silently giggle.

Notably, the project I was joining had *no* customers. Zero. Nobody. Nadie. This was probably the biggest mental adjustment I had to do; the startup I was about to sign a contract with was planning to build a spacecraft nobody asked for. Quite some money and several tens of thousands of man hours of work endeavor were about to be put on something which had no customer, at least just yet. My classic "prime contractor" space brain was short circuiting, because it was expecting the customer to exist and do, you know, what customers do: placing endlessly changing and conflicting requirements, complaining about the schedule, asking for eternal meetings, asking for eternal reviews, being the pain in the neck they usually are. All of a sudden, most of my bibliography of Systems Engineering suddenly turned pointless. Each one of those books would start with: "identify customer needs". These books were taking customers for granted and did not provide any hint on where to get customers from if you did not have them.

When there are no customers, you sort of miss them. But even though customers were still not there, there were lots of ideas. There were scribbles on whiteboards. Some breadboard experiments here and there. There was this strong belief that the project, if successful, would unlock great applications of Earth observation data and grant years of dominance in an untapped market; there was a vision, and passion. So, sails were set toward designing, building and flying a proof-of-concept to validate the overall idea which if it did not work, things would turn more than ugly. It was all or nothing.

Many NewSpace companies around the world work the same way. There is vision, there is commitment, excitement, and some (limited) amount of funds, but there are no revenues nor customers, so their runways are finite, meaning time is very short, and the abyss is always visible in the near horizon. In NewSpace, time to market is king; being late to deliver can mean the end of it. Being slow to hit the market can signify a competitor taking the bite of the cake; i.e. you are out. This is why NewSpace is fascinating: like it is not challenging enough having to build a complex thing like a satellite, it has to be built as fast as possible, as cheap as possible and as performing as possible; i.e. a delicate balance inside the death triangle of cost, schedule and performance. And the most awesome part of it is (which now I know but my 2015 self was constantly wondering) that: it is possible. But, it has to be done the *NewSpace way*, otherwise you are cooked. During those years while I had the privilege of being at the core of the NewSpace enterprise of transforming a scribble into an orbiting spacecraft, I took enough time to look at the whole process from a distance to try to capture the essence of it for others to be able to succeed under similar circumstances. I wanted to understand what the NewSpace *way* is. I wanted to comprehend what were the shortcuts to take, where it was needed to stop, what was to be discarded, when it was time to discard it. I feel many other industries can benefit from the NewSpace way as well. Even *classic* space can. The historic space industry has a perfectionist mindset that everything needs to be perfect before flight. NewSpace challenges all that. In NewSpace, you polish things as you fly.

In hindsight, one factor I did not realize until very much later in the process is how much my attention was centered around the system under design (spacecraft) and how much I (luckily) overlooked the real complexity of what I was about to embark on; I say luckily because, had I realized of this back in that November of 2015, I would have never taken the job offer. All my thinking was put on the thing that was supposed to be launched; but reality is we ended up building a much broader system: a tech organization. A highly creative and sometimes chaotic organization with dependencies and couplings I was initially blissfully ignorant of.

I have interviewed hundreds of people for many different positions throughout those years while I was leading an engineering team. In most cases, the interviewee was coming from a bigger, more established organization. Knowing this, I would love asking them if they were actually ready to leave their comfortable and mature organization to come to a more *fluid* and dynamic (to say the least) environment. In many cases, interviewees were trying to convince me that they were working in a particularly "fluid" department in this multinational corporation and that this department was behaving "pretty much like a startup". A skunkworks group is not a startup, an R&D department may be highly dynamic but it is not a startup either. The main difference lies in the *big system*, which we will describe next.

1.1 The Big System

Years ago, I was reading a magazine and there was an advertisement with a picture of a Formula 1 car without its tires, resting on top of four bricks. The ad was asking "how much horsepower again?". The ad was simple but spot on: a racing car without tires cannot race, it goes nowhere. Connecting that to what we do: systems dependencies are as strong as frequently taken for granted. In racing cars, engines and aerodynamics usually capture all the attention; tires never get the spotlight, although they are essential for the car to perform. Same way in NewSpace we tend to focus on the thing that goes spectacularly up in a rocket with lots of flames and smoke. But, a true space system is way more than that. A spacecraft is nothing without the complex ecosystem of other *things* it needs around it to be able to perform. This is a common pitfall in NewSpace orgs; partial awareness of how the *big system* looks like. I guess it is a defense mechanism as well: complexity can be daunting, so we tend to look away; the problem is when we never look back at it. I will revisit the *big system* concept several times throughout the book. I have seen many NewSpace companies falling in the same trap, finding too late in the process that they have not included important building blocks to make their System/Product feasible. Some references call these components enabling systems. Enabling systems are systems that facilitate the activities toward developing the system-of-interest. The enabling systems provide services that are needed by the system of interest during one or more life cycle stages, although the enabling systems are not a direct element of the operational environment (INCOSE 2015). Examples of enabling systems include development environments, production systems, logistics, etc. They enable progress of the main system under design. The relationship between the enabling system and the system of interest may be one where there is interaction between both systems and one where the system-of-interest simply receives the services it needs when it is needed. Enabling systems strongly depend on perspective: what is an enabling system for me, might be the system-of-interest for the engineer working on that system. Perspective, or point of view, is critical in any complex system; things (architecture, requirements, etc.) look very different depending from where you look at it. Years ago, an early stage NewSpace startup was few weeks away from launching their first proof-of-concept spacecraft, only to realize that they had not properly arranged a ground segment to receive the spacecraft UHF beacon and telemetry. Very last minute arrangements with the Ham Radio community around the world (partially) solved the problem. A spacecraft without a ground segment is like the racing car on top of the bricks from the ad.

To complete the *big system* picture, we must add the organization itself to the "bill" as well; the interaction between the organization and the system under design should not go unnoticed. Is not the organization an *enabling system* after all? If we refer to the definition above (systems that provide services to the system-of-interest), it fits the bill. In the big system, one key thing to analyze and understand is how the way people group together affects the things we design, and also how the technical artifacts we create influence the way we group as people. The engineering witchcraft has two clearly defined sides, which are two sides of the same coin: the technical and the social one. Engineering, at the end of the day, is a social activity. But a social activity needs to happen in natural, spontaneous, non-artificial ways.

If the organization is one more part of the *big system* we must come up with, what are the laws that govern its behavior? Are organizations deterministic and/or predictable? How does the organization couple with the machines they spawn?

There has been a bit of an obsession with treating organizations as mechanistic boxes which transform inputs into outputs applying some sort of "processes" or "laws". Although naive and over simplistic, this reasoning is still thought-provoking. Organizations do have inputs and outputs. What happens inside the box?

1.2 Are Startups Steam Machines?

I would not be totally off by saying that an organization is *kind of* a system. You can assign a "systemic" entity to it, i.e. *systemize* it, regardless if you know how it internally works or not. If you step out from your office and you watch it from a distance, that assembly of people, computers, meeting rooms, coffee machines and information is kind of a thing on its own: let us call it a system. It takes (consumes) some stuff as input and produces something else as an output. Some scholars understood the same, so there has been historical attention and insistence from them to apply quasi-mechanistic methods to analyze and understand social structures like organizations as well as its management. One example is a bit of an eccentric concept called cybernetics. Cybernetics is a term coined by Norbert Wiener more than 70 years ago and adapted to management by Anthony Stafford Beer (1929–2002). Cybernetics emphasizes the existence of feedback loops, or circular cause–effect relationships, and the central role information plays in order to achieve control or self-regulation. At its core, cybernetics is a once-glorified (not so much anymore) attempt to adapt the principles that control theory has historically employed to control deterministic artificial devices such as steam machines or thermostats. Maxwell (who else?) was the first to define a mathematical foundation of control theory in his famous paper *On governors* (Maxwell 1867). Before Maxwell's paper, closed-loop control was more or less an act of witchcraft; machines would be unstable and oscillate without explanation. Any work on understanding closed-loop control was mostly heuristic back in the day, until Maxwell came along; his was mostly a mathematical work. All in all, successful closed-loop control of machines heavily relies on the intrinsic "determinism" of physics laws, at least at the macro level (a quantum physicist would have a stroke just reading that). The macro "model" of the world is deterministic, even though it is just a simplification. Wiener's cybernetics concept came to emphasize that feedback control is an act of communication between entities exchanging information (Wiener 1985). Cybernetic management (as proposed by Beer) suggests that self-regulation can be obtained in organizations the same way a thermostat keeps temperature in a room. Strange, is not it? Like if people could be controlled like a thermostat. It seems everyone was so fascinated about closed-loop control back in the day that they thought it could be applied anywhere, including social systems.

Systems Theory was also around at the time, adding its own take on organizational analysis. Systems Theory uses certain concepts to analyze a wide range of phenomena in the physical sciences, in biology and the behavioral sciences. Systems Theory proposes that any system, ranging from the atom to the galaxy, from the cell to the organism, from the individual to society, can be treated the same. General Systems Theory is an attempt to find common laws to virtually every scientific field. According to it, a system is an assembly of interdependent parts (subsystems), whose interaction determines its *survival*. Interdependence means that a change in one part affects other parts and thus the whole system. Such a statement is true, according to its views, for atoms, molecules, people, plants, formal organizations and planetary systems. In Systems Theory, the behavior of the whole (at any level) cannot be predicted solely by knowledge of the behavior of its subparts. According to it, an industrial organization (like a startup) is an open system, since it engages in transactions with larger systems: society and markets. There are inputs in the form of people, materials, money and in the form of political and economic forces arising in the larger system. There are outputs in the form of products, services and rewards to its members. Similarly, subsystems within the organization down to the individual are open systems. What is more, Systems Theory states that an industrial organization is a sociotechnical system, which means it is not merely an assembly of buildings, manpower, money, machines and processes. The system consists in the organization of people around technology. This means, among other things, that human relations are not an optional feature of the organization: they are a built-in property. The system exists by virtue of the motivated behavior of people. Their relationships and behavior determine the inputs, the transformation, and the outputs of the system (McGregor 1967).

Systems Theory is a great analysis tool for understanding systems and their inner cause–effect relationships, yet it blissfully overgeneralizes by equating complex social constructs as organization to physical inanimate objects. Both Cybernetics and Systems Theory fall victim of the so-called "envy of physics": we seem to feel the urge to explain everything around us by means of math and physics. Society demands this scientific standard, even as it turns around and criticizes these studies as too abstract and removed from the "real world" (Churchill and Bygrave 1989). Why must we imitate physics to explain social things? This involves an uncritical application of habits of thought to fields different from those in which they have been formed. Such practice can find its roots in the so-called Newtonian revolution and later with the scientific method. When we entered the industrial era, the lens of Newtonian science led us to look at organizational success in terms of maintaining a stable system. If nature or crisis upset this state, the leader's role was to reestablish equilibrium. Not to do so constituted failure. With stability as the sign of success, the paradigm implied that order should be imposed from above (leading to top-down, command-and-control leadership) and structures should be designed to support the decision-makers (leading to bureaucracies and hierarchies). The reigning organizational model, scientific management, was wholly consistent with ensuring regularity, predictability, and efficiency (Tenembaum 1998). Management theories in the nineteenth and early twentieth centuries also held reductionism, determinism, and equilibrium as core principles. In fact, all of social science was influenced by this paradigm (Hayles 1991).

The general premise was: if the behavior of stars and planets could be accurately predicted with a set of elegant equations, then any system's behavior should be able to be captured in a similar way; including social systems as organizations. For Systems Theory, after all, stars and brains are still composed of the same type of highly deterministic matter. Systems Theory pursues finding theoretical models of any system in order to achieve basically three things: prediction, explanation, and control. In social sciences, the symmetry between prediction and explanation is destroyed because the future in social sciences is genuinely uncertain, and therefore cannot be predicted with the same degree of certainty as it can be explained in retrospect. In organizations, we have as yet only a very imperfect ability to tell what has happened in our managerial "experiments", much less to insure their reproducibility (Simon 1997).

When we deal with an organization of people, we deal with true uncertainty, true uniqueness. There are not two identical organizations as there are not two truly identical persons. May sound obvious, but worth stopping for a moment to think about it. Too often small organizations imitate practices from other organizations. Imitative practices come in many forms. Among others, firms expend substantial resources to identify and imitate best practices; firms hire consultants and experts to gain access to good ideas and practices that have worked in other firms; firms invest in trade associations to share information; young firms join business incubators and seek access to well-connected venture capitalists in part with the hope to gain access to good practices used by others. On the prescriptive side, firms are exhorted to invest in capabilities that allow them to more quickly and extensively imitate others, to benchmark their practices, to implement "best practices", and to invest in absorptive capacity. The underlying rationale for these activities and prescriptions is that firms benefit from imitation. First, a variety of reasons exist why the act of imitating a particular practice may fail. In other words, a firm tries to copy a particular practice but is unable to do so. Reasons include, for example, cultural distance or a bad relationship to the imitated firm (Csaszar and Siggelkow 2010).

In any case, clearly the hype about Systems Science in organizations dates from the pre-startup era, but it remains more present than we think. Respected authors in management such as Douglas McGregor, who is still relevant nowadays due to his Theory X and Theory Y of management,[1] dedicate a chapter of his "The Professional Manager" to talk about "managerial controls" which he finds analogous to machine control. He states:

> *... application [of feedback loops] to machines is so well understood that extended discussion of them is necessary. In a managerial control system the same principle may be applied to human performance*".

But, aware that such a remark could reach hyperbole levels, he adds:

> "*There is a fundamental difference between the engineering and the human application of the principle of information feedback control loop. Machines and physical processes are docile; they are passive with respect to the information fed to them. They can respond, but only to the extent that the alternative forms of response have been designed into the machine or the process. A more fundamental distinction is that emotions are not involved in the feedback process with machines or technological systems. This distinction, often ignored, is critical.* (McGregor 1967).

And I am not even totally sure about the statement that machines are "docile"; I believe McGregor did not have to deal for example with printers in his time. Then, he timely acknowledges that feedback control with machines lacks emotions, which is precisely why machine-like control could never work on social groups. Our social

[1]Theory X and Theory Y are theories of human work motivation and management. They were created by Douglas McGregor while he was working at the MIT Sloan School of Management in the 1950s, and developed further in the 1960s.

feedback loops are impregnated by emotions, shaped by them. All in all, this outdated school of thought proposes the approach of treating organizations as mere machines composed of interconnected boxes with inputs, outputs and clear boundaries and predictable behavior. Cybernetics, being highly influenced by such ideas, inherits and extends further the mechanistic mindset. Self-regulation, ubiquitous in the cybernetic perspective, suggests that equilibrium is the norm. Is equilibrium a word that would describe a startup? Anyone who has spent even a few hours in a startup would quickly state that equilibrium is not the norm. Startups behave in unstable ways, more like a complex, adaptive, and chaotic *thing*. Is this instability and chaotic nature, which often fosters innovation, and it is also one of its main threats. As ridiculous as it can sound to believe organizations can be treated as machines, current management practice is still deeply rooted in the mechanistic approach (Dooley 1997).

If control theory is applied to organizations as for machines, this could mean that decision-making in organizations could be controlled by computers. This includes the design process, which means a spacecraft could be programmatically designed by an artificial intelligence algorithm from a set of rules or specifications, or from measured market fit. This is what's called algorithm-driven design. How long will design engineers be needed in order to create technical things? How far are we from *the machine that designs the machine*? If organizations were machine-like deterministic systems, algorithm-driven operations would also be possible: i.e. a company being run automatically by a control loop. This algorithm could, for example, analyze sales, analyze the competitors, analyze market share, and decide when to do changes automatically, to meet prerevenues set points, for instance, by launching a new product.

No matter how advanced computers are today, running a company remains a very human-centered activity. Computers help, and they do help a lot, to do repetitive and computing intensive tasks we do not want/need to do ourselves very fast and efficiently, but computers are still not decision-makers. We will dedicate a chapter on Knowledge Management and it will be discussed how capturing and codifying knowledge in ways computers can understand could pave the way for AI-driven decisions at the organizational level in the future.

It is usually said, without great rigorosity, that startups are *chaotic*. Chaos is always associated with disorder. Startups can be also considered organized anarchies. In an organized anarchy, many things happen at once; technologies (or tasks) are uncertain and poorly understood; preferences and identities change and are indeterminate; problems, solutions, opportunities, ideas, situations, people, and outcomes are mixed together in ways that make their interpretation uncertain and connections unclear; decisions at one time and place have loose relevance to others; solutions have only modest connection to problems; policies often go unimplemented; and decision-makers wander in and out of decision arenas saying one thing and doing another (McFarland and Gomez 2016). Well, that pretty much sums up almost every early stage NewSpace organization out there.

A more mathematical definition of chaos says that chaos is about the state of dynamical systems whose apparently random states are governed by deterministic laws, which are very sensitive to initial conditions. We analyzed in the previous section that organizations do not follow deterministic laws as determinism is not found in social systems. In any case, organizations are systems with extremely high numbers of variables and internal states. Those variables are coupled in so many different ways that their coupling defines local rules, meaning there is no reasonable higher instruction to define the various possible interactions, culminating in a higher order of emergence greater than the sum of its parts. The study of these complex relationships at various scales is the main goal of the Complex Systems framework. Think of any organization, which is ultimately a collective of people with feelings, egos, insecurities, with a great variety of past experiences, fears, and strengths. At the same time, most of the interactions in an organization are neither coordinated nor puppeteered by some sort of grandmaster (despite what some CEOs would like...), meaning that there is a great deal of spontaneous interactions among the actors. Organizations fit the complex adaptive systems (CAS) bill very well, and luckily for us there is a reasonable body of research around applying the CAS framework to organizations in order not to be able to predict and control them, but to understand better how they work. A CAS is both self-organizing and learning (Dooley 1997). What is more, the observer is part of the analysis; our flawed and biased perceptions can influence our decisions to alter the scenario in a way that it reinforces our own beliefs.

1.3 Systems Thinking

We might be still far from being able to predict and control organizations in automated manners. As said, managing them remains a very human-centered activity. We can, although, comprehend them better if we think about them as systems, i.e. collection of entities connected together. When we think systemically, we gain a perspective which usually leads to better clarity on what the boundaries are and how the interfaces look like. Essentially, the properties or functions that define any system are functions of the whole which none of its parts has.

There are many different definitions of "system" in the literature, but they are more alike than different. The one that follows tries to capture their core of agreement. A system is a whole consisting of two or more parts, which satisfies the following three conditions:

1. The whole has one or more defining properties or functions.

For example, a defining function of an automobile is to transport people on land; one of the defining functions of a corporation is to produce and distribute wealth for shareholders and for employees; the defining function of a hospital is to provide care for the sick and disabled. Note that the fact that a system has one or more functions implies that it may be a part of one or more larger (containing) systems, its functions being the roles it plays in these larger systems.

2. Each part in the set can affect the behavior or properties of the whole.

For example, the behavior of such parts of the car can affect the performance and properties of the whole. The manuals, maps, and tools usually found in the glove compartment of it are examples of accessories rather than parts of the car. They are not essential for the performance of its defining function, which is yet another example of an output from an engineering process, which enables the system or product of interest.

3. There is a subset of parts that are sufficient in one or more environments for carrying out the defining function of the whole; each of these parts is necessary but insufficient for carrying out this defining function.

These parts are essential parts of the system; without any one of them, the system cannot carry out its defining function. An automobile's engine, fuel injector, steering wheel, and battery are essential—without them the automobile cannot transport people. Most systems also contain non-essential parts that affect its functioning but not its defining function. An automobile's radio, floor mats, and clock are non-essential, but they do affect automobile users and usage in other ways, for example, by entertaining or informing passengers while they are in transit. A system that requires certain environmental conditions in order to carry out its defining function is an open system. This is why the set of parts that form an open system cannot be sufficient for performing its function in every environment. A system that could carry out its function in every environment would be completely independent of its environment and, therefore, be closed. A system is a whole whose essential properties, its defining functions, are not shared by any of its parts.

Summarizing:

A system is a whole that cannot be divided into independent parts without loss of its essential properties or functions.

This hardly seems revolutionary, but its implications are considerable. An organization is nothing but a system with parts and interfaces, whose success is not a function of the sum of its parts; but the product of their interaction. Because the properties of a system derive from the interactions of its parts rather than their actions taken separately, when the performances of the parts of a system, considered separately, are improved, the performance of the whole may not be (and usually is not) improved. In fact, the system involved may be destroyed or made to function less well. For example, suppose we were to bring together one each of every automobile currently available and, for each essential part, determine which automobile had the best one. We might find that the Rolls-Royce had the best motor, the Mercedes the best transmission, the Buick the best brakes, and so on. Then suppose we removed these parts from the automobiles of which they were part and tried to assemble them into an automobile that would consist of all the best available parts. We would not even get an automobile, let alone the best one, because the parts do not fit together. The performance of a system depends on how its parts interact, not on how they act taken separately. If we try to put a Rolls-Royce motor in a Hyundai, we do not get a better automobile. Chances are we could not get it in, and if we did, the car would not operate well (Ackoff 1999).

Something that stands out from Ackoff's remarks is that he seems to assign some sort of negative meanings to the word *analysis*, considering it reductionistic. In fact, the nuance is that Ackoff considers that applying analytic techniques to a working, existing, synthesized system is reductionistic. In that context, his analysis is correct: if you want to analyze a watch by its parts, you will not be able to, since the watch is only a watch with all its parts assembled together. But analysis, and we will discuss this later on, is still an essential part of our problem solving "toolbox" when we design technical objects: we must analyze things, so we can then synthesize them. But it is true that the process is not easily reversible in that sense. What has been already synthesized cannot be (easily) analyzed without losing its global function. Reverse engineering is a case of postsynthesis analysis, and that is the reason why reverse engineering is so hard to do for complex systems.

Thinking in systems provides a perspective, which can greatly impact in early stages of young and small organizations, not only to help visualize and analyze the *big system*, which encompasses all the way to the organization and its business activities, how it interfaces to the market, what are the dependencies, the unknowns, the value chain. Identifying blocks and parts but at the same time understanding how all that functions is of great use when the level of uncertainty and fuzziness is high. Organizations may not be clockwork machinery, but they sure show an internal composition that Systems Thinking can help bring to the forefront.

1.3.1 Habits of a Systems Thinker

Systems Thinking is a habit that can be taught. But it requires active training and sponsorship from all directions. A young organization has the great advantage that the complexity of the *big system* is still reasonably low when things start. Hence disseminating systemic thinking becomes an asset further along the line.

The INCOSE Systems Engineering handbook summarizes the essential properties of a systems thinker (INCOSE 2015):

- Seeks to understand the big picture.
- Observes how elements within the system change over time, generating patterns and trends.
- Recognizes that a systems' structure (elements and their interactions) generates behavior.
- Identifies the circular nature of complex cause-and-effect relationships.
- Surfaces and test assumptions.
- Changes perspective to increase understanding.
- Considers an issue fully and resists the urge to come to a quick conclusion
- Considers how mental models affect current reality and the future.
- Uses understanding of system structure to identify possible leverage actions.
- Considers both short- and long-term consequences of actions.
- Finds where unintended consequences emerge.
- Recognizes the impact of time delays when exploring cause-and-effect relationships.
- Checks results and changes actions if needed: "successive approximation".

1.3.2 Engineering Systems Thinking

At times, Systems Thinking can happen to appear too abstract to engineers. Or too linked to *soft sciences*, or flagged as yet another "thought framework" from the myriads of thought frameworks out there. Systems Engineering is vastly influenced by Systems Thinking, yet this influence remains a bit implicit. To make the connection between the two more explicit, we can discuss now Engineering Systems Thinking.

Dr. Moti Frank was one of the first to identify systems thinking within engineering as a concept distinct from systems thinking within systems science. Russell Ackoff's Systems Thinking approach is highly based on systems science. Through examination of literature and interviews with engineers, Frank derived 30 *laws* of engineering systems thinking (Frank 2000) (reproduced here with permission).

1. In all the projects phases/stages and along the systems life, the systems engineer has to take into account:
    a. The customer organization vision, goals, and tasks.
    b. The customer requirements and preferences.
    c. The problems to be solved by the system and the customer needs.
2. The whole has to be seen as well as the interaction between the systems elements. For this purpose, a circular thinking has to be developed, to replace the traditional linear thinking. A problem should not be solved by just dismantling it to parts, but all its implications have to be taken into account. Each activity in a systems certain element affects the other elements and the whole.
3. Consider that every action could have implications also in another place or at another time. Cause and effect are not closely related in time and space (Senge 1990, 63).
4. One should always look for the synergy and the relative advantages stemming from the integration of subsystems.
5. The solution is not always only engineering one. The systems engineer has also to take into account:
    a. Cost considerations: business and economic considerations (in the development stages the production costs have also to be taken into account).
    b. Reuse or utilization of products and infrastructures already developed and proven (reuse in order to reduce risks and costs).
    c. Organizational, managerial, political, and personal considerations.
6. The system engineer should take as many different perspectives as possible, of every subject or problem, and other aspects have to be reviewed from all points of view.
7. Always take into account:
    a. Electrical considerations.
    b. Mechanical considerations.
    c. Environmental conditions constraints.

   d. Quality assurance considerations.
   e. Benefit indices, such as reliability, availability, maintainability, testability, and productibility.

8. In all development phases, the future logistic requirements have to be taken into account (spare parts, maintenance infrastructures, support, service, maintenance levels, worksheets, technical documentation, and various manuals).
9. When a need arises to carry out a modification in the system, take into account:

   a. The engineering and non-engineering implications in any place and at any time.
   b. The effects on the form, fit, and function.
   c. The delays and the time durations of the modification incorporation.
   d. The system's response time to the changes.
   e. The needs, difficulties, and attitudes of those supposed live with the modification.
   f. That the change could bring short-term benefit but long-term damage.

10. Each problem may have more than one possible working solution. All possible alternatives should be examined and compared to each other by quantitative and qualitative measurements. The optimal alternative should be chosen.
11. Engineering design is not necessarily maximal. Not always should one aspire to achieve maximum performances. At every stage engineering trade-offs and cost-effectiveness, considerations should be considered. One could always improve more. One has to know when to cut and freeze a configuration for production. Excessive pressure in a certain point could cause a collapse at another point. Over stressing one part in the system could weaken another part and thus the entire system. Maximum performance design is expensive and not always results in maximizing entire system performance. The harder you push, the harder the system pushes back (Senge 1990, 58). Faster is slower (Senge 1990, 62).
12. In case of systems malfunction, problem, or failure, repeated structures and patterns should be looked for and analyzed, and lessons drawn accordingly (repeated failure is a failure that keeps returning, after the repairs, until the true malfunction is found and repaired; a repeated non-verified failure is a failure that the user complained about, the technician inspected and could not verify, and the failure reappeared again in the next operation).
13. Look always for the leverage point changes that might introduce significant improvements by minimum effort. Small changes can produce big results but the areas of highest leverage are often the least obvious (Senge 1990, 63).
14. Pay attention to and take into account slow or gradual processes.
15. Avoid adapting a known solution for the current problem that might not be suitable. The easy way out usually leads back in. Today's problems come from yesterday's solutions (Senge 1990, 57).
16. Take into account development risks. In each project, uncertainty prevails on the level of scheduling, cost, resources, scope, environmental conditions, and technology. Therefore, the strategy of eliminating uncertainties has to be taken

e.g. experiments, tests, verifications, analyses, comparisons, simulations, awareness of possible risks, planning ways of retreat, and risk deployment among the partners.

17. It is impossible to run a project without control, configuration management, milestones, and management and scheduling methods. Possible bottlenecks and potential critical paths have to be examined constantly.
18. The operator/user person must be considered as a major part of the system. Hence at each stage, the human element has to be considered. The engineering design should include HMI (Human–Machine Interface) considerations.
19. The engineering design is a top-down design (excluding certain open systems, for which the bottom-up approach is preferable). The integration and tests are bottom-up.
20. At every stage, systemic design considerations should be used (such as decentralized or centralized design, minimum dependency between subsystems, etc.). The systems engineer should be familiar with system malfunction analysis methods and tools.
21. Engineering systems thinking requires the use of simulations. The simulation limitations should be taken into account.
22. Engineering systems thinking requires the integration of expertise from different disciplines. Inasmuch as the systems become more complex and dynamic, one person, as competent as he may be, is inadequate to understand and see it all. Systems thinking, by its nature, requires the examination of different perspectives, calling for teamwork to cover the various perspectives. When setting up the team, proper representation has to be given to all the system's functions. Control and status discussions and meetings as well as brainstorming may have to be more frequent.
23. Try to anticipate the future at every stage. Take into account anticipated technological developments, future market needs, difficulties, problems, and expected changes in the project. (For example: The life expectancy of complex platforms, such as fighter aircraft could reach decades, but the life expectancy of the avionics system is around 10 years on the average. What will be required after 10 years?)
24. Selecting partners and subcontractors could be critical. Before signing agreements, refer to considerations such as the engineering/economic history of the potential partner, manpower (quality, stability, human capital) that he is capable of investing at the projects disposal, division of work and responsibility, and proper arrangements for status meetings, integration tests, and experiments of all kinds.
25. When selecting the software language or software development tools and platforms, make sure that they are usable and supportable, or changeable, throughout the systems life.
26. When selecting components for production take into account their shelf life. Select components whose supply is guaranteed throughout all the systems life. In case of likely obsolescence of components, make sure of having sufficient stock.

27. In order to win a tender, many companies reduce the development price in their offer, assuming that they will be compensated by the serial production and by the cost of modifications (if required). Therefore, in engineering systems thinking, it is recommended not to start development at all, if the serial production budgets are not guaranteed in advance.
28. Always examine the external threats against the system (for example, electromagnetic interference, environment conditions, etc.).
29. Engineering systems thinking resorts to probability and statistical terms, both when defining the system specifications and when determining the project targets (costs, performance, time, and scope).
30. In engineering systems thinking, it is advisable to limit the responsibility assigned to an external factor (such as external consultants), since this increases the dependency on it. Shifting the burden to the intervenor. Teach people to fish, rather than giving them fish (Senge 1990, 382).

1.4 Out of Context Systems Engineering

Systems Engineering is a largely misinterpreted discipline. Or, perhaps the context is misinterpreted, not the discipline itself. Or may be both. For a start, there is no single, widely accepted definition of what Systems Engineering is; any two "systems engineers" are likely to provide different definitions. Tech organizations, big or small, have been forcing themselves to apply Systems Engineering "by the book" just to find out it gave little to none return of investment (investment here understood not only on money but also pain and sweat vs actual tangible results). In cases I experienced first-hand, it only generated heavyweight processes and extensive documents that made work slower and more cumbersome. I had this conversation with one of the sponsors of this SE process introduction, who was quite puzzled (only in private, in public he would still defend it): "if Systems Engineering is the way to go, why does it make everything more difficult than it used to be?"

Let us recap what Systems Engineering "by the book" is. First, probably a quick recap on the history of Systems Engineering. Systems Engineering is the result of necessity, and a reaction to two main problems: complexity and multidisciplinarity, which are sort of two sides of the same coin. As semiconductor and computer technology evolved during the twentieth century, computing became so pervasive that devices stopped being single-disciplined (fully mechanical, chemical, hydraulic, etc.) to be a combination of many disciplines such as electronics, software, and so on. This gave birth to the concept of a technical *system*, where the global function is the product of different components interacting together to accomplish the system's goal, and each one of those components is a combination of multiple disciplines combined. Miniaturization helped to squeeze more and more functionality into the systems architectures, where they started to contain more and more constituent components. And here is when the top-bottom vs bottom-up (analysis vs synthesis) gap became a more like a fissure: it became clear the way to design complex systems required a thorough top-bottom approach, whereas engineers were still "produced" (educated by universities) and trained to be the ones synthesizing the stuff, i.e. to go bottom-up. Universities were providing masons, where the industry was lacking architects. So the top-bottom branch needed some method, and that is what Systems Engineering came to propose. It did so in perhaps a bit of an uncoordinated way, since Systems Engineering never materialized as a formal educational track as other engineering disciplines, leaving it to a bunch of authors to define its practice and methodology by creating an heterogeneous Systems Engineering body of knowledge, which is currently composed of hundreds and hundreds of books, with similar but yet slightly different opinions, approaches and terminologies. Systems Engineering is an industry on itself as well, and a lucrative one for whoever manages to engage with the "big fishes".

Then, World War II and Cold War acted as accelerating factors in many critical military technologies such as weapon systems, radars, spacecraft, missiles, rockets, etcetera, so in the later part of the XX century the defense sector became the main driving force for the need of more integration among systems that had not been integrated together before, giving way to the very complex military systems that are known today as for example C4ISR (Command, Control, Communications, Computers, Intelligence, Surveillance and Reconnaissance). These kinds of "Systems of Systems" are reaching levels of integration no one could have thought about some decades ago. The military background (still one of its strongest sponsors) of Systems Engineering cannot be overlooked and explains most of the content found in the materials. The United States Department of Defense (DoD) has been facing incredibly difficult challenges acquiring those systems, which are without a doubt the most complex technical systems humanity has ever made. What is more, DoD grew increasingly willing to acquire systems that can interoperate. To give an indication of the importance of system acquisition, DoD has even created a university[2] for educating on how to acquire defense systems. This is the Systems Engineering that you will find in most of the books out there. A discipline where the audience is usually a gigantic multibillion dollar budget governmental entity requesting a highly complex mission-critical System (or System of Systems) on programs with lengthy schedules where requirements specification, conceptual design of the architectures, production, and deployment are very defined stages that often span several years or even decades. Just to have some idea, DoD proposed budget for FY2020 is ~ 750 billion dollars.[3]

How is this context anywhere related to small and medium enterprises? Answer is simple: it is not related at all. There is almost nothing you can get out of those books and handbooks that you can apply in our 10, 20, 50, 250 people company, or department. Systems Engineering practitioners around the world have noticed the disconnect between the historic Systems Engineering for giant organizations and the need of Systems Engineering for startups and small organizations. There have been

---

[2]This is the Defense Acquisition University (DAU).

[3]https://comptroller.defense.gov/Portals/45/Documents/defbudget/fy2020/fy2020_Press_Release.pdf.

some attempts to create tailored Systems Engineering standards for small organizations, for example ISO/IEC/TR 29110.[4] When a methodology, process or similar needs a *tailoring* or a *scaling*, that usually indicates the original is just overhead.

But, if we can for a moment forget about the background story of Systems Engineering and think about how we still do technical things, we quickly see we deal with the *top-bottom meets bottom-up*, or analysis vs synthesis problem-solving scheme, regardless if we are designing a CubeSat or an inter-continental ballistic missile. Systems Engineering, in a nutshell, is about shedding light on the top–bottom decomposition and then bottom-up integration of all the bits and parts, which is ultimately common sense: we cannot start picking up the bricks and hope we will end up building a house if we first do not think about what we *need* to build.

Consuming "old school" Systems Engineering material is fine, just if proper care is taken to understand that our context can be very different to the context that material refers to. Good news is that we can still take useful things from the "old" Systems Engineering school of thought: mostly about the life cycle of complex things. The bad news is that every organization, as unique as it is, needs its own flavor of Systems Engineering. No book can capture what your organization needs because those needs are unique. We discussed tailoring before; tailoring is a good practice, but it must be done while embedded in the right context. No external entity/consultant/standard can tailor things for us.

1.5 Sustainable Space by Design

At the height of the Cold War, the USA thought it was necessary to launch 480 million copper needles into orbit, in what was called Project West Ford.[5] It must have seemed like a good idea at the time, considering the context. The United States Military was concerned that the Soviets might cut undersea cables (back then the only way to communicate with overseas forces), forcing the unpredictable ionosphere to be the only means of communication with servicemen abroad. The ring, intended to relay radio signals, became obsolete before it was complete, with the launch of the first telecommunication satellites in 1962. Although most of the individual needles have reentered the atmosphere, clumps of them remain in orbit today, contributing to the serious and growing problem of space debris.

Fast forwarding now to a luckily more peaceful present, the situation in orbit does not look any less ridiculous. Or, it looks even more ridiculous: the space industry is at the verge of a potential new Project West Ford. With semiconductor technology reaching incredible integration densities and computing capabilities, and with prices of these devices dropping by the day, the cost of designing and building a spacecraft is

[4]https://www.iso.org/standard/62711.html.

[5]Project West Ford was a test carried out by Massachusetts Institute of Technology's Lincoln Laboratory on behalf of the US Military in 1961 and 1963 to create an artificial ionosphere above the Earth. This was done to solve a major weakness that had been identified in US military communications.

getting lower and lower; a spacecraft the size of a shoebox can have more computing power than a spacecraft the size of a truck two decades ago. And with more rocket launchers entering the game, and with rocket reusability already a reality, nothing indicates space will get less crowded any time soon. In fact, some estimations indicate the amount of spacecraft in orbit could quintuple by the end of this decade (Ryan-Mosley et al. 2019). But how is the situation today?

According to the Union of Concerned Scientists (UCS), which maintains a database of active satellites in orbit, as of August 2020 (Union of Concerned Scientists 2020), the amount of active satellites in orbit is:

Total number of operating satellites: 2,787

- USA: 1,425
- Russia: 172
- China: 382
- Other: 808

Distributed in the following types of orbits:

- LEO: 2,032
- MEO: 137
- Elliptical: 58
- GEO: 560

In terms of inactive elements orbiting, the numbers are, as of February 2020 (European Space Agency 2020):

- Number of satellites rocket launches have placed into Earth orbit: ~9600
- Number of these still in space: ~5500
- Number of debris objects regularly tracked by Space Surveillance Networks and maintained in their catalog: ~22,300
- Estimated number of breakups, explosions, collisions, or anomalous events resulting in fragmentation: >500
- Total mass of all space objects in Earth orbit: >800 tonnes
- Number of debris objects estimated by statistical models to be in orbit:
    - 34,000 objects >10 cm
    - 900,000 objects from greater than 1 cm to 10 cm
    - 128 million objects from greater than 1 mm to 1 cm

All indicates our dependency on satellites will only increase, going hand-in-hand with our increasing dependency on all kinds of digital services and broadband connectivity. Every new satellite placed into orbit contributes a bit more to the overall probability of collision, which also increases the probability of a cascade effect of runaway debris which could render key parts of the Earth orbit too risky for satellites to operate. Debris-filled orbits could endanger all types of human spaceflight, affecting space exploration and therefore societal development. The economic impact of such scenarios would be of disastrous consequences for millions. NewSpace has the responsibility of designing spacecraft that comply with responsible use of space. Such responsibility needs to be reflected in the design approach and flowed down to all stages of the development process. True sustainability of space operations can only take place if sustainable space is approached "by design". We will define what sustainable space is and what decisions we as engineers must take to ensure space as a resource for many generations to come.

1.5.1 What Is Sustainable Space?

Sustainability is one of those buzzwords which has been so pervasively used that it is lost its true meaning. What does sustainability mean in terms of space?

In 2010, the United Nations Committee on the Peaceful Uses of Outer Space (UN COPUOS) established a Working Group on the Long-Term Sustainability of Outer Space Activities. The Working Group was tasked with producing a set of voluntary guidelines for all space actors to help ensure the long-term sustainable use of outer space; the Working Group's mandate ended in June 2018 at its 61st session. During its mandate, the Working Group agreed on 21 guidelines and a context-setting preambular text that included the following definition of space sustainability (UN COPUOS 2018):

> The long-term sustainability of outer space activities is defined as the ability to maintain the conduct of space activities indefinitely into the future in a manner that realizes the objectives of equitable access to the benefits of the exploration and use of outer space for peaceful purposes, in order to meet the needs of the present generations while preserving the outer space environment for future generations.

The same document defines the Earth's orbital space environment (and radio frequency spectrum) as finite resources. This means, space is no different than any other limited natural resource and as such it must be protected from predation as well as from uncoordinated and uncontrolled exploitation. On the same line, the UN COPUOS report acknowledges that the long-term sustainability of outer space activities is of interest and importance for current and emerging participants in space activities, in particular for developing countries. Overexploitation and pollution are frequent negative externalities of common pool resources, often referred to as the "tragedy of the commons", where the actions of individual users, motivated by short-term gains, go against the common long-term interest of all users. The management of common pool resources, for which market mechanisms are highly imperfect or completely absent, depends crucially on the existence and effectiveness of the rules and institutions (whether formal or informal) to govern their use. For space activities, this translates for example into human activities potentially littering Earth's orbits beyond sustainable limits, creating space debris that could reduce the value of space activities by increasing the risk of damaging collisions and requiring mitigation actions (Undseth et al. 2020).

1.5.2 Challenges to Space Sustainability

Perhaps the biggest challenge for space sustainability is our own increasing dependency on digital services: mainly faster communications and Earth observation data, the latter for both civil and military applications. Digitalization of our lifestyle is the root cause or the principal need to put spacecraft in orbit. The obvious effect: overcrowded orbits. The next two figures depict the increase of traffic in LEO, per sector (Fig. 1.1) and mission type (Fig. 1.2). It is interesting to note the rapid increase of amateur missions from 2010 onward (European Space Agency 2020).

However, the real game changer for the near future will be the deployment of several broadband mega-constellations that are under preparation. In 2018, the US Federal Communications Commission approved SpaceX' plan of almost 12000 satellites, and in 2019, the company submitted filings with the International Telecommunications Union (ITU) for an additional 30000 satellites. In July 2019, Amazon filed an application with the US Federal Communications Commission (FCC) for its 3000 + Kuiper System constellation. In 2019, OneWeb launched the first satellites in their planned 650 + satellite constellation. SpaceX, in its future Starlink constellation, foresees placing first about 1600 satellites at 550 km, followed by ~2800 satellites at 1150 km and ~7500 satellites at 340 km. The Amazon Kuiper System constellation would deploy three "shells" of satellites at 590, 610, and 630 km, while OneWeb intends to use the 1200 km orbit. Other large constellations are being considered in

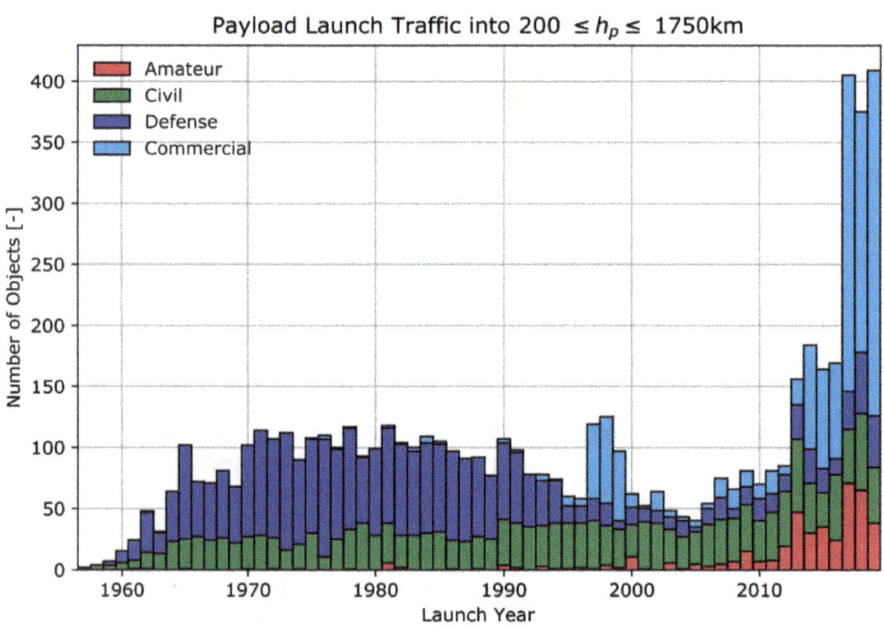

Fig. 1.1 Evolution of the launch traffic near LEO per mission funding (*source* ESA's Annual Space Environment Report, page 21, used with permission)

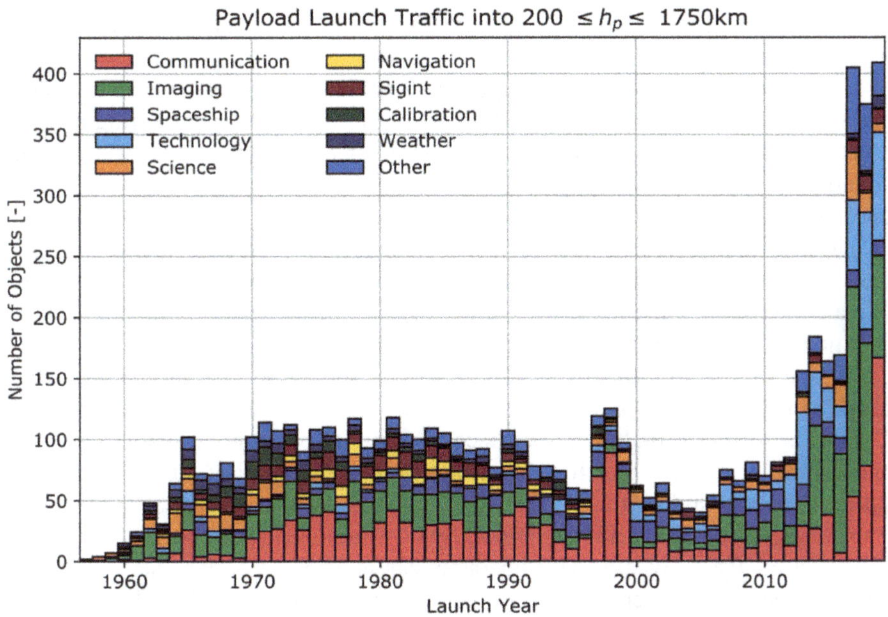

Fig. 1.2 Evolution of the launch traffic near LEO per mission type (*source* ESA's Annual Space Environment Report, page 21, used with permission)

North America and China in particular. With the deployment of these constellations, the number of operational satellites in orbit could double or even triple in the next 5 years. When taking into account, all existing satellite filings, there could be several tens of thousands of operational objects in orbit by 2030. At this pace of orbital density increase, according to multiple modeling efforts, it is not a question of if a defunct satellite will collide with debris, but *when* it will (Undseth et al. 2020).

Issues related to the allocation and use of electromagnetic spectrum is another growing concern for the long-term sustainability of space activities, due to the intensification of space operations and to terrestrial competition (OECD 2019). Radio frequencies, used by spacecraft to communicate with other spacecraft and terrestrial ground stations, are also defined as a limited natural resource (ITU 2015).

The International Telecommunications Union is the organization ensuring equitable access to this natural resource by allocating frequency bands to individual countries and mitigates interference issues by reserving specific bands for specific uses (e.g. fixed satellite uplinks and downlinks). There are growing concerns of interferences from both terrestrial and space networks (e.g. deployment of 5G, the growth of mobile communications worldwide). Concerning the space networks, the sheer size of many planned constellations in low-earth orbit raises particular concerns about orbital interferences. Some satellite operators in the geostationary and medium-earth orbits worry that the increasing crowding of the low-earth orbit could eventually jam the link between higher flying satellites and terrestrial satellite dishes (OECD 2019) cited by (Undseth et al. 2020).

Last but not least, the big challenge for a sustained use of space is collaboration and transparency. Currently, the traffic control or conjunction monitoring of space assets remains based in best-effort, and largely uncoordinated and informal. Space situational awareness (SSA) agencies of different militaries around the world are the ones nowadays tracking, by means of different ground and space-based radars, and categorizing space debris up to some size. Some private companies are now offering tracking services as well, and not for small prices. There is not a single international overarching entity handling space traffic, in equivalence to the International Civil Aviation Organization (ICAO) for air transportation.

In lack of such entities, some organizations stand out as coordinating actors, notably the Combined Space Operations Center[6] (CSpOC) for NewSpace operators. CSpOC is a US.-led multinational space operations center that provides command and control of space forces for United States Space Command's Combined Force Space Component Command. The CSpOC is located at Vandenberg Air Force Base.CSpOC created and administrates a Space Situational Awareness (SSA) website called Space-Track[7] where space operators can register to receive data about space objects. The main objective of Space Track is to enhance SSA data sharing, transparency, and spaceflight safety. Space Track provides space surveillance data to registered users through this public website. CSpOC routinely updates the website with positional data on more than 16,000 objects in orbit around the Earth. Users can also use an available API to pull data from records and automate the retrieval of new data.

Available services are:

- Satellite Catalog (SATCAT) information.
- General Perturbation (GP) positional data for the unclassified space catalog (two-line element sets).
- Satellite decay and reentry predictions.

It is important to note that registering to the SSA database is optional for operators.

Fortunately, there is an increasing number of initiatives in terms of international cooperation for spaceflight safety, for example the recently formed Space Safety Coalition (SSC). Formed in 2019, the Space Safety Coalition publishes a set of orbit regime-agnostic best practices for the long-term sustainability of space operations. These best practices are generally applicable to all spacecraft regardless of physical size, orbital regime or constellation size, and directly address many aspects of the twenty-one consensus Long-Term Sustainability (LTS) guidelines approved by the United Nations Committee for the Peaceful Use of Outer Space (UN COPUOS) in June 2019. SSC has been endorsed by a considerable number of space companies from both classic and NewSpace (Space Safety Coalition, n.d.). In its document, SSC

[6]https://www.peterson.af.mil/About/Fact-Sheets/Display/Article/2356622/18th-space-control-squadron/.

[7]http://www.space-track.org.

expresses as its main concern the need to preserve a safe space environment for future exploration and innovation and the need to limit the creation of new space debris, maximize the information available on both debris and spacecraft, and encourage the development of and adherence to community-wide best practices for all space industry stakeholders, and it urges all space actors to promote and adhere to the best practices described to ensure the safety of current and future space activities, and to preserve the space environment (Space Safety Coalition 2019).

In short, the space industry shows great growth potential for the next 5–10 years. This growth will be mostly impacting Low Earth orbits, so in order to make such a growth viable, the existing and new actors entering the industry will have to closely collaborate toward organizing the exploitation of limited natural resources such as orbit and radio frequency spectrum. NewSpace has the opportunity to impact the lives of millions of people by offering affordable digital services such as connectivity and Earth observation data. By doing so in an ethical way, the access to space will be guaranteed for many generations to come.

1.5.3 The Cost of Space Safety

As it happens with many industries, being ethical and fair with others while doing things can increase costs compared with those who take unethical shortcuts and act more selfishly. This is the reason why many companies worldwide, acting as profit-seeking machines, engage in unethical practices to maximize their gains (by minimizing their costs) at the expense of poor job quality and work conditions, most usually impacting developing countries. In fashion industry, for example, if a company decides to have an ethical production strategy by means of supporting local businesses for their supply chain, with fair pay and work conditions, chances are this fashion brand will not be able to compete to a similar one which chooses to use sweatshops in a developing country where labor is cheap at the expense of very poor and illegal work conditions. Profit-seeking can foster highly questionable practices. And a selfish profit-oriented approach to space will only lead to a massive denial of access sooner than later.

Space, as an industry, is no different than other industries in the sense that cost is king. When costs mandate, companies choose to put to the side everything considered "optional" or nice-to-haves. Costs related to space safety can easily land there. Mostly because the current legislation about space safety does not rigidly impose spacecraft manufacturers to add mandatory space safety on-board capabilities to their satellites, besides a promise that orbital decay will take place in some maximum timeframe; i.e. the famous 25-year rule. Since the introduction of this in the 1990s, many organizations, including the Inter-Agency Space Debris Coordination Committee (IADC) have studied and confirmed the effectiveness of using the 25-year rule to curtail the debris growth in LEO (Liou et al. 2013). Based on nominal future projection assumptions, a non-mitigation scenario (i.e., just abandoning the spacecraft there at the end of its mission) could lead to a LEO debris population growth of approximately 330%

over the next 200 years, whereas a 90% global compliance with the 25-year rule could reduce the LEO debris population growth to approximately 110% over the next 200 years. Reducing the 25-year rule to, for example, a 5-year rule, only leads to another 10% debris reduction over 200 years, which is not a statistically significant benefit. Only a modest increase in de-orbit propellant is needed to reduce the residual lifetime to the ~ 50-to-25-year range. However, decreasing the postmission orbital lifetime from 25 years to a very short time, such as 5 years, will lead to a non-linear, rapid increase in fuel requirements. Therefore, the 25-year rule appears to still be a good balance between cost and benefit. Main problem with this rule is when to draw the line a mission has ended (hence the 25-year clock needs to start ticking) and also enforcement of the rule before launch; as it usually happens, problem is not lack of rules, problem is lack of enforcement. Analyses of global space operations indicate that the biggest issue in LEO is associated with the very low compliance with the 25-year rule (less than 50%) (UN COPUOS 2018).

Then, it can be seen that currently spacecraft are not mandated to carry any equipment (sensor or actuator) related to space safety if they do not want to. Aerospace regulations, in contrast, mandate aircraft to be equipped with something called TCAS (Traffic Collision Avoidance System), also known as ACAS (Airborne Collision Avoidance System). TCAS works by sending interrogations to other aircrafts (more specifically to other aircrafts' transponders). From the time difference between the interrogation and the reply, the distance to the other aircraft is calculated; the reply itself contains the altitude of the other aircraft. The distance and the altitude difference with the other aircraft are tracked to identify a trend. From successive distance measurements, the closure rate is determined. With the closure rate and the current distance (slant range), an approximation is made of the time to the closest point of approach (CPA). The same is done in the vertical plane. If some certain thresholds are met, a Traffic Alert (TA) is raised. When some lower warning level is crossed, an informative Resolution Advisory (RA) is given. The TA is a "heads up" indication, the RA is an instruction that must be followed by the pilot to reduce the collision risk. Originally TCAS did only give Traffic Advisories in the form of an audible warning. There was no avoiding action indicated. With the introduction of TCAS II in the second half of the '80 s the Resolution Advisory made its entrance. In selecting a resolution advisory, there are basically two steps. The first step is to select a sense, either upward or downward. This is based on a calculation of how much altitude difference can be achieved at the CPA by either starting a climb or descent, assuming the target will maintain its vertical rate. The second step is to select a magnitude. The algorithm is designed to be the least disruptive to the flight path, whilst still achieving a minimum vertical separation. When a RA is selected, it is transmitted to the other aircraft. When the other aircraft receives that message, it will only use the opposite sense for its own RA. In the rare case that both aircrafts transmit their RA intent at the same time, the aircraft with the higher transponder address will give in and reverse its RA sense if it is conflicting with the other.

There is no equivalent of TCAS/ACAS in space. Satellites from some particular operator may have automatic collision avoidance systems, but they are non-cooperative meaning that they work for their own private purposes. The value of automated cooperative collision avoidance is also to prevent manual collision avoidance maneuvers to make things worse (for example if both spacecraft maneuver to the same *side of the road*). Interestingly, currently collision avoidance in space is done in a very *ad hoc* and highly human-centered manner. This approach literally requires operators from the incumbent entities calling on the phone or emailing each other to coordinate how the maneuvers will be performed if needed; who moves to which side of the road. Overall, the global space traffic management strategy is in some sort of limbo at the moment. Disagreements on which agency or entity must be responsible for Space Traffic Management has slowed progress down in terms of defining a clear direction (Foust 2020).

In terms of design costs associated with space safety, while data are limited, some operators (in the geostationary orbit) have indicated that the full range of protective and debris mitigation measures (e.g. shielding, maneuvers, and moving into graveyard orbit) may amount to some 5–10% of total mission costs (National Research Council 2011). In LEO orbits, such costs are estimated to be higher, but sources are scarce.

There are operational costs associated with space safety as well. Satellite operators report an increase in maneuvers to avoid collisions with debris. Operators need to take into account different types of data and sources with various formats to plan orbital trajectories. They may receive hundreds of warnings of impending close approaches (conjunction warnings) a year, several of which may be false or inaccurate, creating a significant burden on operators in terms of analysis and data management. What is more, if the conjunction warning is considered critical, a collision avoidance maneuver is conducted, which can take days. This consumes satellite propellant and in addition the payload cannot be used during the maneuvering which means losses for the operators in the form of acquisitions and orders not taken/lost. In 2017, the US Strategic Command issued hundreds of close approach warnings to their public and private partners, with more than 90 confirmed collision avoidance maneuvers from satellite operators.

But, all in all, the current costs of space debris are nothing compared with future prospects. In a worst-case scenario, certain orbits may become unusable, due to continued, self-reinforcing space debris generation. This would have significant negative impacts on the provision of several important government services and would most probably also slow down economic growth in the sector. The social costs would be unequally distributed, with some rural regions more hardly hit, in view of their growing dependence on satellite communications in particular (Undseth et al. 2020).

1.5.4 Design Considerations for Sustainable Space

Collision risk is projected to increase quite significantly in the next decade; launchers are already experiencing challenges to choose launch windows due to congestion

(Wattles 2020). Modeling conducted by the Inter-Agency Space Debris Coordination Committee (IADC) in the 2009–12 timeframe (before the increase in CubeSat launches and the announcement of mega-constellations) predicted an average 30% increase in the amount of low-earth orbit debris in the next 200 years, with catastrophic collisions occurring every 5–9 years, factoring in a 95% compliance rate to mitigation rules (Liou et al. 2013). A 2017 study at the University of Southampton found that adding one mega-constellation of several thousand satellites to the low-earth orbit space environment would increase the number of catastrophic collisions by 50% over the next 200 years (University of Southampton 2017). In a worst-case scenario, researchers suggest that the low-earth orbit could be rendered unusable for future generations, because collisions and the continued generation of debris would become self-sustaining, the so-called Kessler syndrome (Kessler and Cour-Palais 1978). Exactly when and if this ecological tipping-point is reached, is subject to great uncertainty, with current modeling capabilities unable to provide a clear answer. Either way, the economic tipping point, where operations in low-earth orbit become economically unsustainable, may be reached sooner than predicted (Adilov et al. 2018). Human spaceflight operations in the lower earth orbits may also be considered too risky, due to the concentrations of CubeSats and broadband satellites (Undseth et al. 2020, 26).

Reality indicates most of the attention on space safety is allocated once the spacecraft is already in orbit. In other words, the thickest part of the overall analysis revolves on how to mitigate the risks once the objects are up there. The question now is: how can we as spacecraft engineers contribute to a more sustainable use of space from a design perspective? Four areas of interest will be explored (which are greatly intertwined):

- Deorbiting mechanisms:
  - Active and Passive
- Modularity:
  - Both intrasatellite and intersatellite
- Reusability:
  - Can a spacecraft bus serve more than one mission/payload throughout its lifetime?
- Green Engineering:
  - An environmental-aware engineering process, which reduces e-waste, and minimizes the use of hazardous materials.
  - Space safety requires a great deal of orbital agility, which requires extensive use of propellants, which can be hazardous during ground handling and also for the atmosphere during reentry.

1.5.4.1 Deorbiting Mechanisms

If a spacecraft is overall functional, and provided it has enough on-board resources (propellant, power), its disposal is more or less an operational matter: in other words it can be commanded by ground crew to maneuver it either to either a less crowded *graveyard* orbit or force its reentry by decreasing its orbital velocity. Active deorbiting can be challenging since it requires guaranteeing fully, non-degraded performance of all subsystems at their End-of-life (EOL). An alternative could be to incorporate standalone, independent deorbiting modules attached to the spacecraft bus. Such modules would require to be a "spacecraft inside a spacecraft" to be able to fully independently tow a defunct platform to either a graveyard orbit or to make it reenter. As such, this standalone module has the complexity and inherent risks (software bugs, etc.) a spacecraft has.

If a satellite is non-functioning or severely degraded, it is as maneuverable as a rock. For this, passive deorbiting mechanisms are the only possibility, and only for reentry; which also constraints their use mostly for low altitudes. The typical solution for passive deorbiting mechanisms involves using aerodynamic drag one way or the other. Alternatives include using mechanically deployable drag sails which present a big surface area toward the direction of flight to reduce the orbital speed and provoke reentry. Challenges related to deployable sails are: a. reliability of their somewhat complex mechanics, b. handling a tumbling satellite, c. bulky when stowed, heavy, taking up a big part of the spacecraft volume. Another option is to use inflatables. Pneumatic inflatables require a gas source. This can be in the form of a compressed gas or gas producing chemical reaction, similar to a car airbag. Inflatables offer better mass to volume ratio and can be hardened to form solid shells (Chandra and Thangavelautham 2018).

1.5.4.2 Modularity

Modularity will be discussed in much detail further on. Here, we will briefly discuss how architectural decisions and space safety could be matched for a more sustainable use of space. Spacecraft are functional containers (i.e. they are an arrangement of subfunctions aggregated into a global function). As such, the way these functions are combined together is arbitrarily defined by the designers and realized by the architecture. Modularity deals with how such functionalities are coupled together to provide the overarching function. The global function of a system does not care about the physical form factor as long as the function is provided in presence of a pressing need (for example, if I am really tired, I do not care how a chair looks like but just the fact I can sit and rest, or if I am far away from home and all I can find to come back is a three-wheeled bike I will not question that). In the absence of exhaustion or the need for urgent transportation, criteria may vary. On the same line a payload needing to be operated in orbit does not care how the underlying infrastructure to make that happen looks like. In short, what this means is that for a payload to be able to perform its mission, what the payloads ultimately need is a list of functions or services, and the way those functions or services are arranged is irrelevant from a payload perspective (they can, though, be relevant from other perspectives such as cost, complexity, etc.).

Modularity can be:

- Intraspacecraft:
  - Functional elements of the architecture are aggregated in one single physical entity where building blocks are coupled in ways they could be detached as they degrade and deorbit on their own by means of passive deorbiting mechanisms. As the *mothership* loses mass from discarding degraded blocks, the remaining propellant is optimized for active deorbiting to either graveyard orbits or reentry (Fig. 1.3).
- Interspacecraft:
  - Spacecraft functionality is not aggregated in one single physical body but spatially distributed and connected wirelessly by intersatellite links (ISL). In this configuration (also called segmented architecture, as depicted in Fig. 1.4), different functional blocks fly next to each other and couple their functionalities by means of high-speed wireless links. With this configuration, degradation of some functional elements can be handled by deorbiting them and sending new segments to the formation group, extending its lifetime.
  - Each segment of the group is self-contained in terms of safety: it can deorbit itself (passively or actively).
  - Each segment requires almost full spacecraft functionality such as propulsion, attitude and orbit guidance and control, communication links, etc.

Fig. 1.3 A concept depicting a modular spacecraft with detachable subsystems which can be individually disposed (*credit* ReOrbit, used with permission)

Fig. 1.4 A segmented architecture in formation flying. Each spacecraft performs a different function (*credit* CONAE, used with permission)

- One segment could have the role of "group conjunction safety officer" where it could off load ground operations from conjunction maneuvers of the group.
- A distributed architecture like this fly in a physical vicinity, which means automated flight control and precise position determination is a must.

Note that both approaches can be combined. For example, two aging spacecraft could discard their individual degraded subsystems and join efforts as one distributed spacecraft (intra and inter spacecraft modularity can be complementary). For example, if one of them has a malfunctioning downlink subsystem, this spacecraft could send the payload data using the intersatellite link to another spacecraft which has a healthy downlink subsystem. This approach already touches base on reusability which is the next topic in the list.

1.5.4.3 Reusability

In this concept, being perhaps a blend of the intra and inter modularity covered before, the payload is just another block in the modular architecture. This means, if payload fails or reaches its end of life (EOL), it detaches from the bus and deorbits itself. The remaining platform's resources can still be used by sending a new payload with basic resources to couple with the bus remaining capabilities; this coupling can be either physical (which requires physical docking), or by means of wireless communication link. In case of the physical coupling approach, both platform and payload need almost full spacecraft functionalities such as propulsion, attitude and orbit guidance and control, communication links, etc. What is more, a precise autonomous ranging and docking mechanism is needed for payload and bus to be able to attach each other. Reusability is enabled by modularity (Fig. 1.5).

Fig. 1.5 A concept depicting a detachable payload module, which can be deorbited while the bus can be reused (*credit* ReOrbit, used with permission)

1.5.4.4 Green Engineering

NewSpace organizations are central to ensure a sustainable use of space. But what about things before flight? They do not exactly look any better. There is no risk of Kessler Syndrome on the ground but, since the Industrial Revolution, there is a consistently depletion of natural resources and consequent pollution, and it all indicates these trends are accelerating; a kind of growth that will not sustain forever. In "The Limits of Growth", a work which studied the effects of exponential economic and population growth with a finite supply of resources done 50 years ago by means computer simulation, had these conclusions (Meadows et al. 1970):

- If the present growth trends in world population, industrialization, pollution, food production, and resource depletion continue unchanged, the limits to growth on this planet will be reached sometime within the next 100 years. The most probable result will be a rather sudden and uncontrollable decline in both population and industrial capacity.
- It is possible to alter these growth trends and to establish a condition of ecological and economic stability that is sustainable far into the future. The state of global equilibrium could be designed so that the basic material needs of each person

on earth are satisfied and each person has an equal opportunity to realize his individual human potential.
- If the world's people decide to strive for this second outcome rather than the first, the sooner they begin working to attain it, the greater will be their chances of success.

"Limits of growth" created quite some waves when it was published. The main criticism of the report is that it did not take into account our capability to technologically improve in order to make a better use of finite resources. The assumption of its critics that technological change is exclusively a part of the solution and no part of the problem is debatable (Daly 1990).

A truly sustainable use of space starts all the way from ideation phase and concerns both the technological design process of systems and products and also the organization itself. As just any other engineering discipline, space systems engineering needs to consider the environmental impact of its activities throughout the whole life cycle and not only when things are ready to be launched to space.

Sustainability requires adopting new and better means of using materials and energy. Incorporating sustainability as part of the engineering quest is defined as *green engineering*, a term that recognizes that engineers are central to the practical application of the principles of sustainability to everyday life. Increasingly, companies have come to recognize that improved efficiencies save time, money, and other resources in the long run. Hence, companies are thinking systematically about the entire product stream in numerous ways (Vallero and Brasier 2008):

- Applying sustainable development concepts, including the framework and foundations of "green" design and engineering models.
- Applying the design process within the context of a sustainable framework, including considerations of commercial and institutional influences.
- Considering practical problems and solutions from a comprehensive standpoint to achieve sustainable products and processes.
- Characterizing waste streams resulting from designs.
- Applying creativity and originality in group product and building design projects.

A growing environmental concern is the steady generation of massive amounts of e-waste. E-waste or electronic waste is created when an electronic product is discarded after the end of its useful life. The rapid expansion of technology and an ever increasing consumerist culture results in the creation of a very large amount of e-waste every day. Planned obsolescence, including extreme practices as purposely slowing down electronic devices to push users to buy new ones, are contributing to this alarming situation (Reuters 2020).

Some facts about e-waste (ITU 2020):

- E-waste is a growing challenge, matching the growth of the information and communication technology (ICT) industry. There are currently more mobile cellular subscriptions on Earth than there are humans. Since 2014, the global generation of e-waste has grown by 9.2 million metric tonnes (Mt) (21%).

- The fate of over four-fifths (82.6% or 44.3 Mt) of e-waste generated in 2019 is unknown, as well as its impact on the environment.
- E-waste contains substances that can be hazardous to human health and the environment if not dealt with properly—including mercury, cadmium and lead. An estimated total of 50 tonnes of mercury might be contained within the 44.3 Mt of e-waste whose destination is unknown.
- The global regulatory environment for e-waste is rather weak. Currently, only 78 out of 193 (40% of countries) are covered by an e-waste policy, legislation or regulation.
- ITU Member States have set a global ITU e-waste target for 2023 to increase the global e-waste recycling rate to 30% and to raise the percentage of countries with an e-waste legislation to 50%. They have also committed to reducing the volume of redundant e-waste by 50%.

A set of methods to avoid producing irresponsible amounts of waste during product development is what's called Green Engineering practices. Green engineering practices seek to minimize generating and using hazardous materials, and it does so by selecting manufacturing types, increasing waste-handling facilities, and if these did not entirely do the job, limiting rates of production. Green engineering emphasizes the fact that these processes are often inefficient economically and environmentally, calling for a comprehensive, systematic "green" life-cycle approach.

A truly sustainable approach for space (and any other) engineering must include a responsible exploitation of resources from every single aspect related to the activity. From daily company operations, to research, design, production, and disposal of the products, organizations must have a strategy to minimize environmental impact.

References

Ackoff, R. (1999). *Re-creating the corporation, a design of organizations for the 21st century*. Oxford University Press.

Adilov, N., Alexander, P., & Cunningham, B. (2018). An economic "Kessler Syndrome": A dynamic model of earth orbit debris. *Economics Letters, 166*, 79–82. https://doi.org/10.1016/J.ECONLET.2018.02.025.

Anz-Meador, P., & Shoots, D. (Eds.). (2020, Feb). Orbital Debris Quarterly News. *NASA, 24*(1).

Chandra, A., & Thangavelautham, J. (2018). De-orbiting Small Satellites Using Inflatables.

Churchill, N., & Bygrave, W. D. (1989). The Entrepreneurship Paradigm (I): a philosophical look at its research methodologies. *Entrepreneurship Theory and Practice, 14*(1), 7–26. https://doi.org/10.1177/104225878901400102.

Csaszar, F. A., & Siggelkow, N. (2010). How much to copy? Determinants of effective imitation breadth. *Organization Science, 21*(3), 661–676. https://doi.org/10.1287/orsc.1090.0477.

Daly, H. E. (1990). *Steady-state economics* (Second ed.).

Dooley, K. J. (1997). A complex adaptive systems model of organization change. *Nonlinear Dynamics, Psychology, and Life Sciences, 1*(1), 69–97.

European Space Agency. (2020, 2). *Space debris by the numbers*. ESA Safety & Security. https://www.esa.int/Safety_Security/Space_Debris/Space_debris_by_the_numbers

Frank, M. (2000). Engineering systems thinking and systems thinking. *INCOSE Journal of Systems Engineering, 3*(3), 163–168.

Foust, J. (2020, November 3). *Space traffic management idling in first gear*. SpaceNews. https://spacenews.com/space-traffic-management-idling-in-first-gear/

Hayles, N. K. (1991). Introduction: Complex dynamics in science and literature. In *Chaos and order: Complex dynamics in literature and science* (pp. 1–36). University of Chicago Press.

INCOSE. (2015). *INCOSE Handbook* (4th ed.). Wiley. ISBN:978-1-118-99940-0

ITU. (2015). *Collection of the basic texts adopted by the Plenipotentiary Conference*. http://search.itu.int/history/HistoryDigitalCollectionDocLibrary/5.21.61.en.100.pdf

Kessler, D., & Cour-Palais, B. (1978). Collision frequency of artificial satellites: The creation of a debris belt. *Journal of Geophysical Research, 83*(A6), 2637. https://doi.org/10.1029/JA083iA06p02637.

Liou, J. C., Anilkumar, A. K., Virgili, B., Hanada, T., Krag, H., Lewis, H., Raj, M., Rao, M., Rossi, A., & Sharma, R. (2013). Stability of the future leo environment—An iadc comparison study. https://doi.org/10.13140/2.1.3595.6487

Maxwell, J. C. (1867). On Governors. *Proceedings of the Royal Society of London, 16*(-), 270–283. www.jstor.org/stable/112510

McFarland, D. A., & Gomez, C. J. (2016). *Organizational Analysis*.

McGregor, D. (1967). *The professional manager*. McGraw-Hill. ISBN:0070450935

Meadows, D., Meadows, D., Randers, J., & Behrens III, W. W. (1970). *The limits to growth: a report for The Club of Rome's project on the predicament of mankind*. Potomac Associates.

National Research Council. (2011). Limiting future collision risk to spacecraft: an assessment of NASA's meteoroid and orbital debris programs, National Academies Press, Washington, DC. http://dx.doi.org/10.17226/13244.

OECD. (2019). The space economy in figures: How space contributes to the global economy. *OECD Publishing, Paris*. https://doi.org/10.1787/c5996201-en.

Reuters. (2020, March 20). *Apple to pay users $25 an iPhone to settle claims it slowed old handsets*. The Guardian. https://www.theguardian.com/technology/2020/mar/02/apple-iphone-slow-throttling-lawsuit-settlement

Ryan-Mosley, T., Winick, E., &Kakaes, K. (2019, 6 26). *The number of satellites orbiting Earth could quintuple in the next decade: The coming explosion of constellations*. MIT Technology Review. https://www.technologyreview.com/2019/06/26/755/satellite-constellations-orbiting-earth-quintuple/

Senge, P. (1990). *The Fifth Discipline: The art and practice of the learning organization*. New York: Doubleday.

Simon, H. (1997). *Administrative behavior: A study of decision making processes in administrative organizations* (4th ed.). The Free Press.

Space Safety Coalition. (n.d.). *About*. https://spacesafety.org/about/

Space Safety Coalition. (2019, September 16). *Best practices for the sustainability of space operations*. https://spacesafety.org/best-practices/

Tenembaum, T. J. (1998). Shifting paradigms: From Newton to Chaos. *Organizational Dynamics, 26*(24), 21–32. https://doi.org/10.1016/S0090-2616(98)90003-1.

UN COPUOS. (2018). *Guidelines for the long-term sustainability of space activities, UN Committee on the peaceful uses of outer space*. https://www.unoosa.org/res/oosadoc/data/documents/2018/aac_1052018crp/aac_1052018crp_20_0_html/AC105_2018_CRP20E.pdf

Undseth, M., Jolly, C., &Olivari, M. (2020, April). *Space sustainability: The economics of space debris in perspective*. OECD. https://www.oecd-ilibrary.org/docserver/a339de43-en.pdf

Union of Concerned Scientists. (2020, August 1). *UCS Satellite Database In-depth details on the 2,787 satellites currently orbiting Earth, including their country of origin, purpose, and other operational details*. Union of Concerned Scientists. https://ucsusa.org/resources/satellite-database?_ga=2.206523283.1848871521.1598077135-464362950.1598077135

University of Southampton. (2017, April 19). *Biggest ever space debris study highlights risk posed by satellite 'mega-constellations'*. https://www.southampton.ac.uk/news/2017/04/space-debris-mega-constellations.page

Vallero, D., & Brasier, C. (2008). *Sustainable design the science of sustainability and green engineering*. Wiley. ISBN: 9780470130629

Wattles, J. (2020, October 8). *Space is becoming too crowded, Rocket Lab CEO warns*. CNN Business. https://edition.cnn.com/2020/10/07/business/rocket-lab-debris-launch-traffic-scn/index.html

Wiener, N. (1985). *Cybernetics, or control and communication in the animal and the machine* (2nd ed.). The MIT Press. ISBN: 0-262-23007-0

Chapter 2
From the Whiteboard to Space

One man's magic is another man's engineering.
– Robert A. Heinlein

Abstract All things we create with our ingenuity go through a life cycle, even if not formally defined. This cycle is a mixture of divergence and convergence, construction and destruction, order and disorder, in cycles that are visited and revisited, over and over. Nothing *engineered* is automatically nor magically created, but incrementally realized by the execution of a variety of activities, bringing objects from just abstract ideas to an operative device, which performs a job to fulfill a need. To achieve this, a set of factors must be surely taken care of: the business factor, the social factor, and the technical factor. In order to organize the huge task of turning rather fuzzy ideas into an operable space system on scarce resources and without clear requirements, NewSpace organizations must carefully break the work down into smaller pieces they can manage, and then put the pieces back together. The social interactions while designing those systems, and the inescapable connection between the way people interact and the systems they design, shape the journey from beginning to end.

Keywords Life cycles · Work breakdown structure (WBS) · Analysis · Synthesis · Proof-of-concepts · NewSpace business · Pre-revenue engineering

Eagerness to jump into details beforehand is one of the toughest urges to tame for a design engineer. Learning to *hold our horses* and taking the time to think before acting is a skill that takes time to develop; it perhaps comes from experience. Engineering is an evolving, iterative craft; sometimes a slow craft, definitely time dependent. This chapter is about understanding how things evolve from pure nothingness, including the triad of factors that are critical for any engineering adventure: the business, the social, and the technical factors. Factors that are connected to each other and greatly shaped by one valuable (and scarce) resource: time.

All things we create with our ingenuity go through a life cycle, even if not formally defined. This cycle is a mixture of divergence and convergence, construction and destruction, order and disorder, in cycles or waves that are visited and revisited, over

I. Chechile, *NewSpace Systems Engineering*,
https://doi.org/10.1007/978-3-030-66898-3_2

and over. Nothing *engineered* is automatically nor magically created, but incrementally realized, by the execution of a variety of activities, bringing objects from just abstract ideas to an operative device, which performs a job to fulfill a need. For some strange reason, bibliography tends to carve in engineers' brains a discretized image of this lifecycle; a sequence of well-defined stages that have their own names, lengths, and activities. And here lies one of the main misconceptions about life cycles, despite its obviousness: the life cycle of what we create flows in continuous time; we only discretize it for practical reasons. And the way to discretize it greatly varies depending on who you ask, or what book you read. Think about your own life cycle as you grow older: it is a continuous progression. We tend to discretize the typical human life cycle in stages like infancy, adolescence, adulthood, old age, and we expect certain things to happen at certain stages, for example, we expect someone in her infancy stage to go to daycare. Discretizing a life cycle is a technique to make it more manageable, but the system maturity evolves in a continuous manner. There is, though, a clear difference between an artificial system life cycle and a human life cycle, and that difference stems from the fact we humans develop as a single entity altogether (our bodily parts develop altogether), whereas systems can actually be developed separately in constituent parts that have their own lifecycle (which should maintain some level of synchronism with the general system), in a sort of a puzzle. These parts of the puzzle are made to fit and work together through several iterations. Each constituent part can also have an internal composition with parts also developed separately, so the nature of life cycles in a system is highly recursive. Discretizing a life cycle in stages is necessary as a way of organizing work better, but you can choose to discretize it in many different ways. Whatever the life cycle staging method chosen, it is quite important to maintain consistency across the organization which designs and builds systems. Life cycle stage names and scopes become very important as a planning and communicational tool, so commonality on this matter is a great advantage.

During different stages of the engineering design process, many key decisions are taken, and key tasks are performed. Another hidden fact with life cycles and their iterative and evolving nature is that we visit life cycle stages several times through the lifetime of the project, which slowly increases system maturity. Let us state it again: what this means is that we do not visit a life cycle stage only once; we visit it multiple times. It is just that the weight or intensity of some of the activities as we run through it changes as maturity of the work progresses. We may believe we have abandoned a stage for good only to find out at some other later stage that an assumption was wrongly made, forcing us to move the project "pointer" back to the beginning. Here is perhaps an interesting thing to point out: the way time progresses and the way the project "pointer" progresses are related but they do not always follow each other. Time only progresses forward, whereas the pointer can move back and forth, come and go, depending on the Systems Engineering *game*. Like we jump spaces in a board game depending on the gameplay.

A great deal of information is generated as the life cycle evolves. Is it of paramount importance to feed this information back into other incumbent stages for those stages to make better decisions, in a virtuous circle that must be ensured. Although life cycles are artificially split into chunks, the flow between the chunks must be kept as smooth and continuous as possible. Isolated life cycle stages mean a broken project which is only headed to failure.

When does the maturity of the design reach a level to consider the job done? Probably never; later on, we will see how subjective this can be. What we can only hope for is to define meaningful *snapshots*, which produce outcomes that match some specific goal for. For example, a representative prototype or proof-of-concept is a snapshot of a life cycle iteration. Prototypes provide invaluable insight to gain real-world understanding of the fitness of the product, but they can be costly and complex (sometimes as complex as the real thing). Reaching production grade is also a *freeze* of the life cycle work; it means the maturity is such that it can reliably replicate/instance the design and those instances will perform as initially planned. Life cycles are pervasive; everything we create needs time to be done, and we have the freedom to organize the work in whatever way suits us, and declare it *done* whenever suits us; no framework or methodology can dictate this for us. We, as designers, are in full control to decide how to do it. But with great power comes great responsibility: proper life cycle management requires coordination, communication, and control.

2.1 The Business Factor

If the business model is not totally fluid, an organization cannot probably be called a startup. That is of course a bit of a tongue-in-cheek way of saying that the business strategy is yet one more unknown to figure out as the whole project evolves in early stage organizations. The technical side of space is not the only side of it, of course. Space is a business, like any other business, and as such it requires the same thinking and planning as restaurants, or lemonade stands do. There are differences with those, for sure, but to turn an idea or project into a self-sustainable endeavor (as in, making people willing to pay for what you do), the Business Factor must be included as another important piece of the mechanism of growing a NewSpace organization. There are myriads of books about this, and it is by far out of the scope of this book to dive in business topics, but I considered it important to mention this before diving in more technical matters.

NewSpace altered the way of doing space in different ways. Not only about how to design and build the actual things that go to orbit in a cheaper, simpler, smaller, faster way, but it also changed the paradigm of doing business with space; it created a new type of market. Before NewSpace, the way of making revenues with space systems was reserved to a very small group of very heavyweight actors, which were building spacecraft for a limited amount of applications: mostly science and (largely) defense and surveillance, with governmental entities as the sole customers. NewSpace created new business cases, allowing data to reach a broader audience, including smaller private entities, and even particular customers, so serve their various needs for applications such as agriculture, mining, flood monitoring, Internet-of-Things, broadband communications, etcetera. Even though new business models appeared, the way NewSpace companies are stepping up to offer their products to broader audiences remains largely under discussion.

There is something about space that still amazes people, even though the first satellite was launched more than sixty years ago. There is still this sci-fi *halo* about it; it probably has to do with the fact they go up in such a spectacular way on board of those fancy rockets which makes it very hard for the layman or laywoman to take as something usual or normal. So, space business still does not look like selling shoes or insurance. But space *is* a business just like selling shoes or insurance. The sci-fi halo is a bit of a problem for NewSpace, or maybe a distracting factor more than a problem. Everyone wants to go to orbit, everyone wants to be involved in something related to going to space: investors, politicians, engineers, accountants, lawyers, you name it. This is great, since it brings to the industry a great deal of attention without having to make a great effort for it, which other industries would only envy. But the *space halo* can cloud important business questions that any new company planning to be profitable needs to have in mind, from the very beginning:

- What is the product?
- Why would people pay for it?
- Is this company offering a service or?
- How is it different from others doing the same?

The interesting bit is that business propositions (or *models*, even though that is a word I will refuse to use, and later reader will get to see why) totally evolve as the organization and the technical development evolves. It is too naive to believe we will manage to figure all those questions out when the company is founded and that they will remain unchanged forever. What we can only dream for is to define a baseline and then reality will calibrate us, in many different ways: probably the technical complexity of what we are aiming for will call for a change which will impact our business proposition, or our funding will change and force us to change the business strategy, or the market will take a turn and make us rethink the whole thing. In any case, the business questions are as important as the technical questions. Thing is, typically NewSpace organizations are founded by engineers, who are fonder of discussing architectures and design things than value propositions. As the project matures, the business questions will surface on an on, and an immature understanding of the business proposition will eventually become a problem. An immature understanding of the business also means a varying business vision within company walls. There must be a common vision inside those walls. Nothing worse than no one really knowing what the business and product is. That being said, the Product can (and will) change: what the product is today might not be what is going to be in six months. Things change, but it is good the business proposition is communicated and brainstormed properly.

2.1.1 Prototype the Business

There are lots of assumptions at early stages of a business. It is essential to test those assumptions as quickly as possible, since this can impact the decision making on the technical side. Many iterations are needed for this to converge. One way to get started is to make sure there is an internal alignment on what the business proposition is. When you are four, five, seven people overall in the organization, it is a great moment to make sure you are all on the same page about what you are doing. After all, making space systems is a lot of effort and sweat, so better to make sure everybody is aware where all that sweat and tears (of joy, of course…) are going to. This can spark a lot of discussions, and very healthy ones. It is critical to keep the business proposition very closely attached to the technology development; they will feed each other indispensable insight.

It is key to iterate, on and on, and adjust the business approach as it goes. But, when we have literally zero customers, it is not simple to prototype much since there is nobody out there to test the business strategy prototypes with. One way is to do some intelligence and identify a few prospective customers we would love to have in our clientele and reach out. When the project is too early, chances are rejection will be the norm, but that is fine. The intertwined nature of business and technology indicates that, as we gain momentum on the technology side, we increase our chances to engage someone as an early adopter of your proposition. But this must be done with some care; chances of ending up creating a product tailored for one specific customer increase. If early adopters align well to the overall strategy, all is good. In any case, it is important to make sure any early adopters do not influence the business and the technology roadmap in such a way we end up serving them with a highly customized thing no one else will want. For very early stage organizations, this is a tough situation because making revenues is every founder's dream.

2.1.2 The Business Blueprint

Just as blueprints serve as schematic clues for engineers to orient themselves and to communicate to each other while they design, the business strategy needs similar design and development. At very early stages, this takes the form of simple questions and concise answers, in a sort of internal questionnaire which is usually called Business Blueprint. The Business Blueprint provides clues about customers and audience, about how revenue will be made, and about partnerships needed for the project to succeed. The Business Blueprint raises awareness and helps us see the bits and parts that are required to create, capture, and deliver value. Questions may vary a lot for different businesses, and here we will tailor them for a NewSpace enterprise.

2.1.2.1 Customers

1. Question: Who are we serving?

2.1.2.2 Offer (Product Proposition)

1. Question: What product or service will we make and deliver?

2.1.2.3 Revenue Model (and Price)

1. Question: How will we make money?
2. Question: How and what will we charge our customers for?

Note: Thinking about pricing at very early stages can be challenging, since costs are still not fully understood and not properly broken down. One way to go is to assess what might be needed to cover the most important parts of your business, look at what competitors are charging for similar offers. Again, the revenue model will most likely change dozens of times as the company grows; still, important to revisit it over and over.

2.1.2.4 Channels

1. Question: How will we get our product to customers?

2.1.2.5 Value Proposition

1. Question: What customer need(s) are we fulfilling?
2. Question: What is the unique promise our offer provides?

2.1.2.6 Costs

1. Question: How much does it cost to create and deliver our offer?

2.1.2.7 Partners

1. Question: What key relationships will help us create and deliver our offer?
2. Question; What key technologies are out there we could add to our value chain?
3. Question: What big players or other NewSpace actors in the market are worth tapping?

Note: This point is important and often underrated. Bad partnerships at early stages can be very damaging; partnerships for the sake of PR or marketing are very bad ideas. That being said, good partnerships at early stages can really boost the path ahead. One effect NewSpace companies suffer from is the fact they're so small that they can fall way under in terms of priority for established suppliers. For example, if a small company needs an on-board computer from an established supplier X, this supplier X will always pay more attention to bigger customers in their clientele and leave the small guy all the to the end of the queue when it comes to attention, support, etc. To overcome this issue, NewSpace organizations can partner with other similarly small startups and help each other towards a partnership which works for both. There are risks involved with this, of course, since we are talking about two fragile organizations with perhaps immature products. The key is to select partners with clear tech competence and clear strategies and avoid falling into the trap of cheap prices from PR-oriented startups which mostly care about piling up letters of intents and shoot press releases.

2.1.2.8 Team, Talent, and Resources

1. Question: What skills/capabilities/resources will we need?

Young organizations should thoroughly brainstorm the business blueprint frequently, perhaps weekly or so. The questions the blueprint asks revolve around a certain number of concepts or topics to be figured out (Fig. 2.1). A more graphical take can be to lay out all these topics from the blueprint in a sort of cloud or list (more

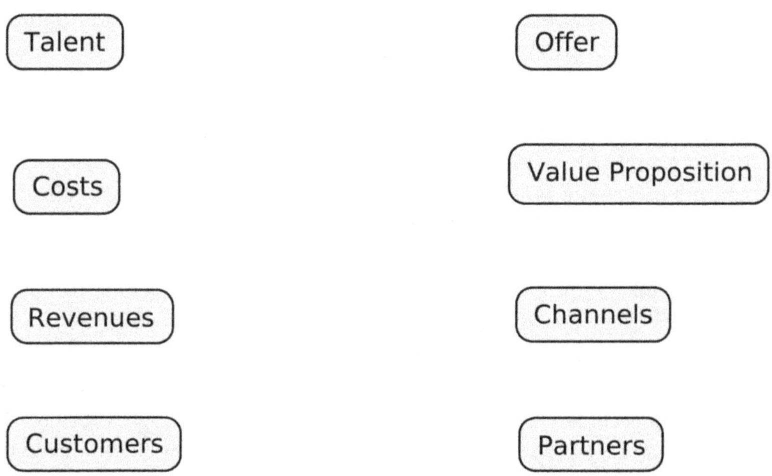

Fig. 2.1 There must be a clear alignment what goes next to each one of these boxes

about this in Chap. 3), and populate short answers next to the boxes, according to the organization's vision and strategy. This can be displayed on a whiteboard in a common hallway to make sure everyone will see it every morning; next to the coffee machine (if there is any) can be a good spot.

2.2 The Social Factor—Connecting Brains Together

The last question of the business blueprint makes a nice connection to this section. Tom DeMarco stated decades ago in his always evergreen *Peopleware*: "The major problems of our work are not so much technological as sociological in nature" (DeMarco and Lister 1999). Understandably, when an early stage small company is running on paper-thin resources and short runways, probably one of the last things it will think about is sociology. The focus will understandably be on coding, architecture, supply chain, schedule, budget, and so on. Still, even if priorities may seem distant from minding sociological factors, the group of people an organization has assembled is a network of brains which need to cooperate to achieve a goal. Teams can be considered, at the end of the day, a cooperative nervous system. The Social Factor is that intangible kind of *magic* that makes a team overachieves even if they have limitations, gaps, or constraints. This social factor, when properly addressed, is what glues a team together and converts it into something else, usually driving startups to achieve what others have failed to achieve. What needs to be done for that alchemy to take place? It is not exactly about finding technical geniuses, but more fundamentally about nurturing some drivers which can definitely increase the probability of making a team bond: equality, diversity, trust, open communication, humor and a very strict sense of aversion to bullshit. All these mixed up make up a potion that can spark a great sense of determination toward a grand goal, no matter how bold the goal might be. Let us quickly elaborate on these drivers:

- **Equality**: Structure is flat. Team has similar compensation and titles. No elites are created in the core team, they are all equals, they feel as equals.
- **Diversity**: Heterogeneity is a team asset and is linked to better team performance (Hackman and Oldham 1980). Diversity also limits the risks of groupthink (Hart 1991), which is a condition where the pressure to keep harmony hinders the ability to evaluate alternatives. Nationality, race, gender, political diversity can enrich the workplace and provide different perspectives for problem-solving.
- **Trust**: It means dependability among team members. It is grown with time and effort, by keeping promises and leading by example, in such a way the team members can feel they have their backs protected. Managers must turn trust into a religion. Managers need to provide enough autonomy to team members and avoid micromanaging and overshadowing. A manager too visible is never a good sign.
- **Open Communication**: No secrecy, no compartmenting of information, no hiding. Feedback is always welcome, even if you would not like it.

- **Shared Aversion to Bullshit**: A team-wide, management-sponsored appreciation for the truth, and rejection of half-truths. Prevalence of evidence vs speculation. Prevalence of emails/messages vs long pointless meetings without clear agendas. It is important to note that the fight against bullshit is a tiring one and requires an active mindset against it, otherwise it will eventually take over. We will cover bullshit in more detail at the end of the book.
- **Humor**: This could sound irrelevant in this list, but it is one of the most relevant ones, if not the most relevant. Startup world can be cruel. At times you can feel all you get from the outside world is hostile disbelief and discredit. Tough times are always around the corner, and a pessimistic attitude will not make things any better. In the darkest hours, cheerfulness and a great sense of humor is an asset, something that should be fostered and encouraged.

Other factors as a quiet office space, amenities, good food and free drinks are always welcome, but those are not team bonding factors. A team that can concentrate in peace (capable of reaching *flow*) is always a good thing, but a truly jelled team will perform even without all that comfort.

In early stage NewSpace organizations, once the dust from the founding *big bang* has settled, the core team is called to duty. This core team is usually small, of maximum five or six individuals. The challenge with the core team is not (only) about finding educated and experienced engineers and putting them in a room; the real challenge is to find people with the right personality and culture fit for the challenge ahead and make them play like a well-tuned orchestra. Bonded, healthy core teams are yet another practical example of the Systems Thinking "mantra": the whole is greater than the sum of its parts.

With flat structures come some challenges related to the decision-making process. A core team that lacks a clear single point of authority (e.g. a Systems Engineer) will have hard times at deciding and eventually get itself into an analysis paralysis state. A single tech manager is instrumental at this stage to decide the steps forward. When the technology maturity is low, every path will look almost equally reasonable to take, every methodology and/or framework will look like the best to adopt, so it is the task of the more experienced manager to guide the team while taking those steps, and often this means showing the way by means of real storytelling from experience, or some proof-of-concept, prototype, analysis, or similar. The manager in charge of a core team who cannot roll up his or her sleeves and go to fight in the trenches, or gather the team to share a story and elaborate why a particular path is wrong should probably find something else to do. But fighting from the trenches does not equally mean being one more of the core team. The manager is the captain, and as such she must remember and remind everyone (by actions, not by words) her authority position and for that is necessary to keep a reasonable emotional distance from the team. This is fundamental for the team members to perceive the manager as the shock absorber and the solver of otherwise endless deliberations. During those times of uncertainty, common sense and simplicity should be policed by everyone in the team, including the manager (more about that later). If common sense is jeopardized in any way, there should be a healthy environment in place for anyone to be able to call it out. The sum being greater than the sum of its parts is no magic: it is achieved by making the team members love doing what they do. Easy to say, but not so simple to do. How do you make team members not only like it but actually *love* it? Engineers are often challenge-thirsty; the feeling of working on things no one else would dare embarking on can be very motivating. Good thing is that NewSpace orgs get easily in trouble, so challenges are probably the only thing that is not scarce around them. The manager should quickly define bold goals to keep the team engaged. For example: "*In two years we will be a major player in the global Debris Removal industry*". Or, more quantitatively: "*In two years' time we will have 20% of the Debris Removal market share*" to use some example. Clear, concise, bold, and reasonably long-term goals are good ways of making a team to create a common vision towards the future. Such vision for *jelled* teams, using Tom DeMarco's term from (DeMarco and Lister 1999), creates strong identities; the team perceives themselves as a pack and will defend and protect other members, come what may. The core team is a very special collective, and it should be composed of people who in general can do many things. Most of these team members will then become IPT leaders or Systems Engineers or similar once the company gets to grow.

Early stage design is more suited for generalists than specialists. This is no wonder; design is too coarse at the beginning for any specialist to be of relevance; at this stage, generalist minds are of more importance. If the generalists can still have one more specialized side where they could do some detailed design, that's a jackpot for the organization.

Core teams can be prone to burnout and turnover, if not properly managed. At times, working for this type of companies can get tough: team's members wear too many different hats; everybody does a lot of different things, change is the norm, budgets are thin; stress levels can go high. It can very easily go into a self-reinforcing feedback loop where budgets, schedule and stress can combine in a way people might stop enjoying what they do and reach psychological saturation. It is important for managers to understand that and make sure that the engineers enjoy waking up in the morning and commuting to work. Managers should continuously scan the team humor and body languages to gauge overall morale. How people walk speaks about their humor, how people behave during lunch provides hints of their mental states. Managers need to remain close to the team, but probably not as close. Team members can be friends and go out together, and it would be great if they do. The manager, on the other hand, should keep a reasonably safe emotional distance to the team, to avoid breaking the trust balance (of course the boss can join the outings, but not all of them), but also to have the right autonomy to make tough calls. One team member perceived by the rest of the team to be too close to a manager could be interpreted in the team as a member with influence to the boss. Mistrust spreads quickly and is very hard to reconstruct. Team trust toward the boss and close friendship are not as connected as it may seem. Team trust toward the boss is more a function of crystal-clear communication, honesty, full disclosure, and a sense of equality, despite the hierarchical position. Bosses who show elitist behaviors, for example by receiving too many benefits the team members (who ultimately are the ones actually doing the work) do not get, end up creating barriers which can be significant to get team's buy-in in difficult times. Managers' arses are the ones permanently on the line, and they deserve a right compensation for it, but a too thick of a gap with the team can severely affect their leadership. Manager's biggest goal is to get the team to jell. Overachieving tech teams never jell by tossing corporate goals to them; they'll go: meh. Competent tech teams jell by giving them challenges and a sense of autonomy and freedom to solve intricate technical problems; that is why drives them. Managers thinking they will get tech teams to jell by creating cringey social happenings or sending the team to go-kart racing miss the point and are just naive. Tech teams jell by doing tech; by coding, designing, figuring things out. They do not jell by keeping them sitting in a video call being lectured. Managers should give them a challenge impossible to solve, or a lost cause. They will die trying to sort it out and become a pack in the process.

As the organization grows, most of the core team members will start leading people (probably IPTs) on their own and gain more autonomy. Eventually the core team will disband in favor of a different team structure that will be put in place to support the company growth and the system architecture evolution. The manager of the core team has the unique possibility of transitioning into an integrating role between the spawned groups; the fact he/she has been working many years and in tough times with the members of the core team (which now lead their own teams) creates a very strong leadership fabric which is a great asset for the organization. This is an instrumental moment for the organization to avoid using the typical lonely middle-manager role, who leads other subsystem managers but has no team (he or she is just a layer between executives and the IPTs). The chance to spawn the management fabric as the architecture evolves brings the unique advantage the trust and bond between the different levels is preserved.

Last by definitely not least, a word on sense of humor. As said, young NewSpace projects are tough environments. An illustrative analogy that has been proven to be quite useful is to think about early stage NewSpace startups like a submarine on a mission. Few environments are tougher than a submarine. Readers may be surprised how much these military machines run on "soft" leadership skills. In submarines, teams are small and constrained, living in uncomfortable quarters for long times. To make it easier for the crew, there is no substitute for cheerfulness and effective story-telling. In fact, naval training is predicated on the notion that when two groups with equal resources attempt the same thing, the successful group will be the one whose leaders better understand how to use the softer skills to maintain effort and motivate. Cheerfulness counts. No one wants to follow a pessimist when things are looking ugly out there. And things do not look good quite often in startups, and in submarines. Navies assiduously records how cheerfulness counts in operations. For example, in 2002, one of Australia's Royal Navy ships ran aground, triggering the largest and most dangerous flooding incident in recent years. The Royal Navy's investigating board of inquiry found that "morale remained high" throughout demanding hours of damage control and that "teams were cheerful and enthusiastic", focusing on their tasks; 'sailors commented that the presence, leadership, and good humor of senior officers gave reassurance and confidence that the ship would survive.' (St. George 2013). Turning up and being cheerful, in other words, had a practical benefit.

It has long been understood that cheerfulness can influence happiness at work and therefore productivity (Oswald et al. 2009). A cheerful leader in any environment broadcasts confidence and capability. In a submarine it is the captain, invariably, who sets the mood of the vessel; a gloomy captain means a gloomy ship, and mood travels fast. Cheerfulness affects how people behave: you can see its absence when heads are buried in hands and eye contact is missing (St. George 2013), and cannot expect to remediate this by taking your team to laser tags or an escape room. You set the tone by showing humor and cheerfulness when things are looking bad and prospects are not nice. Conversely, empty optimism or false cheer can hurt morale. If you choose to be always uber-optimistic, then the effect of your optimism, over time, is reduced. Shit will happen, no matter what; a good sense of humor will not solve it but make it easier. Managers' true nature is seen during storms, not while sailing calm waters.

2.2.1 The Engineering Culture

> It is a great profession. There is the fascination of watching a figment of the imagination emerge through the aid of science to a plan on paper. Then it moves to realization in stone or metal or energy. Then it brings jobs and homes to men. Then it elevates the standards of living and adds to the comforts of life. That is the engineer's high privilege. The great liability of the engineer compared to men of other professions is that his works are out in the open where all can see them. His acts, step by step, are in hard substance. He cannot bury his mistakes in the grave like the doctors. He cannot argue them into thin air or blame the judge like the lawyers. He cannot, like the architects, cover his failures with trees and vines. He cannot, like the politicians, screen his shortcomings by blaming his opponents and hope the people will forget. The engineer simply cannot deny he did it. If his works do not work, he is damned… On the other hand, unlike the doctor, this is not a life among the weak. Unlike the soldier, destruction is not his purpose. Unlike the lawyer, quarrels are not his daily bread. To the engineer falls the job of clothing the bare bones of science with life, comfort, and hope. No doubt as years go by the people forget which engineer did it, even if they ever knew. Or some politician puts his name on it. Or they credit it to some promoter who used other people's money… But the engineer himself looks back at the unending stream of goodness which flows from his successes with satisfactions that few professions may know. And the verdict of his fellow professionals is all the accolade he wants.
>
> Herbert Hoover

Engineering is much more than a set of skills. It consists of shared values and norms, a special vocabulary and humor, status and prestige ordering, and it shows a differentiation of members from non-members. In short, it is a culture. Engineering culture has been well studied by both social scientists and engineers themselves. Researchers agree that there are several distinguishing features of the culture that separate it from the cultures of specific workplaces and other occupational communities. Although engineering cannot sustain itself without teamwork, it can be at the same time an individual endeavor. Engineers routinely spend long hours at a workstation trying to figure things out on their own, forming intimate relations with the technologies they are tasked to create. Because work is so often done individually, engineers engage in seemingly strange rituals to protect the integrity of their work time. For example, it is typical for an engineer to stay at the office or laboratory late into the night or to bring work home with them just so they can work in solace and without distraction (Perlow 1999).

2.2.2 *But What Is Culture?*

Culture is a property of groups, and an abstraction for explaining group behavior (Schein 2004). The definition of "culture" is one of those terms (like quality) that everybody uses in slightly different ways. The following are commonly used definitions of culture.

1. Culture is a set of expressive symbols, codes, values, and beliefs. These are supported by information and cognitive schemas and expressed through artifacts and practices (Detert et al. 2000).
2. Culture is a shared pattern of basic assumptions learned by a group by interacting with its environment and working through internal group issues. These shared assumptions are validated by the group's success and are taught to new members as the "correct way to perceive, think, and feel in relation" to problems the group encounters (Schein 2004).
3. Culture is in the interpersonal interactions, shared cognitions, and the tangible artifacts shared by a group (DiMaggio 1997).

These definitions share the common features of identifying culture through properties, tangible, and intangible, that represent shared thoughts or assumptions within a group, inform group member behavior, and result in some type of artifact visible to members outside the group (Twomey Lamb 2009). These features are influenced by a group's history, are socially constructed, and impact a wide range of group behaviors at many levels (e.g. national, regional, organizational, and inter-organizational) (Detert et al. 2000). Culture can also be considered at smaller levels. For instance, a team may have its own subculture within an organization: heavily influenced by the overall organizational culture but nuanced by the individuals and experiences on a given team. Within a team, much of the tone is set by the team leader and those who have been with the team the longest. Once established, a group's culture is tempered by shared experiences and by the past experiences of those who later join the group, bringing with them new beliefs, values, and assumptions (Schein 2004). In an engineering context, this means a team's culture impacts its creativity, problem-solving, and ability to generate new concepts (Harris 2001). In fact, group norms, one of the characteristics of culture, are key to group performance (Hackman and Oldham 1980). However, efforts to alter group norms can be confounded by culture. New behaviors or processes introduced to a group will fail to catch on if they go against the prevailing culture (Hackman and Oldham 1980). This is because one characteristic of culture is its stability within a group (Schein 2004). The formation of culture begins with the formation of a group, and mirrors the stages of group formation:

forming, storming, norming, and performing (Tuckman and Jensen 1977). Once a group is formed, group norms begin to develop through conflicts, attempts to achieve harmony, and the eventual focus on a mutual goal throughout the execution of which the team matures, adapts, and innovates, constantly testing and updating its behaviors, assumptions, and artifacts (Schein 2004). While culture is a powerful predictor of group behavior, it can also be a barrier to the introduction of new methods, tools, and processes (Belie 2002). However, culture can also be a motivator for change. So-called "cultures of change" empower members to seek out new methods and ideas to solve problems (Twomey Lamb 2009). Organizational culture is a contributor to team success. Because trust is at the base of successful interactions, organizations can emphasize positive team norms and create a cultural context that supports team success by fostering and sustaining intellectual curiosity, effective communications, and the keeping of thorough documentation (Goodman 2005).

2.3 The Technical Factor

We design in order to solve a problem. But there are multiple ways of solving a problem, as there are many ways *to skin a cat*. Regardless of the paths we choose, everything we engineer boils down to a combination of analysis and synthesis, decomposition and integration; this is the true nature of Systems Engineering: facilitating the decomposition of the problem into bits that we realize on their own, and then we integrate all together. The terms analysis and synthesis come from (classical) Greek and mean literally "to loosen up" and "to put together" respectively. These terms are used within most modern scientific disciplines to denote similar investigative procedures. In general, analysis is defined as the procedure by which we break down an intellectual or substantial whole into parts or components. Synthesis is defined as the opposite procedure: to combine separate elements or components in order to form a coherent whole. Careless interpretation of these definitions has sometimes led to quite misleading statements—for instance, that synthesis is "good" because it creates wholes, whereas analysis is "bad" because it reduces wholes to alienated parts. According to this view, the analytic method is regarded as belonging to an outdated, reductionist tradition in science, while synthesis is seen as leading the "new way" to a holistic perspective. Analysis and synthesis, as scientific methods, always go hand in hand; they complement one another. Every synthesis is built upon the results of a preceding analysis, and every analysis requires a subsequent synthesis in order to verify and correct its results. In this context, to regard one method as being inherently better than the other is meaningless (Ritchey 1991). There cannot be synthesis without analysis, as well as we can never validate our analyses without synthesis.

For the tasks we must perform while we design, we must formulate them before executing. There is an old adage that goes: *it is not the idea that counts—it is the execution*. We can poorly execute an excellently thought idea, as well as we can poorly think something we execute perfectly. For anything we want to achieve, it is always about thinking before doing. We cannot expect proper results (proper as in, close to the desired goal) out of no planning or no thinking. It's like taking a bus before you know where you want to go.

Whatever the problem is that we are trying to solve, we must take time to analyze what needs to be done and describe to others involved with us in the endeavor the *whys* and *hows*. Time is always a constraining factor: we cannot take too long in performing analysis, we always need to move to execution otherwise everything just halts. To create technical devices from scratch, like spacecraft, we must start by identifying a need, or an idea, which triggers a set of activities. More knowledgeable and mature companies usually work from an identified need (using techniques as market research, focus groups, customer feedback, etc.), but startups usually work solely on abstract ideas, since they do not have the luxury of a clientele, often not even a single customer. For whatever *thing* we need to do in our lives, be it taking a shower, a train, or having lunch, we go through a sequence of steps where we first define a goal, and then we perform a set of activities to achieve such goal; we do this all the time. Most of the time subconsciously, some other times consciously.

The analysis–synthesis perspective is probably the highest level of abstraction we can put ourselves in order to observe how we design things. Both sides offer a lot of internal content to discuss. As engineers we are put in a quest where we must turn an abstract concept into a device or set of devices that will perform a joint function which will hopefully realize the idea or fulfill the need which triggered all that. Think about it again: something that is a thought in our brains needs to turn into an actual thing which must perform a job in a given environment. What type of device or system will fulfill the need is our decision. For example, picture someone having the idea of providing Earth Observation data every hour of every single point of the globe: does that mean the system to develop needs to be spacecraft? It could be drones, or balloons. Most likely the answer will be satellites, because of operational or cost factors, but still it is a choice we make from a variety of other technically feasible options. The solution space offers different types of alternatives, it is up to us to pick the best up. Having many options on the table tends to make the decision-making process harder compared with having one single option. For example, think about an idea of selling a service for orbital debris removal. In this case, the solution space about the artifacts to choose shrinks dramatically; it must be a spacecraft. It is important, during the analysis phase, not to *jump the gun* into defining technical solutions for the problems we have. Keeping our minds open can lead us to alternatives that are better suited than what past experience could dictate.

Once we decide what type of artifact we must realize for the problem we must solve, either be a spacecraft, a piece of software, or a nuclear submarine, it becomes our design subject; at this stage it is more of a wish than something else (a collective wish to be more precise). We must then dismember it into fictional bits, chunks, or parts that we declare our design subject is made of, even without being anywhere sure what those bits and chunks will effectively work together in the way we initially wished for. Designing requires a quote of betting as well. Eventually, the moment of bringing the bits together will come, and that is precisely the objective of the synthesis part: realizing and connecting the pieces of the puzzle together and checking (verifying) that the whole works as the analytical phase thought it would. In any project, the time comes when analysis meets synthesis: a.k.a reality kicks in, or what we planned vs what we came up with.

Here is one key: analysis works on *fictions* without operative behavior (requirements, models, diagrams, schematics), whereas synthesis works with a more concrete or behavioral reality (either literally tangible objects, or other more intangible ones such as software, etc.). Analysis and synthesis belong to different domains: the former to the intangible, the latter to the tangible and executable which can be operated by a user. During the process, analysis and synthesis feed each other with lots of information. This feedback helps us adjust our analytical and synthetic works by perceiving, interpreting, and comparing the results from our initial wishes and needs. Is the new whole performing as expected? If not, do more analysis, do more synthesis, verify, repeat. If yes, enjoy. The circular nature of the design process clearly shows a "trial and error" essence at its core. We think, we try, we measure, we compare, we adjust (with more thinking), we try again. Engineering activity requires a constant *osmosis* between analysis and synthesis, or, if we want to continue stealing terms from biology: a symbiosis. Both need each other and complement each other. In more classic engineering, the analysis and synthesis activities can run long distances on their own. Extended periods of analysis can take place before thinking about putting anything real together. NewSpace changes that paradigm, by shortening considerably this cycle, and coupling both activities closer together. In NewSpace, you cannot afford to analyze things too much; you need to put the puzzle together as fast as you can, and if it does not look right, you try until it does.

Now, revisiting the "think before doing" bit from some paragraphs before, it is fair to say we perform many tasks without consciously analyzing or planning them. The thinking bit happens without us realizing much. You do not write a plan or a Gantt chart for taking a shower or walking the dog. For such things, the stages of planning and acting are subconscious. We can do many things, repeatedly cycling through the stages while being blissfully unaware that we are doing so. Pouring water in a glass requires all the stages described above, yet we never consciously think about them. It is only when something disrupts the normal flow of activity, that conscious attention is required. Most behavior does not even require going through all stages in sequence; however, most activities will not be satisfied by single actions. Why do we have to plan so much when we design and engineer things, and why do we *not* have to plan so much when we do our everyday stuff? When is it that the analytical planning crosses to the conscious side? Here is when we need to start talking about a few concepts that are central to the way we execute tasks: collectivity, uncertainty, and atomicity (of actions). In short:

- Collectivity: how many actors are involved for doing something dictates the need of coordination and explicit planning. We cannot read minds just yet, so we need to capture in some shareable format our ideas and plans. The smaller the number of actors, the simpler the coordination needed.
- Uncertainty: how much we know about the task we are planning to do. The less we know about an activity, the more conscious planning needs to be done.

- Atomicity: how indivisible a task is. The more complex the composition, the more effort is needed to be understood and managed.

When all these factors combine, planning must be very explicit and clear, otherwise the probability of reaching a desired goal decreases consistently. Take a very simple task or thing to do: you want to take a walk in the park; just you, alone. If this action involves only you and nobody else, all the planning will be mostly introspective; yes, your wish of taking a walk could be reasonably conscious, but you will plan it without really thinking a lot about it, nor leaving a very substantial trace behind. Think about the last time you decided to go for a walk by yourself, and what evidence that action has left behind. Think now about a walk in the park with a friend. The action is the same, nothing has changed, but now you need to coordinate with your friend to meet at the right park at the right time. Your friend might or might not know where the park is, you will have to explicitly send messages, directions, timeline, etc. Think about now the last time you went with a friend for a walk and the trace it has left behind: at least, for sure, some messages, and hopefully memories. Collective actions require conscious analysis and formulation, execution and measure in ways all actors involved can understand what is to be done and converge toward a common goal. Even for tasks we know perfectly well how to perform! Certainty (or the lack of) about an action is the other key factor, which imposes the need to make the formulation and execution explicit and coordinated. If you are trying a new park to walk to, you will have to check the maps, you will have to write down directions, or print something, even if you are going alone. Our lack of knowledge about something increases the need for more detailed planning. As for atomicity, an action is atomic if you cannot divide it any further. Taking a walk in the park by yourself is fairly atomic if you want. But think about for example going to the supermarket: there is nothing "unknown" about it (you know how to do it), there is no collectivity (it's just you), but it is not an *atomic* action, since you probably need to fetch plenty of elements from the shop. You might be able to remember four, five, or seven items top, but anything more than that you will have to create a list for you to remember.

When we engineer things such as space systems, there is a combination of all that: collectivity, uncertainty (lots of) and of course non-atomicity: tasks are composed of other subtasks, sub–subtasks, and so forth, in a flow down structure. Hence, whatever we plan to design for space, to increase its probability of success we will have to tread carefully through a very explicit lifecycle, from idea to realization, mindfully organizing and planning our activities by means of stages and breakdown structures.

Engineering literature is known for calling the same thing many ways, and life cycle stages are no different. In the space industry, NASA has popularized the seven-stage life cycle model (phase A to F) (NASA 2007). Here, we generically define a set of six stages:

- **Conceptualization**

- Identify needs
- Define problem space

- Characterize solution space
- Explore ideas and technologies
- Explore feasible concepts
- Identify knowledge blind spots

- **Development**

- Capture/Invent requirements
- Decompose and break down:

  Capture architecture/system decomposition
  Cost Analysis
  Work decomposition
  Org decomposition
  Align all above under the Work Breakdown Structure

- Capture knowledge
- Integrate, prototype, test.

- **Production (Construction)**

- Replicate the design into multiple instances, which can reliably perform

- **Utilization**

- Operate the instances created

- **Support**

- Ensure Product's capabilities are sustained

- **Retirement**

- Archive or dispose the Product.

There are multiple ways of subdividing the way things are done. It is an arbitrary decision to choose a way of partitioning the schedule. It was discussed before that maturity milestones in system design are basically "snapshots" we take to the work whenever we consider it has met a criterion to be considered a milestone. Reaching production readiness follows the same principle; we iterate the work, we gate the schedule with a review, and if the review is passed then the snapshot can be considered production ready. The word "production" in consumer markets usually means mass replication of the design specification; i.e. creating multiple instances or copies of the design, which will operate according to specifications. In NewSpace, there is no such thing as mass production. Spacecrafts are usually manufactured in small scales, due to their complexity and handcrafted nature. For NewSpace projects, reaching production readiness can be misleading, depending on what the company is offering. If the startup offers, for example, spacecraft platforms off-the-shelf, then the production readiness concerns mostly the space segment. But more complex approaches require making sure that the *big system* (the spacecraft plus all the rest of the peripheral systems that must accompany the space segment to provide the global function) are production ready as well. The "concept" and "development" stages are where the time NewSpace companies spend the most. The development stage can produce an operable system as an output, but this system should still not be qualified as "production" ready.

2.4 The Breakdown Mess

Problem-solving requires us breaking things down into a variety of smaller, more manageable things. These constituent parts of a problem are not isolated from each other, they require some level of connection or linking, in a sort of network. To make things more interesting, the problems we are ought to solve are seldom isolated from the rest of the world. We usually deal with a set of connected problems; what management science calls *messes*. In this decomposition process, we assign to that network of decomposed *bits* a structure that often appears to be hierarchical, where we categorize and group things together. It is like forming a small cabinet in our brains where we have little drawers where we put stuff in, and we take stuff out. But do our brains work in this folder-like manner? The file cabinet approach makes it difficult or impossible to map or reuse the same piece of information in multiple drawers. Each time a change is made to any given file, it must be tracked down and updated in every location in which it exists. This leads to redundancy, with a cluttering of near-identical ideas, and significant work any time a system-wide change is required. It could not be overstated how critical, and ubiquitous hierarchies are in engineering design.

> It is by breaking (as in, decomposing) things down the wrong way that we can break (as in, destroy) projects down beyond repair; i.e. bad breakdowns can break everything down.

Not only the activities and tasks need to be broken down but also the functional and physical organization of the device-to-be as well. Practically every technical object around us is a container: objects composed of other objects, composed of other objects. We commented that during analysis, we work with fictional representations on how things will look like, and one key depiction is the architecture. Architectural diagrams allow for an organized sharing of both business and technical information related to a project and product. Architectural depictions are a cornerstone of the systemic approach: a system is ultimately a collection of relationships, interfaces, form, and function. The essence of architecting is: structuring, simplification, compromise, and balance. Architectures are a powerful graphical tool for representation of project aspects, thus capturing the imagination of the people's mind and streamlining communication. Architecture diagrams are the main consequence of the analytical phase. Analysis requires us to break things down by creating a tree-like set of (parent-child) relationships of objects, which in turn expand to many branches themselves. Breaking down means to apply a "divide and conquer" approach. Challenge is that this analysis phase produces many parallel breakdown structures, and not just one. For example, cost analysis, bill of materials, labor, product, even org charts. The recursive nature of the systems we design doesn't really help; this means there are breakdowns inside the breakdowns, and so on. We don't have the luck of dealing with only one breakdown structure; we deal with many of them in a recursive way. Keeping them all aligned is a continuous activity throughout the project life cycle, which must never be abandoned.

2.5 Work Breakdown Structure

The WBS is probably the one thing that NewSpace must take, or borrow, from classic space. A project without a WBS is a loose project. The WBS is probably the mother of all breakdown structures during the lifetime of a project. When a team needs to develop anything, be it a smartphone, a spacecraft, or a nuclear submarine, one of the first steps after collecting requirements is to come up with the hierarchical discretization on how the work will be done. The WBS is a structure, which contains other structures, like PBS/SBS (Product BS/System BS), CBS (Cost Breakdown Structure) and even OBS (Organizational Breakdown Structure). The WBS glues all those structures together, and it must be kept this way throughout the lifecycle of the project. The WBS also acts as a bridge between Systems Engineering and Project Management. To understand the importance of the WBS in devising space systems, NASA has a specific handbook about Work Breakdown Structures. The handbook states (Terrell 2018, 2):

> A WBS is a product-oriented family tree that identifies the hardware, software, services, and all other deliverables required to achieve an end project objective. The purpose of a WBS is to subdivide the project's work content into more tractable segments to facilitate planning and control of cost, schedule, and technical content. A WBS is developed early in the project development cycle. It identifies the total project work to be performed, which includes not only all inhouse work content, but also all work content to be performed by contractors, international partners, universities, or any other performing entities. Work scope not contained in the project WBS should not be considered part of the project.

The WBS includes not only the technical aspects of the work to be done to develop the space system, but all the rest as well, including logistics, facilities, etc.

In short: Work not specified in the WBS does not exist.

To organize work better, NASA WBS handbook proposes a numbering convention for each level in the hierarchy. Even though this is just a suggestion (you can choose your own), it is a good start to define work items unambiguously. The handbook also limits the amount of levels in the WBS up to seven. The reason for these numbers is not pure magic, but the fact research indicates our capacity of processing hierarchical information beyond depths of seven plus minus two is limited (Miller 1994). Equally important, the naming convention for the system subdivisions is needed: this avoids confusion and solidifies a strong common terminology across the organization. There are numerous terms used to define different levels of the WBS below the topmost system level. An example the reader can use is: subsystem, subassembly, component, equipment, and part. It should be noted that these entities are not necessarily physical (i.e. allocated to a specific instance), but they can also be functional. It is recommended that beyond system level (L2), all the entities are referred in functional ways, also known as functional chains. This way, a functional chain gets its entity in the WBS without locking it to any specific physical allocation. For example, two functional chains are assigned to a spacecraft mission: Attitude Management and On-board Data Handling. A block is assigned to each one of these chains in the WBS, with their own internal structure. Design activities could either decide to allocate both functional chains into one physical Computing Unit (for example using a hypervisor, or simply in two different threads), or allocate these two into two different physical computing units. Using functional chains in the WBS from L3 and below makes sure this decision, whenever taken, will not greatly affect the organization of the work overall. That decision will impact how the pSBS (Physical System Breakdown Structure) or how the As Built or Bill of Materials will map to the WBS, but not more than that. The Work Breakdown Structure must care that the work is done for all functions needed for the System to be successful, and not exactly how those functions are finally implemented.

Project management and other enabling organizational support products should use the subdivisions and terms that most effectively and accurately depict the hierarchical breakdown of project work into meaningful products. A properly structured WBS will readily allow complete aggregation of cost, schedule, and performance data from lower elements up to the project or program level without allocation of a single element of work scope to two or more WBS elements. WBS elements should be identified by a clear, descriptive title and by a numbering scheme as defined by the project that performs the following functions:

- Identifies the level of the WBS element.
- Identifies the higher level element into which the element will be integrated.

One important thing to consider while creating the WBS (which will be addressed in the following chapter) is to keep the WBS strictly mapped to the system under design to reconcile the WBS with the recursive nature of the system. The work grows bottom-up, meaning that lower levels dictate the work needed for the system to be realized. Put in a different way, avoid adding in higher level of the WBS things or work chunks things that will still be needed in lower levels, for example Project Management. If your system is composed of other subsystems, those subsystems will still need Project Management work, so it does not make a lot of sense to add PM at level 2. Graphically.

In the example (Fig. 2.2), both "Project Management" and "Systems Engineering" are put at L2, but this could be interpreted as no PM nor SE presence or work at lower levels, which is inaccurate. A clearer way of stating the work breakdown for the project is to make sure all levels contain their PM/SE effort correctly mapped into it. Ideally, higher levels of the hierarchy should be treated as "abstract classes". They merely exist as containers of lower level entities. But what about during early stages

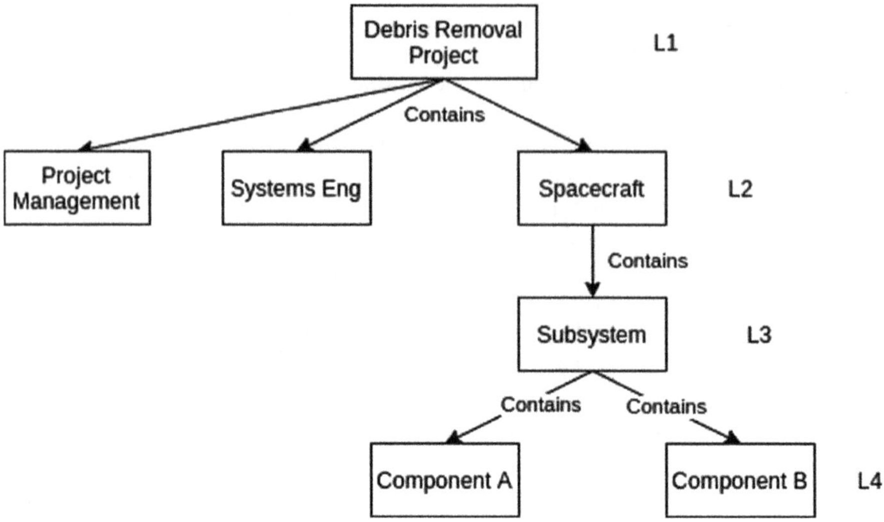

Fig. 2.2 WBS example

of development when the lower level entities do not exist just yet? This brings up a very important factor: The Work Breakdown Structure evolves with system maturity, and system maturity evolves with time, hence the WBS is a function of time. The way to tackle this is as follows: start with the topmost level block (the project) and add to the next level the work needed to create the following level. Let's start the process from scratch. Figure 2.3 depicts the initial WBS.

Your "Debris Removal" project will need PM and SE to understand how to realize this project. Then, PM will work for a while and come up with the idea (obvious one, but bear with me for the sake of the exercise) that a Spacecraft is needed, hence, a new block is added to the WBS (Fig. 2.4).

The creation of a new block at L2 (Spacecraft), will need PM/SE work on its own. It is important this effort is contained and integrated by the Spacecraft block itself. Of course, level 1 PM/SE will talk and collaborate with level 2 PM/SE and so on, but they have different scopes, hence they should be kept separately (Fig. 2.5).

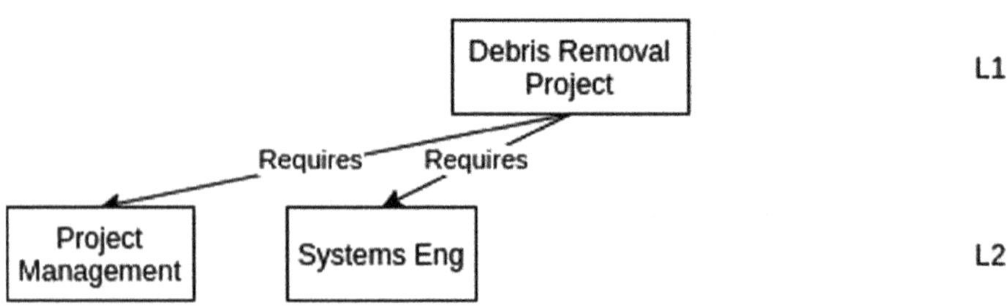

Fig. 2.3 Initial WBS

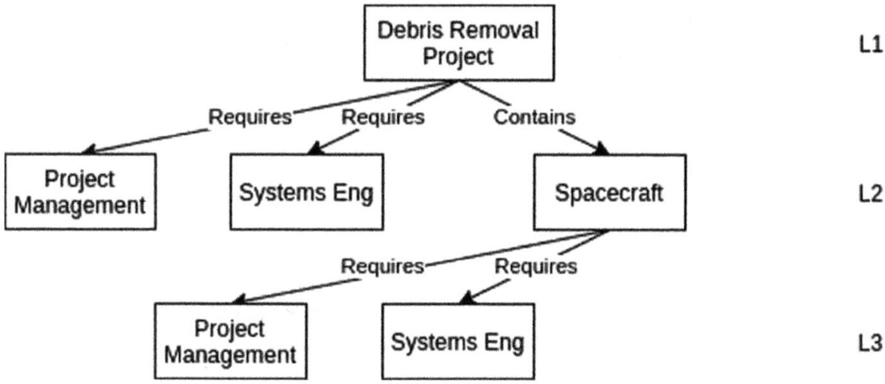

Fig. 2.4 A new block added to the WBS

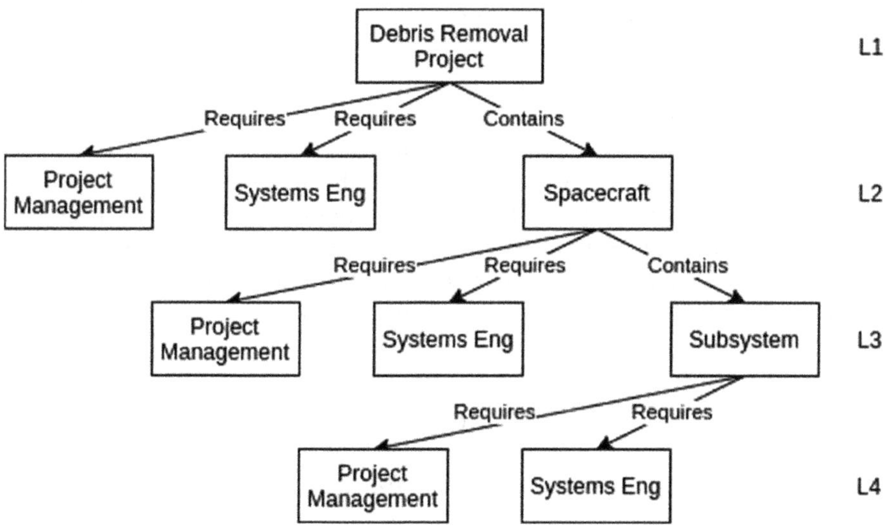

Fig. 2.5 Adding a subsystem to the WBS

Readers may ask (and rightfully so) how can early stage NewSpace projects afford an army of project managers and systems engineers at every level of the WBS? The short answer is: there is no need for such a crowd. Any startup below 30 or 40 people does not need an actual full-time PM nor SE. But the long answer is that PM and SE are understood as "efforts", which can be carried by people wearing multiple hats in the organization, and not necessarily as a specific person or a team allocated just for it.

But in any case, no matter how small the team, it is very useful to have a person defined as the PM and another person as SE of a specific level of the WBS. In extreme cases, one person can still wear these two hats, but the efforts must be differentiated.

Assigned PMs and SEs can be responsible for the activities for a specific level of the WBS and discuss with the PM/SEs of levels below about integration activities. Systems Engineering at every level is capable of spawning or flowing down components for the next level. For the debris removal example, Systems Engineering at level 2, after running analysis along with Flight Dynamics, are in their right to spawn a Propulsion subsystem. This subsystem will require its own PM and SE which will flow things down further and collaborate for integration upstream. It is important to mention that there is an authority flow going downstream as you navigate the WBS. The PM/SE effort at project level (L1) should keep oversight of all the components at that level and below. The PM/SE at L2 (Spacecraft) keeps oversight of that level and below but reports to L1, and so on. The Systems Engineering team of that level is responsible that the component gets properly integrated and verified to be part of the SBS (Fig. 2.6).

As you can see, this is nothing else than analytical work (i.e. flow down). Work gets dissected from requirements and user needs all the way down to components and parts, and the system gets realized (integrated) bottom up. Despite its "classic" smell, the WBS is a great gluing thing to use and to perfect in NewSpace, since it helps organize the work and clearly define responsibilities, despite the team size. Let's redraw it in a more NewSpace-friendly way (Fig. 2.7).

The WBS is just a collection of the work and parts needed to realize a system, but how the life cycle stages are broken down inside the different levels of the WBS is totally open. For example, the PM team in charge of the L2 subsystem "Spacecraft" could use a more classic methodology, whereas the L3 team in charge of Propulsion could use Scrum. Even though it would be advisable to follow similar life cycle approaches across the WBS, it is not mandatory to use the same.

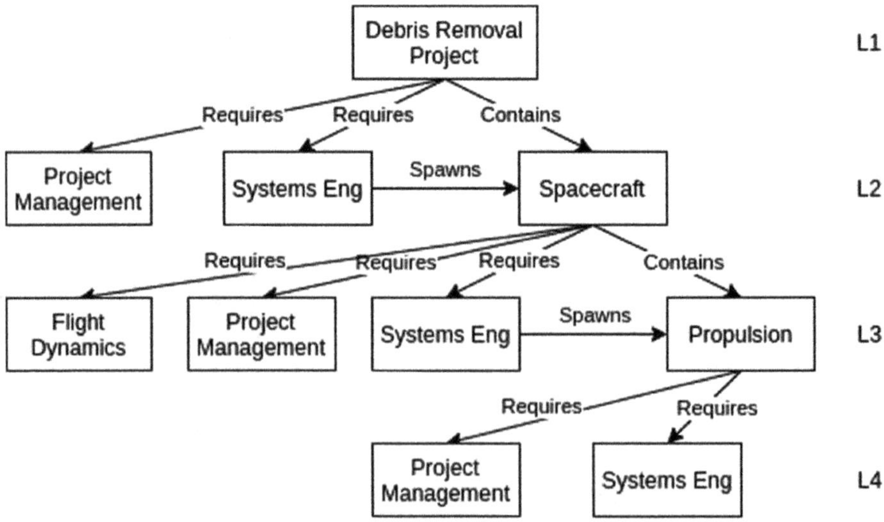

Fig. 2.6 Efforts required to spawn a new element in the WBS

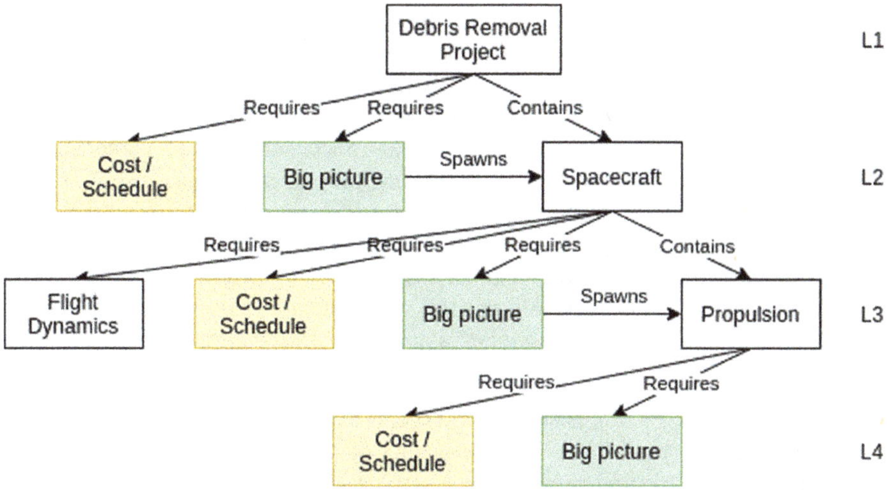

Fig. 2.7 NewSpace-friendly version of a WBS—Every level needs cost/schedule + big picture

We will discuss further ahead how to allocate the integrated teams (IPTs) to realize the different blocks of the WBS.

2.6 Requirements

Requirements are literally sacred in classic space. Years are spent capturing, analyzing and reviewing requirements. Thousands of them are collected in sophisticated and expensive tools. Specific languages have been created or adapted for writing and handling requirements in diagrammatic and hierarchical ways. There surely exists alucrative industry of the requirement. This industry will make it feel we *must* do requirements, and they would like us to do nothing else if possible. But requirements have a different meaning and status in NewSpace. For a start, for one to be able to have requirements, a sort of customer figure must exist in the project. A customer with wishes we could data mine and interpret, giving a great deal of information to help streamline our design efforts. Small space startups usually do not have the luxury of having a customer to interrogate; so, what happens with requirements in a no-customer situation?

Some possible scenarios: 1. Totally ignoring them or 2. Faking requirements to make their absence feel less uncomfortable, with very little practical use because those fake requirements are usually shallow. For example, I once had the chance to skim through a "Requirements Specification Document" of a very early stage NewSpace project and this document had requirements like "The spacecraft shall be able to take images". What is the use of documenting such obvious things? Writing that kind of requirement is pointless but jumping directly to design with zero requirements is dangerous. The risk of doing that is that the organization's scarce resources could be employed to synthesize a totally wrong solution, or to overlook some important factor, feature or performance measure which could come back later to bite them. For example, "minimum time between acquisitions" in an Earth-observation spacecraft might not be a requirement at the proof-of-concept stage but become a critical technical performance measure later. So, it is healthy to repress the instinct to start designing out of gut feeling and take some time to do some thorough analysis, in the form of open source intelligence, or to observe/collect what competitors do, and how they do it. Requirements analysis cannot take forever unfortunately, otherwise the system never gets to market. Anything helps at this stage to capture what the system is ought to do. Also, role-playing when someone in the team can act as a *pretend* customer. In summary.

- Understand the market: it is money well spent to invest in reaching out to potential customers and extract what kind of needs they have: price targets, scope, what they are currently not getting from existing players, etc. Get figures from the market you are trying to address: its growth, size, etc.
- Do some competitive intelligence and benchmarking: Run your own intelligence gathering exercise where you scan the market you're planning to join, choose three or four of your closest competitors, and compile a few requirements based on their offers. Reach out to big players who are doing similar things; chances are they will be willing to help.
- Identify key technical performance measures your system must meet and create a shortlist of the most meaningful requirements, which can fulfill those technical performance metrics.

It is enough to capture the topmost 10 or 15 technical performance metrics and requirements for a design to be able to kick off. These master requirements can still (and will evolve) because things are very fluid still. Requirements spawn other requirements, and requirements have their own attributes and kinds.

What other types of requirements are there? There are different types, and many ways of categorizing them, depending on what bibliography you have in your hands. One reasonably simple categorization of requirements is Kano's model. This model is a theory for product development and customer satisfaction developed in the 1980s by Professor Noriaki Kano, which classifies customer preferences or requirements into three major groups:

- Must-be requirements
- Performance requirements
- Attractive requirements (Fig. 2.8).

2.6.1 Must-Be Requirements

The "must-be" requirements are the basic needs for a product. They constitute the main reason that the customer needs the product. That is, they fulfill the abstract need

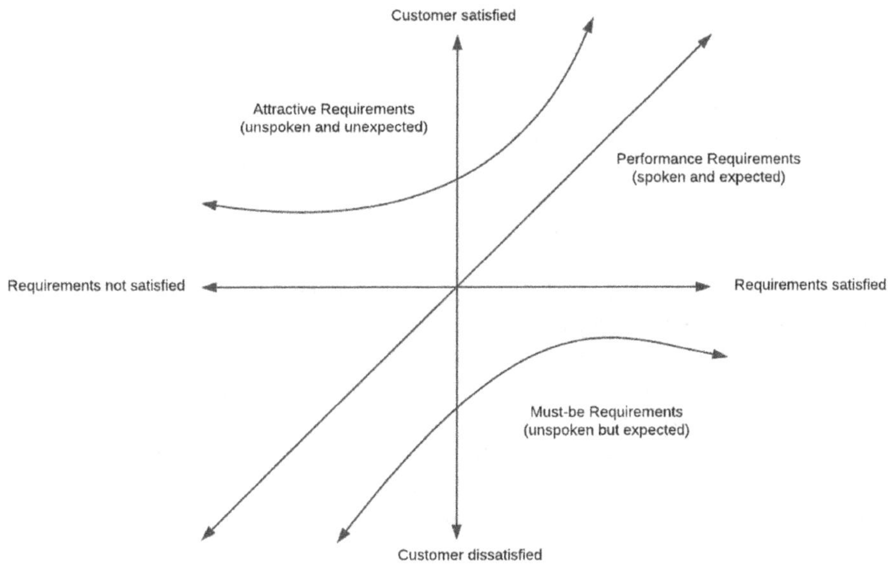

Fig. 2.8 Kano's model for requirements

of the customer. The customer expects these requirements, and the manufacturer gets no credit if they are there. On the other hand, if these requirements are not fulfilled, the customer will be extremely dissatisfied (the customer will have no reason to buy the product). Fulfilling the "must-be" requirements alone will not satisfy the customer; they will lead only to a case of a "not dissatisfied" customer. Must-be requirements are unspoken and non-measurable; they are either satisfied or not satisfied. They are shown in Kano's model by the lower right curve of the figure.

2.6.2 Performance Requirements

Performance requirements are spoken needs. They can be measured for importance as well as for range of fulfillment levels. The customers explicitly tell us or answer our question about what they want. Performance requirements include written or spoken requirements and are easily identified and expected to be met. Usually, customer satisfaction increases with respect to the degree to which these requirements are fulfilled. Also, performance requirements serve as differentiating needs and differentiating features that can be used to benchmark product performance. The diagonal line in Kano's model depicts these requirements.

2.6.3 Attractive Requirements

Attractive requirements are future oriented and usually high-tech innovations. These requirements are unspoken and unexpected because the customer does not know they exist. Usually, these requirements come as a result of the creativity of the research and development effort in the organization. Because they are unexpected, these creative ideas often excite and delight the customer and lead to high customer satisfaction. These requirements are shown by the curved line in the upper left corner of the figure. It should be mentioned here that these requirements quickly become expected. But once requirements, either artificially created or realistically captured from a real customer, are in some reasonably good shape, it starts the breakdown activity. The team needs to subdivide work in smaller, more manageable chunks in order to solve the problem in a more effective way.

NewSpace projects usually start with a partial understanding of the problem and/or the market. It is normal they will assume or simplify many things which will prove to be different as the project matures. The design needs to be done in a way it can easily accommodate for new requirements found along the way, or for changes in the current set of requirements.

2.7 Life Cycle Cost

How can we possibly know the costs of what does not exist? Clearly, we cannot. Still, oftentimes, when we develop technical things, we are asked to precisely reckon the total cost of what we have just started to design. This usually leads to off estimations which fire back later in the process. And we try, frequently failing miserably. In the Engineering practice, there is the recurrent need to crystal ball. But, without wanting to challenge what it has been part of the job for ages, let us try to understand how cost predictions can be done; let us stop using the term prediction and start using estimation instead. Too often, we become fixated on the technical performance required to meet the customer's expectations without worrying about the downstream costs that contribute to the total life cycle cost of a system (Faber and Farr 2019).

There are multiple factors that can affect our cost estimations:

- Constantly changing requirements
- Uncertain Funding
- Changing technology
- Change in regulations
- Changes in competition
- Global markets uncertainties.

For very early stage cost analysis, scope is king. A company embarking in a new project must evaluate what to include and what not to include and cost estimations, and this is not a simple task.

Let us go by example: a company is having as an idea selling a debris removal service. It is a LEO spacecraft with the capability of physically coupling to a target satellite and either move it to a more harmless orbit or force both to re-enter the atmosphere together. The company is seeking funds; hence it engages in the process to understand how much money it would cost to develop such a thing. What is the scope of the cost analysis? Is it:

(we are excluding facilities costs here for simplicity)

- Just the spacecraft (material costs)?

  But wait, someone needs to decide what to buy

- Spacecraft + R&D labor?

  But wait, once it is designed, someone needs to assemble it together and verify it

- Spacecraft + R&D labor + AIT (Assembly, Integration and Test) labor?

  But wait, once it is assembled, needs to be launched into space

- Spacecraft + R&D labor + AIT labor + Launch?

  But wait, once it is in space, someone needs to operate it

- Spacecraft + R&D labor + AIT labor + Launch + Operations?

  But wait, to Operate it, tools and systems are needed

- Spacecraft parts + R&D labor + AIT labor + Launch + OPS + Ground Systems?

You can see how the scope only creeps. Each one of these items is most likely non-atomic, so they have internal composition, up to several levels of depth. Take the spacecraft itself: its functional and physical architectures are poorly understood or defined at early stages. The *make vs buy* strategy hasn't been properly analyzed. And yet, on top of this very shallow level of understanding, engineers are supposed to define budgets which also impact for example funding. Risks here are:

- Overestimating the cost: will scare investors, funders and potential customers away. Will make the margins look too thin.
- Underestimating the cost: being unable to deliver the project without cost overruns, risking the project to be cancelled, but also creating a bad reputation.

What to do then?

There are three generally accepted methods for determining LCC (Faber and Farr 2019, 13) and (NASA 2015):

1. Engineering build-up: Sometimes referred to as "grassroots" or "bottom-up" estimating, the engineering build-up methodology produces a detailed project cost estimate that is computed by estimating the cost of every activity in the Work Breakdown Structure, summing these estimates, and adding appropriate overheads.

2. Analogy: an estimate using historical results from similar products or components. This method is highly heuristic, and its accuracy depends on the accuracy of cost estimations for those other similar products. The analogy system approach places heavy emphasis on the opinions of "experts" to modify the comparable system data to approximate the new system and is therefore increasingly untenable as greater adjustments are made.
3. Parametric: based on mathematical relationships between costs and some product- and process-related parameters.

The methodology chosen for cost estimation cannot remain rigid throughout the project lifecycle. The analytical methodology is better suited at early stages when the uncertainty is maximum, whereas the build-up methodology is more accurate as the WBS is matured and more insight about the overall system composition is gained.

2.8 Proof-of-Concepts

While we iterate through the life cycle of our designs, is it important to get things tangible as fast as possible and test our assumptions early. This is usually done by means of prototypes, or proofs-of-concepts. In its most essential form, a prototype is a tangible expression of an idea. It can be as simple as a cardboard sketch or as complex as a fully functioning system like an aircraft. Prototypes can show different levels of fidelity. Low-fidelity prototypes provide direction on early ideas. These can happen fast (sometimes just minutes to produce). High-fidelity prototypes can help to fine-tune architectures, assess areas as manufacturability, assembly tooling, testability, etc.; these are more typical in engineering, and they can be costly and complex. High-fidelity prototypes can in cases show the almost full subsystem composition and, in some cases, also the same physical properties and functionalities as the real thing. For example, the aerospace industry reserves big hangars to host full-size aircraft functional testbeds, which are called "iron bird" or "aircraft zero". These testbeds are composed of most of the subsystems an aircraft is made of, laid out matching the general body shape and physical lengths, but without the fuselage. By means of close-loop simulation environments connected to setup, manufacturers can test a high amount of the complex subsystems present in aircrafts such as fuel systems, hydraulics, electrical systems, harness, etc. An Iron bird is basically an aircraft, but it cannot fly. It is a framework in which major working components are installed in the relative locations found on the actual airframe, arranged in the skeletal shape of the aircraft being tested. With the components out in the open, they can be easily accessed and analyzed. In the years leading up to a new aircraft's first flight, changes made during the development phase can be tested and validated using this valuable tool. Aircraft components that function well in isolated evaluations may react differently when operating in concert with other systems. With the aircraft's primary system components operational on the Iron Bird (except the jet engines, which are simulated), they are put through their paces from an adjacent control room.

There, within a mini cockpit, a pilot "flies" the testbed on a simulator through varied environmental conditions while engineers rack up data on the flight. Interestingly, an Iron Bird's work is not done once the aircraft is certified and deployed. These testbeds are still in operation after aircrafts reach production, where they can be used to provide insights into specific issues that arise or to test new enhancements before they are introduced on in-service aircraft. Even in the age of advanced computer simulations, Iron Birds (or Aircraft zeros) maintains a vital role in commercial aerospace testing protocols. They may never fly, but each Iron Bird is the precursor to an aircraft that does.

At the same time, aerospace builds prototypes which *do* fly, but are not full replicas of the production specimens. For example, flying prototypes are not equipped with cabin subsystems but instead equipped with test equipment to evaluate the in-flight performance.

Even though digital mockups[1] have gained a lot of traction in recent years, the value of laying out subsystems in real physical space with real harness and time constants still provides invaluable insight that computer simulations cannot accurately replicate. Big aerospace companies can afford such facilities, but NewSpace usually cannot. NewSpace companies cannot afford spending money and resources building expensive replicas which will not fly. Hence, the typical NewSpace approach is to build two setups: one that is usually called a Flatsat or Tablesat, and the actual setup which will see orbit. The latter cannot still be called a "production" thing, but still a prototype since most of its components are intermediate iterations. The former (Flatsat) is a functionally similar setup which can sit on top of a table (hence its name). A flatsat is usually composed of non-ready-for-flight pieces of hardware, development kits and even breadboards. The main purpose of the flatsat is to provide an environment for the software guys to grow their software on top of an environment, which comes reasonably close to the real environment (Fig. 2.9).

Recalling the *enabling systems* concept from early chapters, a flatsat is an enabling system for the prototype. What is more, there are a myriad of other bits and things surrounding the prototype which are also prototypes on their own: telemetry parsers, viewing tools, scripting engines, command checkouts, and a long etcetera. A collection of smaller prototypes combined into a big aggregated prototype.

A prototype that successfully works in orbit is a massive hit for a NewSpace project. But it is also a bit of a curse. If things work great, chances are the prototype will be quickly dressed as a product. Moreover, a product *fog* will soon cover all the other bits and parts which were rudimentary minimum-viable things, and quickly turn them into 'production grade' things when they are not ready for that. All these early stage components are suddenly pushed to 'production'.

Nailing a working prototype does not mean development is over. In fact, development may very well continue on the flying thing. NewSpace development environments reach all the way out to orbit. Things are fleshed out in orbit, debugged in orbit. For early stage NewSpace orgs, it is of paramount importance to get to space

[1]Digital MockUp or DMU is a concept that allows the description of a product, usually in 3D, for its entire life cycle.

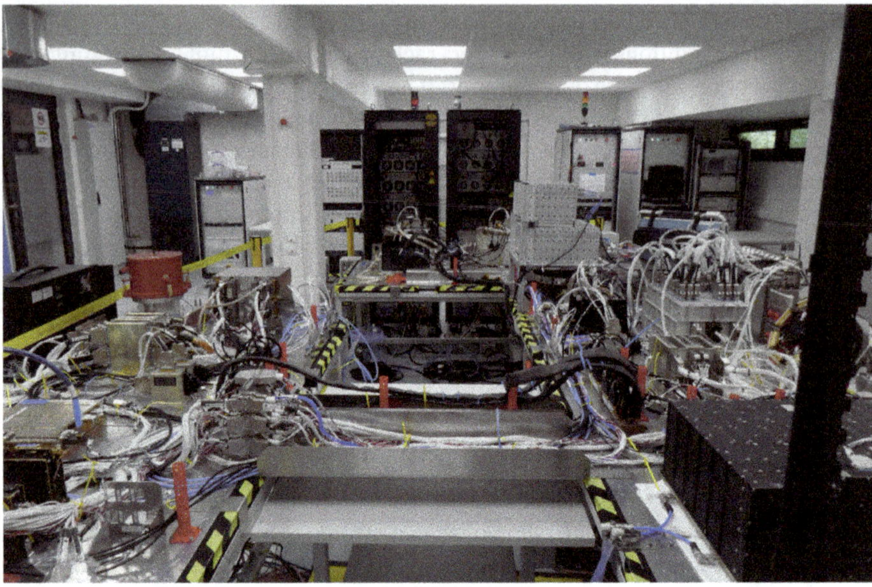

Fig. 2.9 EM-Flatsat and EGSE SmallGEO (*credit* OHB AG, used with permission)

as quickly as possible, to show they can deal with what it takes to handle a space system end to end. Unlike classic space, it does not have to be fully completed to get to orbit and be able to work. The important bit during the development stage is to pay attention to the must-haves; the rest can be done in space.

2.9 Time Leakage

In his classic *The mythical man month*, Fred Brook (Brooks 1995): "How does a project get to be a year late? One day at a time." On the same note, Hofstadter's Law states: "It always takes longer than you expect, even when you take into account Hofstadter's Law". An entire book could be written about how massively error-prone we are when we estimate the time our tasks will take. Task estimation is deeply affected by multiple psychological factors: we are usually overoptimistic (we think it will take less than it will take) and reductionistic (we overlook dependencies that our task will need to progress). Most of us view the world as more benign than it really is, our own attributes as more favorable than they truly are, and the goals we adopt as more achievable than they likely to be (Kanehman 2011). We underestimate uncertainty by shrinking the range of possible uncertain states (by reducing the space of the unknown); the unexpected always pushes in a single direction: higher costs and a longer time to completion (Nassim Taleb 2007), never the other way around. A way to reduce uncertainty in task estimation can be done by relying on heuristics and data; if you have done something similar in the past, you can use that experience as a calibration factor for your estimation on how long something else will or could take; we rarely do things for the first time ever, unless in the very early stages of our professional career. Time is a central part of what we do. Designing technical systems is organized around projects, whose lifecycle takes a fair amount of time to go through its stages, from infancy to retirement. Duration of those stages is never fixed, but they seem to continuously stretch due to delays.

It is because of the highly collaborative and interdependent nature of what we do that delays are frequent. But delays are not indivisible units.

Delays are the aggregation of small, tiny ones that are usually taken as acceptable or inoffensive, when they are not. Just like a million microseconds are needed to elapse to get a second in your watch, a million project micro-ticks are needed to elapse for one project tick to elapse. When one hears of a schedule slippage in a project, it is hard not to imagine that a series of major calamities must have taken place. Usually, however, the disaster is due to ants, not tornadoes; and the schedule has slipped imperceptibly but inexorably. Indeed, major calamities are easier to handle; one responds with major force, radical reorganization, the introduction of new approaches; the whole team rises to the occasion. But the day-by-day slippage is harder to recognize, harder to prevent, harder to make up (Brooks 1995). Yesterday a key actor was unavailable, and a meeting could not be held. Today the machines are all down, because of network maintenance. Tomorrow the latest software patches will not be deployed, because deployments are not allowed on that specific day. Each one only postpones some activity by a half-day or a day, and the schedule slips, one day at a time.

2.10 System Failure

Failure is seldom atomic, i.e. it never happens in a unitary, indivisible action, unless something extremely catastrophic, and not even so. Failure is, more generally, a trajectory or sequence of events toward an outcome, which represents some type of a sensible loss. Just as a particle resting on a frictionless surface at point A needs a driving action (force) and an unobstructed path to reach to a point B, failure requires a driving action and a cleared path to reach to the point where loss occurs. A particle needs time to travel from A to B, just as failure needs time to build up; it is never an instantaneous nor, as said, a single-factor phenomenon. Think about football for example, where team 1 plays against team 2. Any goal scored from team 1 is a failure for team 2, and vice versa. From this perspective, the forcing action is the offensive pressure from one of the teams against the other; Team 1 pushes continuously team 2 to fail, until it eventually happens. For the defending team, in order to prevent failure, it is crucial to add as many barriers as possible, to create the right number of obstacles for the force the attacking team is exerting. This requires taking decisions about how to locate and move such impedances, and such decisions heavily rely on information. In sports, such information is mostly visual. Also, in sports offensive and defensive roles can easily change places. But not so much when we operate systems such as organizations or systems; we can only adopt defensive roles against failure; it is the external environment against the system and we cannot take any offensive action against the environment hoping that will put the forcing action to a stop; an aircraft cannot do anything to modify the physics laws which governs its flight, nor an organization can do much to alter the behavior of the markets it's embedded in. In these cases, we can attack the effects, but we cannot touch the causes.

Failure prevention is a game where we can only play defense; failures are insistently wanting to happen, and we need to work against it. In tennis, there is something called unforced errors. Unforced errors are those types of mistakes that are supposedly not forced by good shots of an opponent. The people in charge of recording the statistics try to determine if, when a player makes a mistake, whether that mistake was forced or not forced by a good play by his or her opponent. But if the player seems to have balanced himself or herself for the shot and to have enough time to execute a normal stroke (not abbreviated one) and still makes a mistake, that would count as an unforced error. As it can be seen, the definition of an unforced error is purely technical, and wrong. There is no mention of physical fatigue, tactical decision-making, and *mental game*. Every tennis player is continuously under the exertion of a forcing action which is pushing her/him to fail. Every error is, therefore, forced (Tennis Mind Game, n.d.). In tennis, whoever makes the least amount of error wins, not the one who makes the most spectacular shots.

When we deal with complex systems, failure-driving *forces* are multiple, persistent, and adaptive. When we operate those systems, we rely on measurements to construct the understanding for assessing faults and failures. But what if the measures are wrong? Then our understanding is flawed, detaching from actual reality and tainting our decisions, eventually opening an opportunity for failure forces to align and find a trajectory towards loss. Errors and failures are troublesome. They can cause tragic accidents, destroy value, waste resources, and damage reputations.

An organization like a NewSpace company, at the highest level of abstraction, operates what we have called before the *big system*, which is an aggregation of systems. All in all, operating this System involves an intricate interaction of complex technical systems and humans. Reliability can only be understood as the holistic sum of reliable design of the technical systems and reliable operations of such systems. Any partial approach to reliability will be insufficient. Organizations, particularly those for whom reliability is crucial, develop routines to protect them from failure. But even highly reliable organizations are not immune to disaster and prolonged periods of safe operation are punctuated by occasional major failures (Oliver et al. 2017). Scholars of safety science label this the "paradox of almost totally safe systems", noting that systems that are very safe under normal conditions may be vulnerable under unusual ones (Weick et al. 1999). Organizations must put in place different defenses in order to protect themselves from failure. Such defenses can take the form of automatic safeguards on the technical side, or procedures and well-documented and respected processes on the human side. The idea that all organizational defenses have holes and that accidents occur when these holes line up, often following a triggering event—the Swiss cheese[2] model of failure—is well known. In the Swiss cheese model, an organization's defenses against failure are modeled as a series of barriers, represented as slices of cheese. The holes in the slices represent weaknesses in individual parts of the system and are continually varying in size and position across the slices. The system produces failures when a hole in each slice momentarily aligns, permitting a trajectory of accident opportunity, so that a hazard passes through holes in all the slices, leading to a failure.

There is some "classic" view of automation, which points out that its main purpose is to replace human manual control, planning and problem-solving by automatic devices and computers. However, even highly automated systems, such as electric power networks, need human beings for supervision, adjustment, maintenance, expansion, and improvement. Therefore, one can draw the paradoxical conclusion that automated systems still are human–machine systems, for which both technical and human factors are important. Quite some research has been done on human factors in engineering, which only reflects the irony that the more advanced a control system is, so the more crucial may be the contribution of the human operator (Bainbridge 1983). It might also be implicit that the role of the operator is solely to monitor numbers and call an "expert" when something is outside some safe boundaries. This is not true. Operators gain crucial knowledge when they operate in real conditions that no designer can gain during the development stage. Such knowledge must be captured and fed back to the designers, continuously.

Automation never comes without challenges. An operator overseeing a highly automated system might start to lose his/her ability to manually react when the system requires manual intervention. Because of this, a human operator of a space system needs to remain well-trained into manual operations in case automations disengage for whatever reason. If the operator blindly relies on automation, under the circumstances that the automation is not working, then the probability of making the situation even worse increases considerably (Oliver et al. 2017). The right approach is about having human operators focusing more on decision-making rather than systems management. In summary:

- Reliability can only be achieved holistically by means of combining complex systems with reliable operations through human–machine interfaces, which consider the particularities of such systems.
- All defenses to prevent failures might have flaws. It is crucial to avoid that flaws (holes in the cheese) align, providing a trajectory to failure.
- Automation is beneficial to optimize data-driven repetitive tasks and streamline human error, but the possibility to run things manually must always be an alternative and operators must be trained to remember how to do so.

[2]https://en.wikipedia.org/wiki/Swiss_cheese_model#Failure_domains.

References

Bainbridge, L. (1983). Ironies of automation. *Automatica, 19*(6), 775–779. https://doi.org/10.1016/0005-1098(83)90046-8.

Belie, R. (2002). Non-technical barriers to multidisciplinary optimization in the aerospace community. In *Proceedings. 9th AIAA/ISSMO Symposium on Multidisciplinary Analysis and Optimization*, Atlanta, GA.

Brooks, F. P. (1995). *The mythical man-month: Essays on software engineering* (20th anniversary ed.). Addison-Wesley.

DeMarco, T., & Lister, T. R. (1999). *Peopleware: Productive projects and teams* (2nd ed.). Dorset House Publishing Company.

Detert, J., Schroeder, R., & Mauriel, J. (2000). A framework for linking culture and improvement initiatives in organizations. *Academy of Management Review, 25*. https://doi.org/10.2307/259210

DiMaggio, P. (1997). Culture and cognition. *Annual Review of Sociology, 23*(1), 263–287.

Faber, I., & Farr, J. V. (2019). *Engineering economics of life cycle cost analysis* (1st ed.). CRC Press.

Goodman, J. (2005). Knowledge capture and management: Key to ensuring flight safety and mission success. In *Proceedings AIAA Space 2005, Long* Beach, CA.

Hackman, J., & Oldham, G. (1980). *Work redesign*. Addison-Wesley.

Harris, D. (2001). Supporting human communication in network-based systems engineering. *Systems Engineering, 4*(3), 213–221.

Hart, P. (1991). Irving L. Janis' victims of groupthink. *Political Psychology, 12*(2), 247. https://doi.org/10.2307/3791464.

Kanehman, D. (2011). *Thinking, fast and slow*. New York: Farrar, Straus and Giroux.

Miller, G. A. (1994). The magical number seven, plus or minus two some limits on our capacity for processing information. *Psychological Review, 101*(2), 343–352.

NASA. (2007). *Systems engineering handbook*. NASA.

NASA. (2015). *NASA Cost estimating handbook*, v4.0.

Nassim Taleb, N. (2007). *The black Swan: The impact of the highly improbable*. New York: Random House.

Oliver, N., Calvard, T., & Potočnik, K. (2017, Jun 9). Cognition, technology, and organizational limits: lessons from the Air France 447 Disaster. *Organizational Science, 28*(4), 597–780. https://doi.org/10.1287/orsc.2017.1138

Oswald, A., Proto, E., & Sgroi, D. (2009). A new happiness equation: Wonder + happiness = improved productivity. *Bulletin of the Warwick Economics Research Institute, 10*(3).

Perlow, L. A. (1999). The time famine: Toward a sociology of work time. *Administrative Science Quarterly, 44–57,* xxx. https://doi.org/10.2307/2667031.

Ritchey, T. (1991). Analysis and synthesis: On scientific method – based on a study by bernhard riemann. *Systems Research, 8*(4), 21–41. https://doi.org/10.1002/sres.3850080402.

Schein, E. H. (2004). *Organizational culture and leadership* (4th ed.). San Francisco: Jossey-Bass.

St. George, A. (2013, June). Leadership lessons from the Royal Navy: the Nelson touch. *Naval Historical Society of Australia*. https://www.navyhistory.org.au/leadership-lessons-from-the-royal-navy-the-nelson-touch/

Tennis Mind Game. (n.d.). Unforced errors in Tennis—Are they really not forced? *Tennis Mind Game*. https://www.tennismindgame.com/unforced-errors.html

Terrell, S. M. (2018). *NASA Work Breakdown Structure (WBS) handbook* (p. 20180000844). AL, United States: NASA Marshall Space Flight Center Huntsville.

Tuckman, B., & Jensen, M. (1977). Stages of small-group development revisited. *Group and Organization Management, 2*(4), 419–427.

Twomey Lamb, C. M. (2009). Collaborative systems thinking: An exploration of the mechanisms enabling team systems thinking. PhD Thesis at the Massachusetts Institute of Technology.

Weick, K. E., Sutcliffe, K. M., Obstfeld, D., & Straw, B. M. (1999). Organizing for high reliability: Processes of collective mindfulness (R. I. Sutton, Ed.) Research in organizational behavior. *Elsevier Science/JAI Press., 21,* 81–123.

Chapter 3
Knowledge Management and Understanding

The greatest enemy of knowledge is not ignorance; it is the illusion of knowledge.
—Daniel J. Boorstin

Abstract Startups have the unique opportunity to define a knowledge management (KM) strategy early in their lifetime and avoid the pain of realizing its importance too late down the road. NewSpace enterprises must grow as a body of knowledge to gain the understanding needed in order to create successful technical systems, and usually such knowledge grows largely unstructured. Multidisciplinary projects are composed of people from different backgrounds, with different terminologies and different past experiences, which impose challenges when it comes to sharing information throughout the design process. This chapter addresses the methods and tools such as diagrammatic reasoning, model-based systems engineering, concept maps, knowledge graphs, and hierarchical breakdown structures and how they can facilitate or complicate the design of space systems.

Keywords Knowledge management · Knowledge graphs · Concept maps · MBSE · Model-based systems engineering · Block diagrams · Breakdown structures

Readers may ask: Why a knowledge management chapter in a systems engineering book? Nobody has time for that! I happen to disagree. Or, I would like to contribute to change that mindset. There is always time for knowledge management. But the main reason why I included this chapter is because startups have the unique blessing (and curse) of being early stage. Being early stage means that they have the great advantage of standing right at the *big bang* of many things in the organization, including information and knowledge creation. Things only expand and get more complex with time, and so does information and knowledge. The main goal of this chapter is to raise awareness on how the organizational knowledge is created, where it resides, and how to protect it. In the beginning, practically everything is a fuzzy cloud of loose concepts and words which need to glue according to some strategy.

I. Chechile, *NewSpace Systems Engineering*,
https://doi.org/10.1007/978-3-030-66898-3_3

Such strategy may mutate multiple times, and this mutation will most of the time respond to better insight gained during the journey.

Early-stage organizations must deal with the massive challenge of building together a product of reasonable complexity, and at the same time grow the knowledge to achieve the *understanding* needed in order to create the products they want to create. New product development always requires overcoming many learning curves. Creating a product does not guarantee per se market success. Since life is too short to develop products nobody wants, we must also learn how the audience of our products react to them and calibrate our actions toward maximizing their satisfaction (and by consequence our profits). Selling anything is ultimately an endeavor of learning and adjusting. The need for knowledge management is both tactical and strategic, and it covers a very wide spectrum that goes well beyond the technical knowledge needed to create a system: from understanding business models, creating the right culture, dealing with daily operations, and doing the accounting; there are learning curves everywhere, intertwined, highly coupled. Knowledge evolves with the life of an organization, and its creation and capturing is not trivial. Fostering knowledge sharing through the interactions between its actors is essential to any practical implementation of knowledge management (Pasher and Ronen 2011).

But what is knowledge? What is the difference between *understanding* and *knowledge*? It's very easy to assume that these two words have the same meaning, but there is some nuance distinguishing them. Understanding can refer to a state beyond simply knowing a concept. For example, we can know about spacecraft, but we might not have a comprehensive understanding of how they work. It's possible to know a concept without truly understanding it. To design successful systems, it is not enough just to know a concept, but have the understanding that allows the organization to apply the concepts and skills in a collaborative way to solve even the most unfamiliar problems. Gathering knowledge and building understanding takes time, and knowledge becomes a strategic asset for small organizations. Learning and understanding is not just the accumulation of knowledge, but the capacity to discover new insights in changing contexts and environments. What is more, knowledge accumulated in unstructured ways across the organization creates risks due to the fact that it becomes difficult to access. Organizational knowledge does not sit on files or documents. It resides in people's brains, but more importantly in the social networks across the organization. Perhaps the biggest tragedy of space startups going bankrupt (after the obvious job losses) is how the organization's knowledge network vanishes away.

There are vast amounts of ontologically unstructured information in organizations, even in mature ones. For startups, understandably, knowledge management comes way later in their hierarchy of needs; when you are burning money, your problems are somewhere else. This is a pity, since startups have the unique chance of defining a knowledge management strategy at probably the best time possible: when things are still small, and complexity has not yet skyrocketed. Minding knowledge management when there are four people compared to when there are 400 is a different story.

Knowledge needs to be constructed; i.e. grown in time. From basic terminology up to complex concepts, a solid common "language" across the organization can act as a foundation to organize and structure knowledge. Different engineering disciplines manage different jargons and vocabularies.

During meetings, way more often than it should, you most probably got the impression that two colleagues were using the same word with a different meaning. Same words mean different things to different people. We all have our internal dictionaries, so to speak, and we carry them everywhere we go; such dictionaries are populated with words we collected throughout our life and are shaped by our education and largely by our life experiences. When we engage in professional exchanges during our professional practice, we have our dictionaries open in our heads while we try to follow what's being discussed. This mechanism is very automatic, but the mechanism to double check if our dictionary is stating the same as our colleague's is not so automatic, it seems. Frequently, many decisions are taken based on these types of misunderstandings. Or, which is not less serious, partial understanding. Using dictionaries as analogy requires a bit further elaboration, to illustrate the problem I am trying to describe here.

A dictionary is an alphabetically ordered list of words with some listed meanings. How are those meanings defined and who defines them? How is it decreed that a word means something and not something else? For a start, dictionaries are created by historians. To add a word in a dictionary, the historian (in the role of the editor) spends a fair amount of time reading literature from the period that a word or words seemed more frequently used. After this extensive reading, all the occurrences of those words are collected, and a context around them is built. The multiple occurrences are classified into a small subset of different meanings and then the editor proceeds to write down those definitions. The editor cannot be influenced by what he thinks a given word ought to mean. The writing of a dictionary, therefore, is not a task of setting up authoritative statements about the "true meanings" of words, but a task of recording, to the best of one's ability, what various words have meant to authors in the distant or immediate past. Realizing, then, that a dictionary is a historical work, we should read the dictionary thus: "The word *mother* has, most frequently been used in the past among English-speaking people to indicate -a female parent". From this we can safely infer, "If that is how it has been used, that is what it *probably* means in the sentence I am trying to understand". This is what we normally do, of course; after we look up a word in the dictionary, we re-examine the context to see if the definition fits. A dictionary definition, therefore, is an invaluable guide to interpretation. Words do not have a single "correct meaning"; they apply to groups of similar situations, which might be called areas of meaning. It is for definition in terms of areas of meaning that a dictionary is useful. In each use of any word, we examine the context and the extensional events denoted (if possible) to discover the point intended within the area of meaning (Hayakawa 1948).

While we grow our internal *dictionaries*, we not only assign meaning to words according to different abstract situations, contexts, actions, and feelings but also establish relationships between them. Dictionaries say nothing about how words relate to each other, it is just an array of loose terms, if you want. So, what we carry is not, strictly speaking, only a dictionary, but a sort of a *schema* (Pankin 2003); schemas are like recipes we form in our brains for interpreting information, and they are not only affected by our own past experiences but also shaped by culture. Self-schema is a term used to describe knowledge we accumulate about ourselves by interacting with the natural world and with other human beings which in turn influences our behavior toward others and our motivations. Because information about the self is continually coming into the system as a result of experience and social interaction, the self-schema will be constantly evolving over the life span (Lemme 2006). The schema concept was introduced by British psychologist Sir Frederic Bartlett (1886–1969). Bartlett studied how people remembered folktales, and observed that when the recalling was inaccurate, the missing information was replaced by familiar information. He observed that they included inferences that went beyond the information given in the original tale. Bartlett proposed that people have *schemas*, or unconscious mental structures or recipes, that represent an individual's generic knowledge about the world. It is through schemas that old knowledge influences new information. Bartlett demonstrated the uniqueness of our schemas as well: no two people will repeat a story they have heard in the exact same way. The way we create those recipes is totally different for each one of us, even under the exact same stimuli. When we communicate with someone, we are just telling each other a combination of what we have felt and learned in the past. Even with a high degree of present moment awareness, how we communicate is often a reflection of the information we have previously consumed.

The entities which are members of a schema are not tightly attached to it. For example, a chair is a concept in our minds that we know very well (because we happened to understand a while ago what chairs are, their function, etc.). But a chair is a concept that is a member of many different schemas. For example, if you go to a restaurant you expect the tables to have chairs (your restaurant schema tells you so). In the same way, you do not expect to find a chair in your refrigerator (your fridge schema is telling you so). As you can see, the chair concept is "on the loose" and we connect it to different schemas depending on the context or the situation; membership is very fluid. On the other hand, the systems we design and build are quite rigid containers, like boxes inside boxes or matryoshkas. What resides inside a box cannot be inside some other box; membership is strict. The way our brains work and the things we build are structured in a different manner.

But how do our inner schemas connect and relate to others? Can organizations grow their collective schemas? Can schemas affect the way an organization collectively learns? They play a part.

- Schemas influence what we pay attention to: we are more likely to pay attention to things that fit in our current schemas.
- Schemas impact how fast we can learn: we are better prepared to incorporate new knowledge if it fits with our existing worldviews.
- Schemas affect the way we assimilate new information: we tend to alter or distort new information to make it fit to our existing schemas. We tend to bend facts in our own service.

As engineers, we all tend to work around a domain or area, according to our education or what our professional career has dictated for us. Each domain, whatever it is, revolves around a certain set of information that makes the domain understandable and tractable, such as concepts, properties, categories, and relationships; if you want to perform in that domain, you need to have at least some grasp. The technical term for this is ontology. It is a fancy, somewhat philosophical term; if you google it you can easily end up reading about metaphysics, but make sure you just turn back if you do; when we say "ontology" we mean more probably a bit more a computational ontology than a philosophical. Traditionally, the term ontology has been defined as the philosophical study of what things exist, but in recent years, it is used as a computational artifact in any computer-based application, where knowledge representation and management are required. In that sense, it has the meaning of a standardized terminological framework in terms of which the information is organized. It's very straightforward: an ontology encompasses a representation, formal naming and definition of categories, properties, and relations between the concepts, data, and entities that substantiate one, many, or all domains of discourse. Ontologies are all around us. Product management is an ontological activity by definition. Think about how products are usually listed in trees, showing how features and attributes define the way such products relate, or how subfamilies are described. Another concrete example is software engineering. The object-oriented programming paradigm has deep ontological roots, since every object-oriented program models the reality by means of classes and the way they relate, along with attributes, properties, features, characteristics, or parameters that objects, and classes can have. It happens to go unnoticed, but the object-oriented paradigm is very ontological; it must capture and ultimately execute the behavior of the underlying concepts and entities involved in a particular domain, problem, or need.

Engineering in general involves and connects multiple fields and domains, and every field creates their own schemas and ontologies to tame complexity and organize information into data and knowledge. As new schemas are made, their use hopefully improves problem-solving within that domain. This means that every field tends to create its own terminology and jargon. Easily, organizations end up dealing with a network of dissimilar domain-specific schemas around a project, and often those go out undocumented. You surely have witnessed a meeting where for example a software person presents something to a technically heterogeneous audience. While she uses very specific software jargon and acronyms, you can feel how nobody outside software understands a thing what is going on, but they're probably too shy to ask.

Years ago, I was in one of those meetings, and after an hour of slides about software architecture and whatnot, one person from the audience raised his hand and asked:

> What does API stand for?

3.1 Diagrammatic Reasoning and the Block Diagram Problem

There are as many block diagrams types as there are engineers. The reader might have surely assisted many times to a presentation or similar situation where a colleague showed a collage of colored boxes and arrows and went to a lengthy explanation around the diagram, assuming the audience dug what those blocks and arrows meant automatically. Diagrams can help to produce new insight and help create new knowledge, that is not in discussion. But the way we extract such insight from those diagrams is a different experience to each one of us. It is us, the consumer of those diagrams who can transform the boxes and arrows into insight; we make sense out of some pixels arranged in some arbitrary way on a screen or with ink on a paper.

Think for a moment about how graphical the practice of design engineering is. Images play a major role. Engineering information is often conveyed in different types of graphical depictions and images. Just as cartographers have created visual representations of areas to orient explorers around the world for centuries, we, engineers, create visual aids as we go through the system life cycle to orient ourselves and our teams when things are too abstract and intangible. Engineering design heavily relies on diagrammatic reasoning, and this type of reasoning mostly deals with flat diagrams we create to transform the depth of the system into a surface. Our screens or printed paper becomes projection space to assist our thinking. Diagramming also needs to be highly schematic; designs can be represented in various ways, as simple boxes with arrows or some other more complex types. As in all diagrams, a schema is the key for any eventual consumer to extract the same information that the diagram generator decided to convey when it was drawn.

We, engineers, are highly visual people. If you don't fully agree with me, here's an experiment you can try on your next meeting, if you happen to be bored (which is probably going to be the case): as the meeting tediously progresses, and if no one is presenting or drawing on the whiteboard, just stand up and scribble some rudimentary and simple diagram (but loosely connected to the discussion, so it doesn't look too out of the blue) on the board. Now see how everybody will start consistently staring at that figure for the rest of the meeting, even when the discussion has passed the point where the rudimentary drawing was relevant. We desperately search for visual clues when we discuss things.

There is a fundamental problem with all block diagrams around the world, no matter the disciplines: they are only fully understood by those who drew them. Let's emphasize this again with the first postulate of block diagrams:

1. A block diagram is only fully understood by its creator.

Of course, this also applies to block diagrams in this book. There is always a non-zero "loss of meaning" between diagram creation and consumption; block diagrams are a lossy way of transmitting information between a producer and a consumer. The more complex a block diagram gets, the thicker the semantic barrier and therefore the meaning loss. The previous postulate can have a corollary, which is:

2. When a block diagram is shown to an audience, there are as many interpretations as people in the audience.

As if this was not enough, diagrams are mostly manually drawn, so they quickly become a liability and subject to obsolescence, which leads me to another postulate:

3. A block diagram is obsolete as soon as it is released.

Manually created diagrams need to be maintained, and they can easily disconnect with respect to the design activity they were meant to represent in the first place. Block diagrams are instantaneous depictions of the continuously changing underlying subject they attempt to describe. To illustrate how semantics and diagrams relate, let's start with an example of a diagram which will describe the way telemetry data flow to and from spacecraft to ground (Fig. 3.1):

Is it a good diagram? Certainly not. What is wrong with it? Well, it uses the same "box" symbol for different things (a spacecraft is an object, whereas "Transmit Telemetry" sounds like a function). What is more, the same line symbol is used for different kinds of relationships. This example captures the essence of the block diagram problem: users or consumers cannot reconstruct the meaning without a pre-specified schema. All block diagrams try to say something; the problem is that we need clues to understand what the boxes represent and how the lines associate the boxes.

How can we improve it? Let's try this (Fig. 3.2):

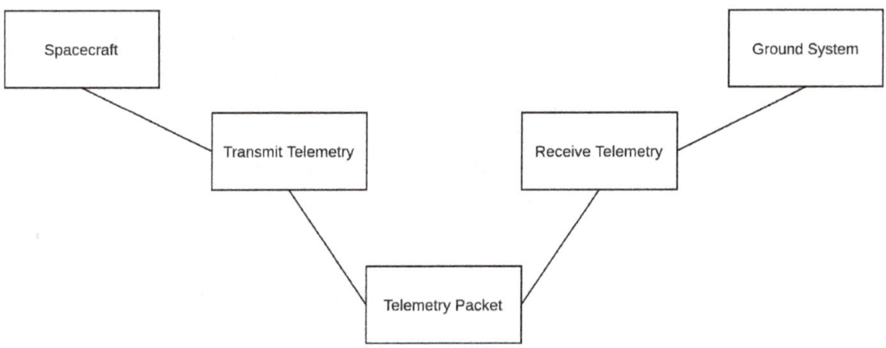

Fig. 3.1 A block diagram representing how some space concepts relate

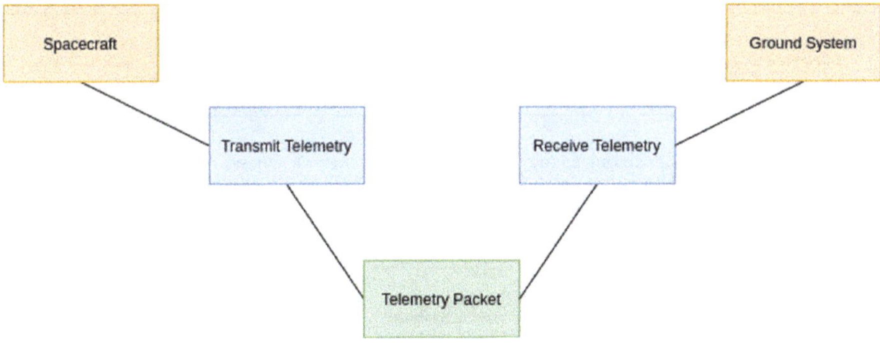

Fig. 3.2 Adding more meaning to the previous model, does it help?

Some color coding was added. “Spacecraft” and “Ground System” have the same color, indicating there is something that relates them, same as “Transmit Telemetry” and “Receive Telemetry”. Does the color coding by itself explain what they share? Does it improve the conveyance of meaning in any way? Not so much, since the color distinctions are merely suggested; they should be made explicit. Rather than just hinting at distinctions with shapes or colors, it is more straightforward if contextual information (meta-information) can be added just as text. This way, there is no hidden convention, but an explicit one. Let’s do so by adding types or classes to be applied to model elements. This again is quite arbitrary, and very domain-dependent: Systems engineers need to convey different things and talk about different types of relationships and objects compared to, say, doctors. The key factor here is that meaning can be defined in many ways, depending on application, domain, or choice of the person drawing the diagram. Is red wine a different type from white, or is it merely a property of wine? It depends on what you want to say about wine. In the domain of systems engineering, we need to clearly represent things like component, interface, function, requirement, work package, product, process, objective, message, and so on. Let’s add this information to the model. For now, every element has one type, denoted like this: « type » and one name, which identifies an individual of that type (see Fig. 3.3).

You can start to extract more meaning from this new graph; it is becoming easier to understand what it is trying to say. Still, the relationships between these boxes remain all the same, let’s create typed relationships (Fig. 3.4):

Stop for a moment and compare this last graph to the first one in this section. See how what was at first a set of meaningless boxes and arrows has transformed into a series of schematic symbols with some meta-information which makes it easier for us to understand what it is trying to say. It is important to police against block diagrams which do not clearly (explicitly) specify what its boxes and arrows mean; an arrow without legends connecting two boxes just means they are related somehow, but who knows why.

Note from the figure that arrows are now directed. This graph is ready to start answering questions, such as:

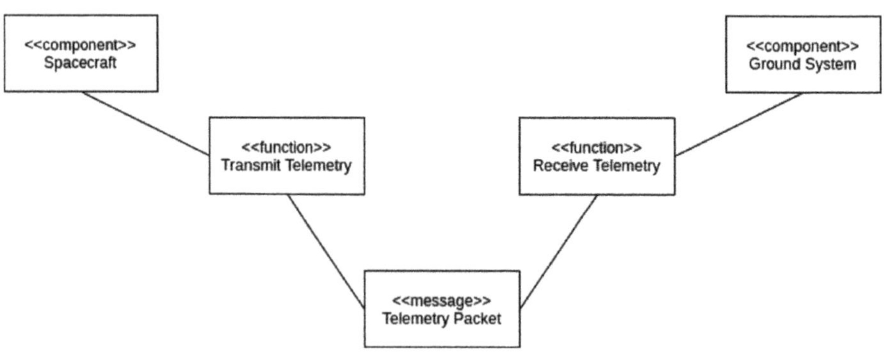

Fig. 3.3 Adding types to boxes

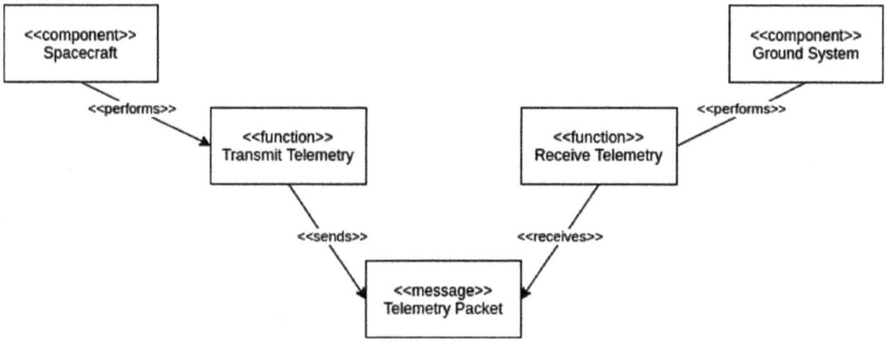

Fig. 3.4 Adding types to relationships

- What function does the Spacecraft component perform? Answer: It transmits telemetry.
- What is common between "Spacecraft" and "Ground System"? Answer: They are both components.
- What messages does the function "Transmit Telemetry" send? Answer: A telemetry packet.

You can ask questions related to function but also related to dependencies, to understand the impact of change. For example:

- What component design may be affected if the definition of telemetry packet changes?

As said, block diagrams must answer questions. The questions may be about the system itself:

- What is it?
- How does it work?
- Is the performance adequate?
- What happens if something breaks?

The questions may be about the model:

- Is it complete?
- Is it consistent?
- Does it support required analyses?

The questions may be about the design artifacts:

- Are all required documents present?
- Does each document contain all required content?

When a diagram is only composed of boxes and arrows with textual information about how all of them relate, it turns into a concept map. Concept maps will be discussed in more detail later in the next section and further in the chapter.

3.2 Concept Maps

A concept map or conceptual diagram is a diagram that depicts suggested relationships between concepts. It is a graphical tool used to organize and structure knowledge. A concept map typically represents ideas and information as boxes or circles, which it connects with labeled arrows in a downward-branching hierarchical structure. The relationship between concepts can be articulated in *linking phrases* such as "causes", "requires", or "contributes to". Words on the line, referred to as *linking words* or *linking phrases*, specify the relationship between the two concepts. We define concept as a perceived regularity in events or objects, or records of events or objects, designated by a label. The label for most concepts is a word, although sometimes symbols such as + or % can be used, and sometimes more than one word is used. Propositions are statements about some object or event in the universe, either naturally occurring or constructed. Propositions contain two or more concepts connected using linking words or phrases to form a meaningful statement. Sometimes these are called semantic units, or units of meaning. Another characteristic of concept maps is that the concepts are represented in a hierarchical fashion with the most inclusive, most general concepts at the top of the map and the more specific, less general concepts arranged hierarchically below. The hierarchical structure for a particular domain of knowledge also depends on the context in which that knowledge is being applied or considered. Therefore, it is best to construct concept maps with reference to some question we seek to answer, which we have called a focus question. The concept map may pertain to some situation or event that we are trying to understand through the organization of knowledge in the form of a concept map, thus providing the context for the concept map (Novak and Cañas 2008).

To structure large bodies of knowledge requires an orderly sequence of iterations between working memory and long-term memory as new knowledge is being received and processed (Anderson 1992). One of the reasons concept mapping is so useful for the facilitation of meaningful learning is that it serves as a kind of template or scaffold to help to organize knowledge and to structure it, even though the structure must be built up piece-by-piece with small units of interacting concept and propositional frameworks (Novak and Cañas 2008).

In learning to construct a concept map, it is important to begin with a domain of knowledge that is very familiar to the person constructing the map. Since concept map structures are dependent on the context in which they will be used, it is best to identify a segment of a text, a laboratory or field activity, or a problem or question that one is trying to understand. This creates a context that will help to determine the hierarchical structure of the concept map. It is also helpful to select a limited domain of knowledge for the first concept maps. A good way to define the context for a concept map is to construct a focus question, that is, a question that clearly specifies the problem or issue the concept map should help to resolve. Every concept map responds to a focus question, and a good focus question can lead to a much richer concept map. When learning to construct concept maps, learners tend to deviate from the focus question and build a concept map that may be related to the domain, but which does not answer the question. It is often stated that the first step to learning about something is to ask the right questions. To start a concept map, it is usually useful to lay out the loose concepts in what is called a “parking lot”. This way, we can observe the boxes without links and start analyzing how these relate according to the context we want to describe. For example, in Fig. 3.5, we laid out the four typical concepts in a space organization:

The way these four things connect greatly depends on the business proposition. One try could be (Fig. 3.6):

A different approach could be (Fig. 3.7):

It quickly appears that the number of concepts laid out might not be enough for a meaningful explanation on what we are trying to describe (the description is too coarse). So, let’s add some more concepts to it, while keeping the original ones with a distinctive green color for awareness.

From the Fig. 3.8, some knowledge can already be extracted:

- In this organization, there seems to be three main products: Spacecraft, raw data and services on top of the raw data. This is not explicitly said, but one can infer those are products since that’s what the organization sells.
- Raw data and services can only exist if a spacecraft exists, since the spacecraft generates the raw data.
- There is no mention if the organization designs and builds the spacecraft or not; it is left unspecified.
- There is no mention if the organization operates the spacecraft or not; it is left unspecified.

Fig. 3.5 Concept maps start with loose concepts in a *parking lot*

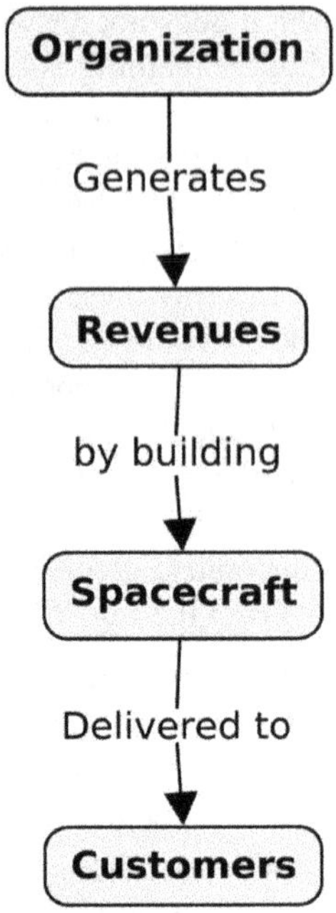

Fig. 3.6 One first attempt to relate the concepts laid out before

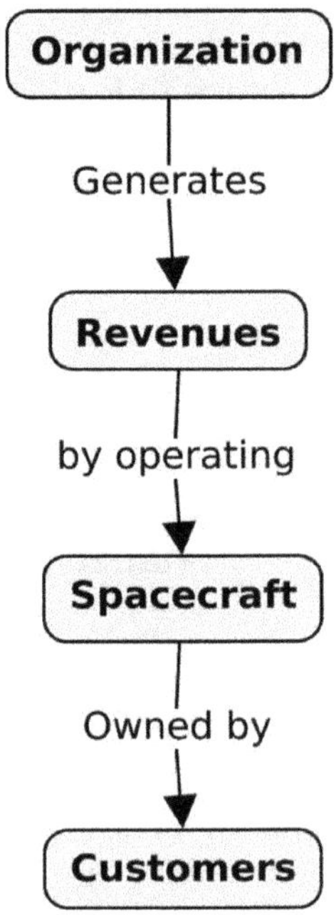

Fig. 3.7 Another way of linking the four concepts

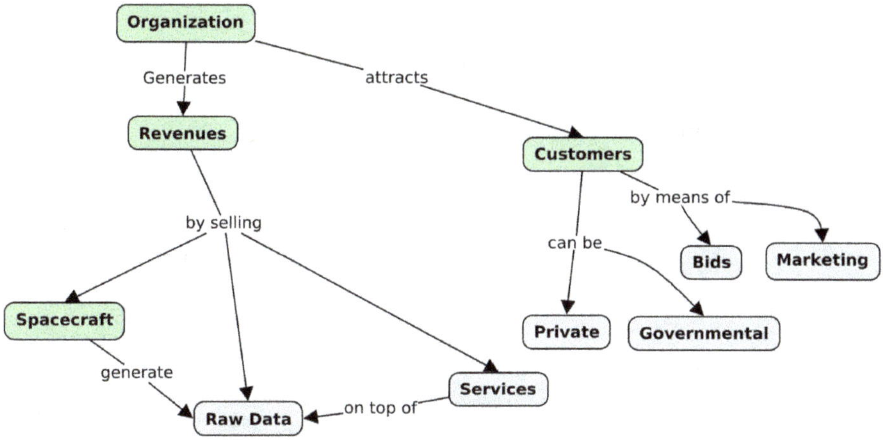

Fig. 3.8 Extending the concept map for a description

It will be discussed a bit further on with knowledge graphs how the missing information can be interpreted in different ways.

Note how a single link changes considerably the scenario:

The red link in Fig. 3.9 states that "Customers Operate the Spacecraft". See the difference now if we move the link to the organization (Fig. 3.10):

There are multiple ways of explaining the same thing, and concept maps are no exception. For example, we could add some more entities for the sake of clarity, like a more explicit "product" entity and make the link between customers and products explicitly with the relationship "consume". Diagrams must be readable and tidy. A

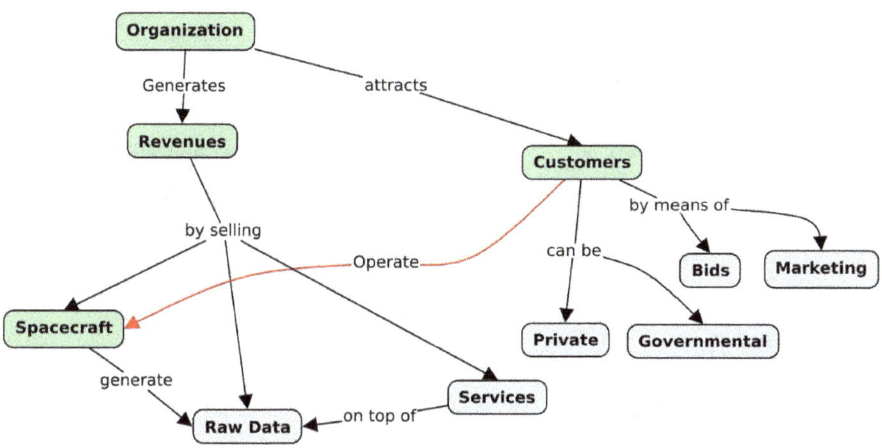

Fig. 3.9 Concept map where customers operate the spacecraft

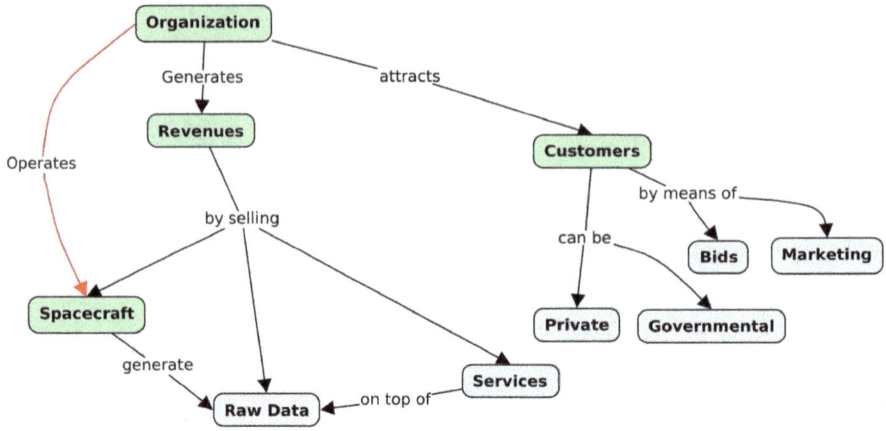

Fig. 3.10 Concept map where the organization operates the spacecraft

good rule of thumb is not to add "common sense" to the layout; i.e. concepts or relationships that are in some way obvious (Fig. 3.11).

Concept maps are great diagrammatic tools for early-stage organizations where ideas are still fluid. Something that needs to be considered is that concept maps are not explicitly hierarchical; some hierarchy could be added to them by means of adding the right links, for example "is A" or "is Composed of" links.

Concept maps help solidify concepts that might not be fully understood, yet across the organization they turn implicit assumptions into explicit statements. Concept

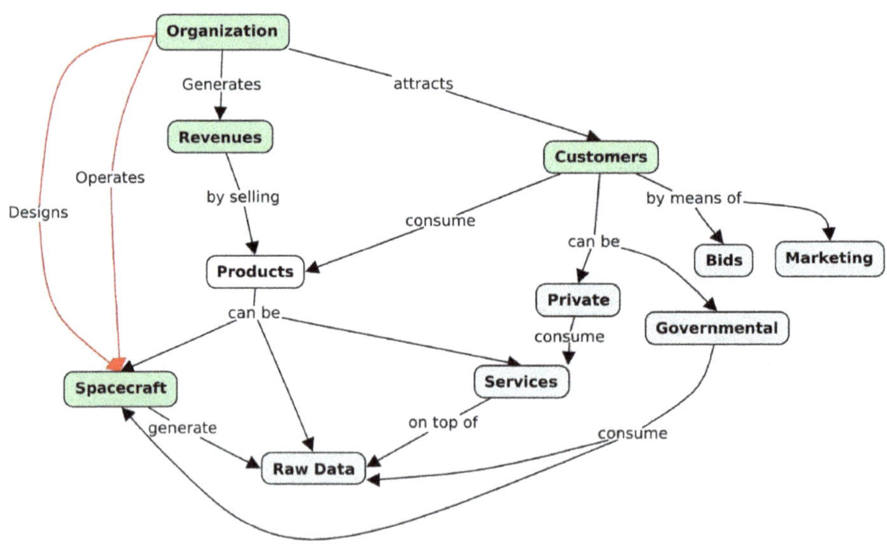

Fig. 3.11 Rearranging the concept map and making the product entity explicit

maps can spark discussions and debate: if they do, it surely indicates a concept needs rework. At very early stages of a NewSpace organization, the documentation should mostly be composed of concept maps!

Concept maps in this section were produced using CmapTools version 6.04. CmapTools is free software from the Institute of Human and Machine Cognition (http://cmap.ihmc.us/), and it runs on Linux, Windows, and Mac.

3.3 Hierarchical Structures

Breakdown structures and hierarchies are related concepts, but they are not synonyms. When we break things down (work, a system, a product, or an organization), we define a set of elements or components a certain object is made of, but to decide how those components relate to each other is another story.

In hierarchies lie one of the main "secrets" of successful (or dreadful) engineering. Hierarchies are all around us, and they are a trap in some way. Any design activity (such as engineering) requires arranging the constituent elements of a system together, but it also requires arranging the data and information generated in this process. Hierarchies are inescapable when it comes to arrange anything non-atomic. It is not enough to identify the building blocks; it is also needed to identify how those building blocks will interact in order to provide the overall function of the object. Hierarchies are always shaped by the context. System design is about multiple hierarchical arrangements interacting together, notably the system-of-interest internal arrangement, and the organization. This has been the subject of research and it has been called *mirroring hypothesis*. In its simplest form, the mirroring hypothesis suggests that the organizational patterns of a development project, such as communication links, geographic collocation, team and firm membership, correspond to the technical patterns of dependency in the system under development (Colfer and Baldwin 2010). According to the hypothesis, independent, dispersed contributors develop largely modular designs (more on modularity later), while richly interacting, collocated contributors develop highly integral designs. Yet many development projects do not conform to the mirroring hypothesis. The relevant literature about this mirroring effect is scattered across several fields in management, economics, and engineering, and there are significant differences in how the hypothesis is interpreted in the different streams of literature. This was also captured by Conway's law (an adage more than a law): "organizations which design systems are constrained to produce designs which are copies of the communication structures of these organizations" (Conway 1968). Some different research notes that organizations are boundedly rational and, hence, their knowledge and information processing structure come to mirror the internal structure of the product they are designing (Henderson and Clark 1990). These theories propose that the way people group and interact toward creating a technical artifact is reflected not only in the design and architecture synthesized but also in how information and data is structured. This implicitly suggests a unidirectional influence from the team structure to the system under design. But how about in the opposite direction? Can the system architecture also exert forces to re-shape the way designers' group together? In any case, we are talking about two hierarchies (the group's and the system's) interacting with each other. When it comes to assign relationships between entities in a hierarchical structure, there are several ways. Let's use a typical example. Let's consider folders in a hard disk drive. Every time we start a new folder structure in our external drive, we probably struggle with the same problem:

- Should you create a parent folder called Software and then create a folder called Flight Computer in that?
- Or should you create a folder called Flight computer and then create a folder called Software in that?

Both seem to work. But which is right? Which is best? The idea of categorical hierarchies was that there were "parent" types that were more general and that children types were more specialized versions of those types. And even more specialized as we make our way down the chain. But if a parent and child can arbitrarily switch places, then clearly something is odd with this approach. And the reason why both approaches can actually work depends on perspective, or what some systems engineering literature call viewpoints. Systemic thinking, at the analytical level, is a construction we use to understand something that otherwise is too complex for us. We can choose to observe a complex object using different systemic lenses when we analyze it, depending on our needs. But once we synthesize the system, the structure becomes more rigid and less malleable. Following the example above, once you choose the folder structure, that structure will impose constraints in the way you can use it. A file in a specific folder is right there in that folder and not somewhere else. The file sits there because of the structure chosen, and the way that file relates to the rest as well. Let's continue the discussion on what type of hierarchies are there before we elaborate on this and how it impacts knowledge management.

There are two main types of relationships in hierarchies: classification/categorical or compositional/containment. Categorical hierarchies are very much related to knowledge management—relating ideas, concepts, or things into types or classes. Categorical hierarchies express specialization from a broader type to more specialized subtypes (or generalization from the subtypes upstream). Another way of relating things or ideas together is using a containment hierarchy, by identifying their parts or aspects; containment hierarchy is an ordering of the parts that make up a system—the system is "composed" of these parts. In a nutshell, containment hierarchies describe structural arrangement (what's inside what). Containment hierarchies also show what is called emergent properties: behaviors and functions that are not seen at the subpart level. An aircraft, for example, can only show its emergent properties when assembled altogether; you can't fly home sitting on a turbofan turbine. If you look at the real world, you'll find containment hierarchies everywhere; categorical hierarchies have rare real-world analogy. The real world is filled with containment hierarchies. All artificial systems out there are basically containers, like your car, your laptop, a chip, a smartphone, or a spacecraft; if you look around, you will perhaps quickly realize that all the things we engineer are just different types of containers. This is the main factor why categorical hierarchies seldom work for representing physical reality; things around us are mostly boxes inside boxes. But containment relationships are somewhat rigid: if something belongs to one box, it cannot belong to some other box, because the structure dictates it so. System structure does not allow many "perspectives" about it once it's established. The main challenge lies in the fact that structuring our work, information and thinking around system design is a dynamic activity; things can group differently according to the situation and the context. Imagine we want to run a preflight subsystem review of the flight computer, so we want to list everything related to it for the review. In that sense, we need the "Software" folder to appear under "Flight computer", along with any other thing that flight computer contains, since we want to review it as a whole. But what if the review is a system-level software review? If that is the case, we want to list all the software bits and parts of the system. In that context, "Flight computer" appears under "Software".

The way engineering teams organize is frequently discipline-oriented. We as professionals specialize in things or topics by means of our education: such as engineering, accounting, law, or medicine, and so on. And inside those professions, we also tend to group with people from our same "flavor" or specialty. So, when companies are small, it is understandable software engineers hang around together, and mech engineers do the same. These are *categories* in the taxonomical sense of the word. Then, it should be no surprise that young organizations, in the absence of an active command, naturally group categorically. Ironically, this is the seed of what is called functional organizational structure. It is very typical to find R&D groups structured in such categorical (functional) ways, keeping these reservoirs of people subdivided by what they know to do, with one functional leader supervising them. Categorical hierarchies and function are at odds but for some reason they are called.

There is even a more interesting caveat. We analyzed in the previous section that WBS, as all the other breakdown structures, is a function of time. The system has a totally unknown structure (architecture) in the beginning; the design process is supposed to come up with it. The architecture evolves from very coarse blocks into more detailed subsystems and parts as the design progresses. Until the component structure and requirements are well understood, it can be challenging to allocate a team for it, since the overall understanding is poor. System components will require attention from many different functional areas (a flight computer will require software resources, electrical resources, mechanical, thermal, and so on) and if the functional resources remain scattered at their home bases, the mapping between component and team is broken since the design work for such component will undoubtedly require collaboration which will span multiple groups, boundaries, and jurisdictions. Responsibility and authority are severely affected. A way to tackle this is to define a matrix organization. This approach is not new, and many companies in the aerospace industry have been moving/co-locating specialized resources to work on specific subsystems and projects, in an attempt to ease the friction between system and group hierarchies. Then, specialized/functional engineers would move from their home bases to different projects or subsystems, and once these projects are done, the engineers would come back to their functional homeland until the cycle is restarted. This

Fig. 3.12 Dual authority dilemma of matrix orgs

approach has several known drawbacks, and some hidden ones. Duality of authority is probably the classic, where the practitioners report to both their functional and project/subsystem leads (Fig. 3.12).

Lack of org design consistency creates hybrid organizations. In these situations, a particular department chooses one org layout (say, functional), whereas another one chooses matrix. This is perhaps the worst-case scenario possible, since subsystems and components will most likely require input from across the organization one way or the other. Without "big system" level harmonization of org philosophies, the friction in hierarchies increases and becomes a source of constant maintenance. For multidisciplinary endeavors, a sound approach is to implement integrated product teams; this differs from the matrix organization in the sense it eliminates the "functional" home base. The engineering department has a pool of *capabilities* (which are managed and assured by executive leadership) and those capabilities are allocated to components and subsystem projects as the needs arise and as the architecture evolves. More information about IPTs are given in Sect. 3.7.

3.4 Projects, Products, Systems (And Role Engineering)

A classic dilemma for organizations which develop systems is how to organize the teams better as the system evolves from whiteboard to production. The fact the engineering world is prone to both evangelism and terminology vagueness does not make things much easier. One of the latest confusions is with the word *product* and the word *ownership*, which are used in very light manners and for a multiple variety of different things, often sparking antagonisms. One of those antagonisms is the "function vs product". This one is not an option in this book, as I have probably insisted enough: to avoid damaging frictions between the social structure and the system structure, the work breakdown structure must dictate how teams are formed according to the system's hierarchy. More accurately, the system's hierarchy dictates how the WBS forms and from there it spills to the teams' structures. Failing to do so will reveal the *hierarchy problem* in full power and make everything go downhill from there. Functions must only exist to be fed to multidisciplinary groups which own some element of the WBS.

Another antagonism is the "product vs project"; or "product vs system". Hours and hours are spent to argue and discuss which one is better: projects or products. Just google "product manager vs project manager" to see a big dog chasing its own tail. If you ask a project evangelist, of course "everything is a project"; ask the same to a product guy, he will say everything is a product. For some reason projects have been lately flagged as old-fashioned, and too *waterfall-ey*. This only reveals an alarming level of ignorance, since a project, just the term itself, does not specify in which way the tasks are handled versus time. But there are more infectious misconceptions about projects that are parroted by many. One of the most surprising definitions found about projects is that they are "temporary, one-off endeavors", with "clear beginning and an end", whereas products are more *permanent* endeavors that constantly *evolve* to fulfill customer's needs. Another hilarious one: "A project manager's job is to ensure that the project is completed on time and on budget. The product manager's job is to ensure that the right product gets built". Like if a project manager would go and build the wrong thing as long as it is on time and budget. Again, dictionary says:

- **Product**: "*a thing that is the result of an action or process*".
- **Project**: "*a set of coordination activities to organize and plan a collaborative enterprise to realize a system or product*".

In fact, as soon as we are dragged to the discussion, it can be quickly realized that the whole argument is more about role naming and role evangelism than about the actual definitions of project and product. The definitions are pretty clear: a product is an outcome; a project is a process. We stated before that design process expands to infinity if we allow it. Products are, then, arbitrary snapshots of our designs, at arbitrary times, which are interesting enough for a customer to pay for it. The concept is straightforward: to create a good product we must organize the multidisciplinary design process in a project which will handle dependencies, budget, and scope, following a system architecture defined by systems engineering. Budget for the project is overseen by the project but also acts as a constraint (or feedback loop) for systems engineering to constraint the solution space. From this budget, a cost for the product is obtained which defines as well the price the customer will end up paying for (adding profit, margins, etc.). Graphically (Fig. 3.13):

For product, in this context, we refer to *something* technical that a customer needs; this customer is usually an external entity willing to engage in a commercial relationship in order to acquire (pay for) such a product. In contrast, any technical artifact the organization internally needs for its own use is more precisely a system, not a product. Calling it a product is misleading since it removes the technical and multidisciplinary nature it surely carries. A pizza and a Boeing 787 are both products, but only one is a system. A server for ingesting and storing a company's spacecraft payload data is a subsystem, not a product. It is a product from the manufacturer's perspective, but not from the end user company's perspective. There is no connection

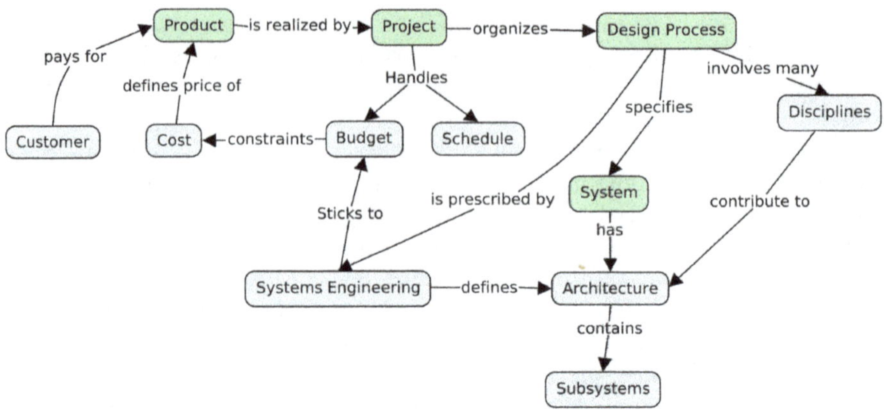

Fig. 3.13 Relationship between product, project, system and the multidisciplinary design process

between the term product and the internal complexity of the object it represents. The term *system*, on the other hand, quickly brings a mental idea of internal composition and some assumed minimum level of complexity.

Systems engineering oversees specifying and prescribing the design process while the project, as an overarching process, coordinates the collaborative effort toward a desired outcome. The technical products we create always need an underlying project. On the other hand, projects do not always create products as outputs. In the technology domain, a project can generate a disparate set of outcomes: systems, studies, investigations, etc. And in life in general, even more disparate outcomes: I recently half-destroyed a wall at home to hang a bookshelf. It was a project for me, from end to end. A long, painful project. I had to measure things, cut the wood, get the fixation hardware, and execute. I cannot call that bookshelf hanging from the wall a product, can I? In short, the term product is just an abstraction to manage a bit easier the outcomes of a project, but as any abstraction it hides details which might be good to keep explicit when it comes to creating complex systems. Picture a fancy dish out of a kitchen in a luxurious restaurant where the *Chef* (the systems engineer) works with the *Sous chef* (project manager) to coordinate with the *Chefs de Partie* (mechanics, electronics, software) to realize her recipe. Restaurants don't usually call their fancy dishes *products*; such a term can opaque the delicate handcrafted nature of cuisine. The term product carries connotations which can be misleading if not properly used.

Having said this, and taking into account that products could be called *plumbuses*[1] and nothing would really change, all the rest about the product versus project dilemma (which actually is not a dilemma as we just discussed) is irrelevant *role engineering*, or the urge to pay more attention to titles and roles than the actual work needed. These types of pointless arguments are a function of startup size, just because the probability of an evangelist coming through the door increases with headcount. Such arguments

[1]Plumbus: Everyone has one at home, so there is no need to explain what a plumbus is.

turn simple discussions into long and philosophical debates and steer the whole point away from what is important (such as minimizing time to market, have a better market fit, increase reliability, etc.) to peripheral questions about how people are called and who *owns* what (there seems to be a growing interest to assign *ownership* lately in engineering design). The product versus project argument is nothing else than the result of proliferating *role engineering*, which stems from role-centered methodologies. These methodologies propose an inverted approach: (a) Let's define the roles first; (b) Second, let's discuss how these roles fit the work; (c) If the work does not fit the roles, let's change the work to make it fit the roles. Roles are so rigidly specified that the main objective becomes to comply with the role scope. This is the one long-lasting effect of transplanting methods and roles from single-discipline domains (such as software, i.e. Scrum) to multidisciplinary enterprises. The fact that multiple disciplines need to be orchestrated to create reliable technical products is what differentiates it from single-discipline engineering, for example, software. This is not, per se, a critique to Scrum. The only observation here is that the system architecture and the work breakdown must always drive the way teams are formed, selecting the best from each discipline and creating tightly knitted integrated teams. The task of assigning roles and the names we put in our badges and email signatures must be an effect, and not a cause. We must know what to do before knowing who will do it and what role will this person have, and not vice versa.

Product and project are complementary concepts, not antagonistic. Even symbiotic if you want: you cannot expect any results (i.e. a good product as an outcome) if you do not engage in a collaborative enterprise to make it happen; i.e. a project. Can projects and products coexist then? Mind-blowing answer: yes, they can, and they must. To consider the word "project" as outdated or "too waterfall" is just evangelistic parroting, or just blissful ignorance.

3.5 Product Identification, Management, and Modularization

We may be witnessing a sort of product *beatlemania* these days, and this perhaps stems from the fact that product is a very slippery term in tech. Those industries where the product is easily identifiable and with clear boundaries, like pizza, count with the great advantage of having clarity and unambiguity. Many other industries, such as space, have hard times identifying what the product *is* and what the product *is not*. Everything can be a product if you are vague enough. The definition of product and its boundaries can change dozens of times throughout the life of an early-stage enterprise.

If we stay away from the hysterical "freestyle" use of the word *product* discussed in the previous section and we stick to the actual meaning of it (*product* as something a customer wants to pay for and an instrument to make profits), then product

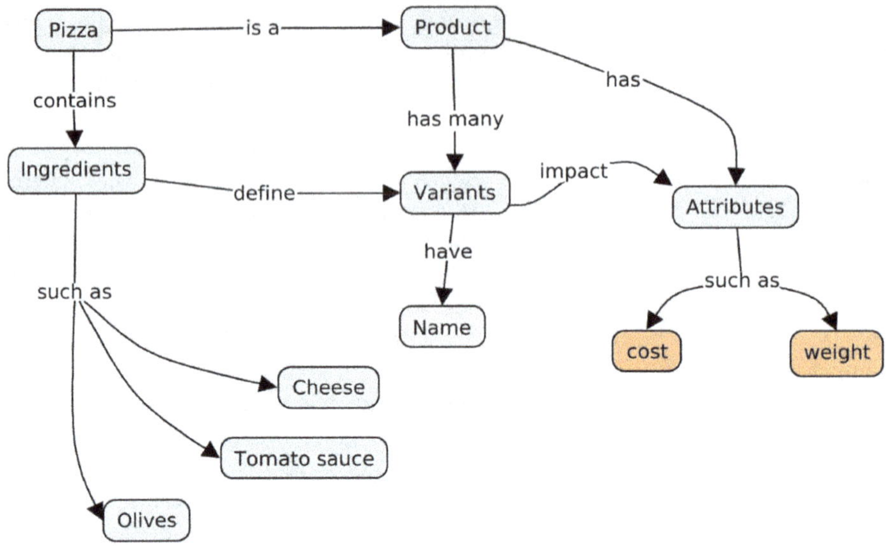

Fig. 3.14 Product management for pizza

management is actually an extremely needed activity, also for early-stage organizations where products may be still fluid. Having a solid understanding across the board on what the product is and iterating through different options and attributes is not less than essential from day one. Once prototypes or proofs-of-concept reach minimum-viable-product (MVP) stage and can be sold to customers, even as early adopters, the organization needs to align on how to call it, what variants to consider versus what not to consider, what attributes are open for customers to modify, and so on. Product management in tech is a combinatorial and experimental activity where multiple combinations of constituent parts can lead to multiple different "flavors" or variants of a product. We said before that pizza is perhaps the paradigm of simple product management, so we can use it as an example in a concept map (Fig. 3.14). It is hard to imagine the owners of a pizza place sitting down to brainstorm what the product is. The product is pizza. People are willing to give their money to put that in their stomachs and feel satisfied. There is no product identification needed; things are crystal clear.

Product management for a pizza place is about combinations. The pizza people must sit down and define what variants will be offered versus what will not. As depicted in the concept map above, different variants of pizza are defined by combining a finite combination of ingredients. The variants must be distinctively named (as an overarching abstraction from the ingredients list). The variants modify or impact product attributes, such as cost and weight. Product management is also about constraining what will *not* be offered. For example, the pizza place could decide not to deliver pizzas and only work with table service. Some attributes can be "optimized" according to customer needs, and product management must define which degree of freedom the products will have to adjust those attributes. Different customers will have different optimization wishes. Some customers may want to get the cheapest pizza possible (optimizing for minimum cost), whereas some other customers may have a different criterion. Product management must define constraints and limits to all these combinations. The pizza concept map is pretty much universal for any other product, and it clearly depicts the highly combinatorial nature of product management. Product management runs with the disadvantage of easily getting lost in the sea of combinations of "ingredients" and variants, going in circles and getting awfully abstract. The product management activity must go incremental and iterate, testing products with early adopters and adjusting the product line according to that.

Leaving the pizza analogy behind, the following concept map captures the product management relationships for a NewSpace company which sells spacecraft. In orange color, there are attributes which can be altered depending on the selection of different on-board elements.

From the Fig. 3.15, some resemblance with the pizza concept can be seen; actually, this concept map is the pizza one modified. Some things to extract from the graph; spacecraft attributes to be optimized can be many. A spacecraft can be optimized for specific application (optical, radar, etc.), for mass (to make the launch cheaper), etc.

One of the key decisions which must be considered in product management, with close collaboration with systems engineering, is modularization. Modularization is the process of subdividing a product into a collection of distinct modules or building blocks with well-defined interfaces. A modular architecture is a collection of building blocks which are used to deliver an evolving family of products. A modular system

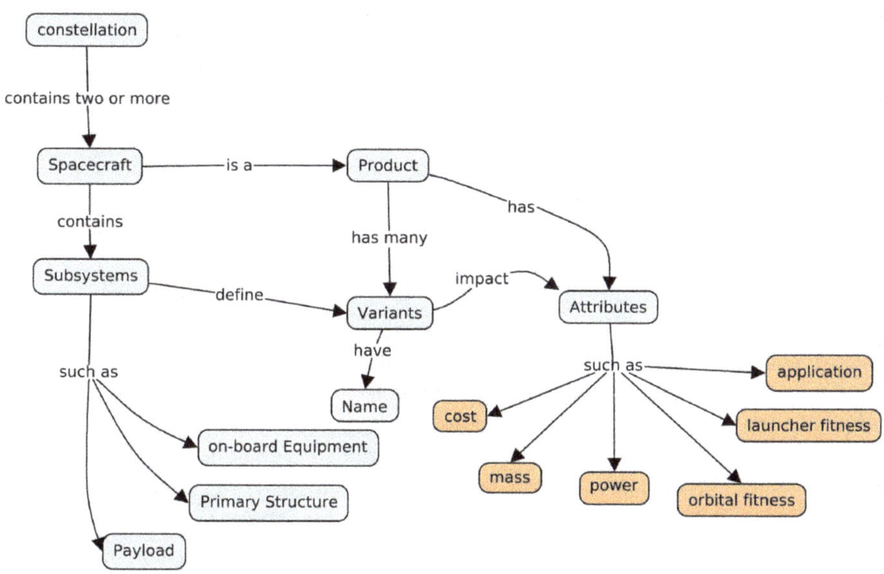

Fig. 3.15 Concept map depicting the relationship between product, attributes, and variants

architecture divides the whole system into individual modules that can be developed independently and then be plugged together. New products can emerge from adding or re-combining modules, as when adjusting to new market demands, new technological developments, or the phasing out of declining product options. A modular system can be imagined as the opposite to an all-in-one, one-size-fits-all, or "uniquely designed for a specific situation, but not adjustable" approach to product design (Fuchs and Golenhofen 2019). More on modular design is discussed in Chap. 5.

3.6 Ownership

Ownership is another buzzword that has found its place quite up in the rankings as one of the most parroted words around the workplace. Scrum greatly contributed to the spread of this *ownership* fashion by explicitly introducing the term "owner" in a role: the product owner. The product owner represents the product's stakeholders and is the voice of the customer. He is responsible for delivering good business results (Rubin 2013). He is accountable for the product progress and for maximizing the value that the team delivers (McGreal and Jocham 2018). The product owner defines the product in customer-centric terms (typically user stories), adds them to the product backlog, and prioritizes them based on importance and dependencies (Morris 2017). The product owner is a multifaceted role that unites the authority and responsibility traditionally scattered across separate roles, including the customer, the product manager, and the project manager. Its specific shape is context-sensitive: It depends on the nature of the product, the stage of the product life cycle, and the size of the project, among other factors. For example, the product owner responsible for a new product consisting of software, hardware, and mechanics will need different competencies than one who is leading the effort to enhance a web application. Similarly, a product owner working with a large Scrum project will require different skills than one collaborating with only one or two teams (Pichler 2010). The product owner has reached the deity status in Scrum circles, with sort of magical superpowers. Nowadays, the product owner is usually the answer for every question. Thing is, the highly centralized nature of the product owner role collides with the natural collaborative DNA of multidisciplinary engineering design. Rubin (2013), for instance, states that "the Product Owner is the empowered central point of product leadership. He is the single authority responsible for deciding which features and functionality to build and the order in which to build them. The product owner maintains and communicates to all other participants a clear vision of what the Scrum team is trying to achieve. As such, the product owner is responsible for the overall success of the solution being developed or maintained". There seems to be as many product owner definitions as authors there are. Not a lot is written about the relationship between product owners and the system hierarchy. Pichler (2010), for example, refers to "Product owner hierarchies", and states that these vary from a small team of product owners with a chief product owner to a complex structure with several levels of collaborating product owners. This somehow indicates that there can be a group or team of product owners coordinated by a chief. How this hierarchy maps to the system hierarchy is not specified.

Having reviewed the software-originated product owner concept, how is ownership interpreted in systems engineering and multidisciplinary design contexts?

As we engineer our systems, we become somewhat entangled to them. Since systems are the result of our intellects, our ideas and thinking are reflected in them, and our intellectual liability grows as the system's life cycle progresses. We might not own the *property* the systems we design are made of (which is owned by the company), but we *own* the intellectual background story behind them. All the decision-making, the trade-offs, the mistakes; all is wired in our brains. Perhaps not in a very vivid manner, but that information we do possess (i.e. own) and it creates ties from us to the system. That's the ownership sense of systems engineering. As perpetrators of our designs, we become the ultimate source of truth for them, throughout its life cycle. This intellectual liability makes us responsible for what we devise with our ingenuity: we know the design the best, probably nobody else does at the same depth. The knowledge about the design resides so deeply in us that even though some of that knowledge could be transferred, the fact we have designed it puts us in a privileged position in terms of understanding. This intellectual advantage does not come for free: we are the ones to respond if the component or element, for whatever reason, does not work as expected.

In short:

1. Responsibility on the space systems we design as engineers ends when they re-enter the atmosphere and evaporate.

It does not change if we work on systems that do not go to space. Still, the things that sit on the ground are enabling and/or supporting assets in orbit, and because of that the statement above remains valid, but can be slightly re-written as:

2. Responsibility on the things we create ends when the systems our work supports re-enter the atmosphere and evaporate.
3. Put slightly different: design responsibility ends when the system or object we have created has officially reached its irreversible end-of-life.

And the same goes if we work on something which goes on top of something that enables or supports something else that goes to space. That's the beauty of the systemic approach: our work is a building block in a bigger architecture whose total output is more than the sum of its parts; this means that our work, whatever it sits in the hierarchy, is not less important than anything else; if our (sub)system or component does not work as expected, the whole system underperforms. Responsibility and accountability only fully decay and disappears when the system is completely retired.

3.7 Integrated Product Teams (IPT)

Perhaps the bridge between the product owner concept referred before and the sense of *ownership* in systems engineering is the integrated product team. Here, the use of the term "product" refers to anything multidisciplinary that must be created, either for external or internal use. As the architecture of the system forms and blooms, and the work is broken down accordingly, small multidisciplinary teams can be incrementally allocated to synthesize different blocks of the architecture. This is what is called an integrated product team (IPT). IPTs are small multidisciplinary groups populated with resources from different disciplines/functional areas, who are tasked to run studies, investigate, or design and build. Although IPTs have some origin in the "big space" or "classic" systems engineering, which we described in some chapters above, it is not a concept which cannot be adapted to small organizations. IPTs do not have to be created upfront, and they probably cannot be created upfront when the organization is very small, but they can be incrementally allocated as the system architecture matures. In fact, every NewSpace org begins with an IPT, which we can call stem[2] IPT, and then this mother-of-all-IPTs spawns children IPTs as the architecture spawns its children subsystems. This way, as the organization grows both its design and social complexity, the way people group, and the system architecture remain aligned, because the system architecture drives it. Conway law (again, not a law, but just an observation) has been largely misinterpreted as something we have to live with, like if the system architecture is hopelessly a victim of how people group. What is more, Mel Conway's observation is more than 50 years old…it dates from 1968! The context of systems and software design in 1968 was different than today; today teams communicate in ways that were unimaginable in 1968. Perhaps teams were sending letters to communicate decisions in 1968, no wonder things were growing apart and architectures were reflecting this "insular" team effects. A software architecture in 1968 has probably little to do with a software architecture in 2020. That is something every NewSpace organization must understand very well: there are degrees of freedom on both ends. Companies must stop relying on very outdated observations as ultimate truths. There is coupling between org and system, that is out of discussion (since one creates the other, the coupling is more than obvious), but both ends can be adjusted accordingly. It can be true that some group of people have been working together for some time and it would make sense to keep them together since they seem to bond well. The IPT definition process must take all these topics into account. An IPT is not just an ensemble of disciplines, no matter the affinity between them. An IPT needs to work as a solid unit, or as DeMarco and Lister (1999) state, *jell*, and for that all the sociological factors a healthy team needs must be taken into account. Systems engineering must ensure there is integrity not only between IPTs across the organization but also within the IPT itself. But IPTs must be autonomous and self-contained, and the systems engineering should refrain from micromanaging it. Hence the leadership of an IPT must be carefully populated

[2]Here I borrow the word from botany. It is the main trunk of a plant which eventually develops buds and shoots.

with individuals who are systems thinkers but also great communicators, coaches, and leaders. IPTs must be teams and not cliques (DeMarco and Lister 1999), and this also includes the stem, precursor IPT. Given that NewSpace organizations are usually populated by young people, it is of great importance as the members of the stem IPT are trained in leadership, systems thinking, and refine their soft skills since they will become the IPT leaders of tomorrow.

IPT is fully responsible for a specific subsystem or component of the system hierarchy, throughout the component's entire life cycle. For a spacecraft, this also means assembly, integration, and verification (AIV). Then, the IPT is responsible not only for designing the subsystem but also to put it together and make sure it works as specified before by integrating it to the system. Once the component is integrated, the system-level AIV team takes over and remains in close contact with the IPT for a seamless integration process. To ensure the verification of the subsystem is reasonably independent (the IPT verifying its own work could be prone to biases), the system AIV runs their independent verifications and ensures that results match with what the IPT team stated.

Making the IPT intellectually responsible for the design process and life cycle of that component of the hierarchy decentralizes the decision-making. Of course, the IPT cannot go for great distances without collaborating and agreeing with other IPTs and with the systems engineer. By being self-contained, IPTs help break the dual-command problem typically found in matrix organizations, where the leadership is typically split between the function and the "verticals" (i.e. components/subsystems). The IPT maps directly to a block in the system hierarchy and straight to the WBS, creating a unit of work that project management can gauge and supervise as a solid entity with clear boundaries and faces, streamlining monitoring and resource allocation. Ideally, an IPT exists while the subsystem exists, up to operations and retirement (remember the intellectual responsibility burns in the atmosphere with the re-entering spacecraft). In small organizations like startups, keeping an IPT together can happen once the subsystem that has been launched can be prohibitively expensive, since basically the IPT would turn into a support team. Here, two things are important: (a) The IPT has surely accumulated a great deal of knowledge about the subsystem delivered. The organization must ensure this knowledge is secured before disbanding the team; (b) If the subsystem delivered will be re-spun or a new iteration or evolution is planned in the foreseeable future, it is better to keep the IPT formed instead of eventually forming a new one. Systems engineering is essential in IPT reuse as the system design evolves.

IPTs can also be allocated for research and development, advanced concepts, or innovation projects by creating *skunkworks*[3] flavored IPT with loose requirements and an approach which is closer to what some research calls *improvisation* principles. Improvisation (understood as improvisation in music and art) is defined as a creative

[3]A skunkworks project is a project developed by a relatively small and loosely structured group of people who research and develop a project primarily for the sake of radical innovation. The term originated with Lockheed's World War II *Skunk Works* project.

act composed without a deep prior thought; this includes creative collaboration, supports spontaneity, and learns through trial and error (Gerber 2007).

As mentioned before, autonomy is key for IPTs. They should not be looking to higher management permission for their decision-making. They should, however, properly communicate and inform their actions and decisions to others, mainly other IPTs directly interfacing with it, but also the system-level integration team, and of course project management. Autonomy must not be mistaken as a license for isolation. An isolated IPT is a red flag that systems engineering must be able to detect in time. The value of the integrated teams is that they can focus on their local problems but the whole must remain always greater than the sum of the parts. This means the true value of the IPTs is in their interactions, and not in their individual contributions. A point to pay special attention is that system complexity can eventually spawn too many IPTs running in parallel if the architecture has too many components; for small R&D teams, this can impose a challenge to track all the concurrent work and create conflicts with people being multiplexed to multiple IPTs at the same time, breaking their self-contained nature and mission. It can be seen that the interaction between the system under design and the way we group people around it is unavoidable: too granular architectures can be very difficult to handle from a project perspective. Another aspect to consider with IPTs is not to confuse autonomy with isolation. IPTs are meant to be aligned with the system life cycle and structure, so they need to keep the interfaces to other subsystems healthy and streamline communication.

Finally, a mention on IPT and multitasking. We said ad nauseam that IPTs must be responsible for a block of the architecture and realize that block end-to-end, from the whiteboard to operations. Truth is that small orgs cannot afford to have one multidisciplinary team fully dedicated to one block of the architecture and absolutely nothing else. So, IPT members usually time-multiplex their tasks (they contribute to more than one subsystem, project, or component at a time), breaking the boundaries or "self-containment" IPTs that are supposed to have. This problem is inherent in small organizations which start to take more tasks than their current teams can support; a natural reaction for every growing business. In startups and even in scaleups, the federated "1 team does one and only 1 thing" approach is a great idea which makes management way leaner, but it requires a critical mass of workforce that is seldom available. In these situations, technical complexity directly drives headcount; in simple words: it does not easily scale. Managers react in two ways to this reality: (a) They acknowledge ballooning the org is a problem in itself and they find ways of organizing the time-multiplexing of tasks for the time being as they incrementally grow the team to meet the needs; (b) they ignore it and start bloating the org with people to reach that critical mass needed for the federated approach to work, which creates not only a problematic transient (a lot of people coming in and trying to get up to speed, draining resources from the existing people) but also a steady-state issue: huge teams, lots of overhead, and a cultural shake. Moreover, massive spikes in team sizes and uneven distribution of management for that growth foster the creation of small empires of power inside the organization which can become problematic on their own; i.e. small organizations spawn inside the organization.

3.8 The Fallacy of Model-Based Systems Engineering (MBSE)

Having discussed diagrammatic reasoning, and block diagrams, let's then dive into a practice which heavily relies on both to convey different kinds of information about a system-of-interest.

"All models are wrong, but some are useful" says the phrase. All thinking is model-based. Thinking involves not only using a model to make predictions but also to create models about the world. The word *model* is probably one of the least accurately used terms in engineering, so let's start by precisely defining it. What are models after all? It is for our need to understand the world that surrounds us that we create simpler representations of a reality that appears as too complex for us from the outset. A model is always a simplified version of something whose complexity we do not need to fully grasp to achieve something with it or to reasonably understand it. We devise scaled versions of things we cannot fully evaluate (like scaled physical mockups of buildings), visual representations of things we cannot see (atoms, electrons, neutrons), or a simplified equivalent of just anything that we don't need to represent in its full extent which help us to solve a problem. All models lack some level of fidelity with respect to the real thing. A model stops being a model if it is as detailed as the thing it represents.

One of modeling foundations is abstraction. I'll restrain myself from the urge of pasting here Wikipedia's definition of abstraction. In short, we all abstract ourselves from *something* in our everyday lives, at all times, even without thinking we do it: the world around us is too complex to be taken literally, so we simplify it ferociously. This enables us to focus on the most important aspects of a problem or situation. This is not new, we all knew it but yes, it has a name. Modeling and abstraction go hand in hand. Ok, but what is to *model*? To model is to create simpler versions of stuff for the sake of gaining understanding. But are *modeling* and *abstraction* synonyms then? Not quite. Modeling is an activity which usually provides as a result of a conceptual representation we can use for a specific goal, whereas abstraction is the mechanism we use to define what details to put aside and what details to include. Abstraction is also a design decision to hide things from users we consider they do not need to see. Drawing the line on where to stop modeling is not trivial, as one can easily end up having a model as complex as the real thing. This has been depicted in Bonini's paradox[4]: too detailed models are as useful as a 1:1 scale map of a territory. Look to Jorge Luis Borges' "On the exactitude of science" to see Bonini's paradox written in a beautiful way:

> … In that Empire, the Art of Cartography reached such Perfection that the map of a single Province occupied a whole City, and the map of the Empire a whole Province. In the course of time, these Disproportionate Maps were found wanting, and the Colleges of Cartographers elevated a Map of the Empire that was of the same scale as the Empire and coincided with it point for point. Less Fond of the Study of Cartography, Subsequent Generations understood that such an expanded Map was Useless, and not without Irreverence they abandoned it to

[4]https://en.wikipedia.org/wiki/Bonini%27s_paradox.

> the Inclemencies of the Sun and of Winters. In the deserts of the West, tattered Ruins of the Map still abide, inhabited by Animals and Beggars; in the whole Country there is no other relic of the Disciplines of Geography.

We all create models, every single day of our lives. We cannot escape models since our brains are modeling machines. When we write the grocery list, we have in our hands or in our phones an artifact which is a simplification from the actual list of real items you will end up carrying in your basket. You leave many details out from this artifact: colors of packaging, weights, even brands for many items (an apple is just an apple after all), and so on. Many things that you don't remotely care about. The grocery list is a model in the strict sense of the word. Now **is** the list the model? It is not. Just as the word "apple" in your grocery list does not mean the real thing, the models we create, either in our heads or on some sort of diagram, are symbols that represent the real thing. And this representation is an incomplete description of the real thing. So, let's get to the point on this:

1. Everybody models, every day.

Then, engineers also model. No matter the discipline, engineers cope with complexity by creating simpler symbolic representations of whatever they're trying to create. These symbolic representations might or might not correlate physically to the actual thing. A CAD or a solid for a mechanical engineer is a physically representative model: the shapes on the representation will match the shapes on the real thing. A schematic diagram for an electrical engineer is a model with loosely physical matching; but a PCB design is a physically representative model. Plain block diagrams written in whiteboards are also models, with or without physical resemblance. Block diagrams are the communication tool of choice when ideas need to be conveyed rapidly. In all these cases (mechanical, electrical, etc.), the modeling activity is fully integrated in the design process, and the practice is very much adopted without questions: nobody remotely sane would design a PCB without knowing what are the building blocks or decide that on the go; in the same way nobody would start stacking bricks for a house without a prior idea on what needs to be built. In all these cases, the models are part of the design *ritual* and you can't circumvent them in order to get to production; I call this *single path* design: the way things are done follows a very clear line among designers, no matter experience or tastes. They will go from A to B doing the same, tooling aside. The implementation effort progresses hand-in-hand with this modeling activity; the more you model, the closer you get to production. Modeling is an inescapable part of the design process. If you ask an electrical or mechanical engineer if he or she is aware they're modeling, they'll probably say no. It is so embedded in the way they work that they don't think about it, but have internalized it. For software, the situation seems to be different.

Software industry has a very distinctive approach to modeling, where the modeling activity stands out, with its own modeling languages. Notably, UML being by far the most popular. But there is some continuous debate in the software world about the need of modeling as a pre-step before implementation, but also about the language itself. Main criticism toward UML is for not giving the right vocabulary for expressing ideas about software architecture. On the one hand, it is too vague to describe ideas; on the other hand, it is excessively meticulous up to the point where it is just easier to implement something directly in object-oriented language rather than draw a UML diagram for it (Frolov 2018). Try now to find if there is any discussion about schematic diagrams being or not necessary for electronics to be implemented. Something is different in software.

Unlike other disciplines, software chose to make modeling very explicit in the design process and tried to formalize its practice. This seems to have created an alternative path for getting from A to B, which breaks the *single-path* design paradigm I commented before. And we as engineers take the least impedance path, just as current does. In software, implementation does not progress as formal (UML) modeling progresses; implementation starts once formal modeling ends. Someone could argue that explicit modeling with diagrams should be considered part of implementation, but that again is open for interpretation. Strictly speaking, you can implement software without drawing a single diagram, while you cannot implement electronics without drawing a single diagram. The key is the formality of the software modeling. Since we cannot help to model as the humans we are, then software engineers also model (if you can call them humans), regardless if they use a formal language or not. Is source code a *model* then? Well, it turns out it is. Models do not need to be boxes and arrows. There's this phrase that goes: "the diagram is not the model", which I have always found confusing. The way I understand the phrase should be:

2. Models do not need to be diagrams

It *can* be a diagram but does not necessarily *need* to be. A piece of code is a model, and its syntax are the artifacts we have at hand to model what we want to model. Someone could argue UML is easier to understand and follow than source code. But is it? If you really, really want to know how a piece of software works, you will sooner than later end up browsing source code. The UML diagrams can be out of sync from the source code since its integrity could have been broken a long time ago without anything crashing; software will continue running, totally unaware of the status of the diagrams. In an ideal world, only source code completely generated from UML models would keep the integrity. But experience shows almost an inverse situation: most of the UML diagrams I have witnessed have been created *from* existing source code. Modeling with UML can reach the point where it becomes easier to implement something directly in object-oriented language rather than drawing a UML diagram for it (Frolov 2018). If it is easier to describe the problem in source code and the diagram proves to be "optional", what is the value of spending time drawing the diagram? Years ago, I had the (bad) luck of attending a course on UML for embedded systems. Course started ok, but suddenly, the instructors started to mix up diagrams together (adding for example state machines in sequence diagrams but also taking many liberties in some class diagrams). I remember pointing that out and even getting a bit triggered about it. The absolute whole point of getting a consistent semantics in diagrams is about being strict on it. If we are not strict, then it just becomes a regular block diagram where all the postulates I listed above apply. So far, this has been mostly about software, but how does all this relate to systems engineering?

Systems engineering deals with multidisciplinary problems, encompassing all disciplines needed to design and build a complex technical system. Explicit modeling has not been a stranger in systems engineering. Model-based systems engineering (MBSE) is the umbrella term for all-things-modeling in the SE world. MBSE was on the hype maybe around 2010 or 2012; everybody was trying to do it. You could feel that if you were not doing MBSE, you were just an outsider; you were old-fashioned. Small- and medium-sized companies were trying to imitate big companies and adopt MBSE methodologies; they were buying training and books and trying to be cool. MBSE promised to change a paradigm: from the document-centric to a model-centric one, where engineers could revolve around "the model" and create documents from it if needed. In the MBSE context, the model is a collection of diagrammatic (graphical) descriptions of both the structure and the behavior of a system by using some language. For creating this model, MBSE decided to use a graphical language, and the choice was, interestingly, SysML, which is sort of a systemic flavor of UML. First problem: SysML *inherits* (pun intended) all the problems UML carries: a semantics that is accessible for some few. Second problem: a multidisciplinary system is made of multidisciplinary teams, which are forced to understand a language which may be largely unfamiliar to them; hard for non-software engineers. It is at least arguable that MBSE claims to streamline communication by choosing a language only a small subset of actors will interpret. Third problem: tooling. Most MBSE tools used to create nice diagrams are proprietary. There are some open source options here and there, but they don't seem to match the commercial counterparts. There are books, courses, training, and there is quite a nice industry around MBSE. Fourth problem: formalizing modeling in systems engineering creates an alternative/optional path which teams might (and will) avoid taking; engineers will always choose the least impedance path. If they can go on and design without MBSE, then why bother? Because management asks for it? Fifth problem: MBSE incentivizes a sort of analysis paralysis[5]: you could model forever before starting to get anything done. There are multiple cases you can find around of different MBSE initiatives where people have been modeling for years, piling up dozens of complicated diagrams, without anything real ever created, reaching very close to the paradoxical Bonini's boundary in terms of detail. MBSE loop goes like this: the longer you model, the higher the fidelity of the model you get, which (in theory) should increase understanding for all stakeholders involved. The longer you model, the more complex the model gets, the less understandable it becomes, and the longer it takes for actual engineering implementation to start. Sixth problem: MBSE diagrams are seldom "digestible" by external tools to extract knowledge from them. They become isolated files in a folder which requires human eyes and a brain in order to interpret them. Hence, their contribution to knowledge management is limited.[6] MBSE is a formal modeling

[5]https://en.wikipedia.org/wiki/Analysis_paralysis.

[6]Some initiatives such as ESA's Open Concurrent Design Tool (OCDT, https://ocdt.esa.int/) are trying to tackle this.

activity which belongs to a path engineers will not take if they don't naturally feel they must take it. Perhaps the important thing to notice about MBSE is: all systems engineering is model-based, regardless of what diagrams or syntax we use. We are modeling machines since the moment we wake up in the morning. Modeling is inescapable; we all do it and will continue doing it at the engineering domain levels in very established ways, without explicitly thinking about it. Systems engineering for any organization should be aware of this and adopt an approach of accompanying and facilitating this activity without trying to force a methodology which might only create an optional, suboptimal path in the design process, adding overhead and increasing costs. In short:

1. All Engineering is Model-Based, thus:
2. All Systems Engineering is Model-Based Systems Engineering.

3.9 Knowledge Graphs

We discussed concept maps a few sections before. But concept maps have to be manually drawn and kept manually updated as the concepts and knowledge evolves. Concept maps are born and meant to be consumed by our human brains. Startups must capture and share knowledge at all cost; a manual approach such a concept map is great compared to nothing. But the best way of capturing knowledge is in formats that computers can process in order to help us connect dots we might not be able to see ourselves with our bare eyes. This section is about discussing more computer "friendly" ways of capturing knowledge.

A concept map is nothing else than a graph, and graphs are representations computers can work better with. In a knowledge graph, nodes represent entities and edges represent relationships between entities. A knowledge graph is a representation of data and data relationships which is used to model which entities and concepts are present in a particular domain and how these entities relate to each other. A knowledge graph can be built manually or using automatic information extraction methods on some source text (e.g. Wikipedia). Given a knowledge graph, statistical methods can be used to expand and complete it by inferring missing facts. Graphs are a diagrammatic representation of ontologies. Every node of this graph stands for a concept. A concept is an entity or unit one can think about. Essentially, two nodes connected by a relationship form a fact. Ontologies can be represented using directed graphs, while some say even more specifically that they are directed acyclic graphs (DAGs). If you need a concrete example of DAGs, note that a PERT diagram is just another case of them. So, existing edges indicate known facts. What about missing edges?

There are two possibilities:

- Closed World Assumption (CWA): non-existing edges indicate false relationships: for example, since there is no isA edge from a Chair to Animal, we infer that a chair is not an animal.

- Open World Assumption (OWA): non-existing edges simply represent unknowns: since there is no *starredIn* edge from Joaquin Phoenix to Joker, we *don't know* whether Joaquin Phoenix starred in Joker or not.

KGs typically include information in hierarchical ways ("*Joaquin Phoenix is an actor, which is a person, which is a living thing*") and constraints ("*a person can only marry another person, not a thing*"). Details and hierarchies can be left implicit; i.e. you must have a proper amount of insight in order to discard some concepts that are "obvious". But what is obvious for you may not be for someone else. Bonini's paradox kicks again: you can describe to ridiculous levels of details a concept or a set of concepts and how they relate, but soon you will realize it is as complex as the actual thing you're describing; it stops making sense. A knowledge graph is also simplification, and as such it, for clarity, leaves certain details aside. Thing is, again: what is obvious for you and you can discard can be a key concept for someone else, and vice versa. And here lies one of the main problems with knowledge management in organizations: if concepts are not captured and formalized in some way, a great deal of meaning is left to interpretation. And there are two sides of this issue. One is related to concepts that are somehow isolated in me, for example, what is my own understanding of the world as such, which might (and will) differ from my colleagues, and that is fine. But when it comes to concepts that are collective and needed for the organization to progress for example something as simple as "what is the company's main product?", meaning cannot be left to interpretation.

Nodes of a knowledge graph (or ontology) are connected by different kinds of links. One important kind of link is called IS-A or isA link. The nodes and isA links together form a rooted directed acyclic graph (rooted DAG). Acyclic means that if you start at one node and move away from it following an IS-A link, you can never return to this node, even if you follow many IS-A links. *Rooted* means that there is one single "highest node" called the root. All other nodes are connected by one IS-A link or a chain of several IS-A links to the root. For the image above, the node "hardware" is having a lot of inbound "isA" links, meaning that this node is a root, or a parent category. From software, a root is usually an abstract class, meaning that there is no "instance" of that concept, but serves as a parent category for real instances of things. You can feel how all this is very similar to an object-oriented approach; even the diagrams are quite similar. But here we're talking about just conceptualization of knowledge as a result, not about software. Now, if an isA link or edge points from a concept X to a concept Y, then that means every real-world thing that can be called an X can also be called a Y. In other words, every X isA Y. Let's try it: every flight computer isA [piece of] hardware, and every hardware isA[n] entity (square brackets added for grammar). Makes sense. To be fully consistent with my own writing, I mentioned previously that categorical hierarchies are sort of "anti-pattern", and of little usefulness to describe reality, and I stand on my point when it comes to system design, because, as I said, it is better to keep hierarchies mirrored with the physical system hierarchy (system description is, in effect, an ontology). But when it comes to specifying knowledge graphs, categorical hierarchies are a bit inescapable, because that's how our brains construct knowledge; categorizing helps us gain knowledge about something. Taking the case of the flight computer again, we kind of *know* (because of naming chosen and because of our past experiences) that it is a piece of hardware, and we could just skip specifying the "hardware" root/parent, but for someone who is lacking context or experience that relationship can be useful. When it comes to define ontologies, there has to be some reasonable tailoring of the hierarchy according to potential users/audience, and some over specification (if it doesn't clutter the graph) does not do much harm. Also, nodes can contain information attached to them. This information includes attributes, relationships, and rules (or axioms); axioms express universal truths about concepts and are attached to concepts (e.g. If X is the husband of Y, then Y is the wife of X). An attribute is like a simple number or variable that contains additional information about that concept. I chose, though, a bit of a different approach and defined an *attribute* as an entity as well; some others choose attributes to be just numbers attached to a node. For me, an ontology must be explicit enough also about attributes and their hierarchies. In the above example, *mass* is an attribute *had* by both hardware and the vehicle.

There is some disagreement whether only the attributes are inherited or also the values of attributes. But, if values are inherited, they may be "overridden" by attributes at lower nodes. Example: I could assign the value "1" to the attribute *mass* at the entity *Hardware* in my ontology. However, the child entity called Flight Computer (which isA hardware) could specify the value "2" for its mass and this would be the value that is used. Most researchers say that we inherit semantic relationships down. Some ontologies have "blocking mechanisms" to stop inheritance. I will come back to this when we see the code. A relationship (or an edge) is a link (arrow) that points from one concept to another concept. It expresses how the two concepts relate to each other. The name of a relationship is normally written next to the relationship link. Other links/edges (relationships) are commonly called "semantic relationships". This is to avoid confusion with isA relationships. Our human judgment in defining ontologies is important because data could be described in an infinite amount of ways. Machines are still unable to consider the broader context to make the appropriate decisions. The taxonomies and ontologies are like providing a perspective from which to observe and manipulate the data. If the element of interest is not being considered, then the knowledge graph won't provide any insight. Choosing the right perspective is how value is created.

Organizations do not spend much time analyzing and capturing what they know. And the key here is that what organizations know is not a monolith body of knowledge but a mixture of both tacit and explicit knowledge. The great deal of "tacit" knowledge is very difficult to capture in frames or schemata. If an organization has a considerable amount of manual work (integrating a spacecraft together is still a handcrafted endeavor), capturing and codifying that type knowledge in any format is very troublesome; for example, go try to learn to play guitar reading a book. Tacit knowledge can be transferred by mentoring and first-hand practice.

The author had a job many (many) years ago which consisted of repairing fax machines. The job was not great, but it was good pocket money while I was studying at the university. Repairing is maybe too big of a word for it. The task was rather simple: faxes could get dirty after some use, some dark marks could start showing in the sent documents, so a clean-up was needed to fix the issue. I hadn't ever seen a fax machine in my life before, so when I opened it (or my mentor opened for me) I was astonished: an intricate system composed of mirrors, plastic gears, and tiny motors was revealed in front of my eyes. Challenge was that there was one mirror which was the cause of 99% of the problems, and to access that mirror and be able to clean it with a cloth, half of the thing had to be torn apart. During my "training", my mentor quickly disassembled it in front of me, expecting that I would remember all the screws, along with what was supposed to come out first, what was supposed to go next. Then, he wishfully sent me to the field to make customers happy. I still remember the first customer I went to: it was a private home of a journalist (or similar), and the guy was in a rush to have the fax fixed for him to be able to send his work; there was (apparently) some deadline he needed to meet. Of course, he never met the deadline because I never figured out how to reach the damn mirror. Guy hated me, and rightfully so. So, after that traumatic experience, I went back to my employer's office, took one defective fax which had been laying around, sat down one afternoon and assembled and disassembled the thing probably a hundred times (no joke). While I was doing that, I was taking notes (phones with cameras were not a thing), with rudimentary diagrams and flowcharts. I never failed cleaning that mirror again. I trained myself into it. After a short while, I didn't need the notes anymore. Training is exactly that: knowledge development versus time. The fax machine experience made me think: what would it take to make it possible to pick a random person from the street and get this person to successfully clean that mirror right the first time, and all subsequent times? If I prepare the best instructions ever, and get this random person to be able to do it right the first time, would that mean the person *knows* how to do it? I figured that because disassembling the fax implied basically unscrewing screws and de-snapping snap-fits (two things that we probably know from before), disassembling and assembling back something like a fax could be turned into a sequence of steps. If that sequence of steps would be properly accompanied by photos and visual cues, it should not be crazy to think someone totally unaware of the task could be able to nail it. Assembling and integrating a spacecraft contains a great deal of fax machine-like parts and bits. These assemblies are aggregations of activities people (with the right background) already know how to do: screwing, gluing, torqueing, and snapping. It is possible, then, to capture and describe the way of tearing assemblies apart and/or putting them back together using step-by-step sequences, provided they are composed of familiar activities. Is this knowledge? Not quite. Being able to perform a task following instructions is not knowledge. After all, that's how GPS works in our phones with its "turn right/turn left" approach. Provided we know how to drive a car or walk, we can get to places we had no previous clue how to get, just by following those simple instructions. Same way the famous Swedish furniture supplier can make you assemble a cupboard even if you haven't ever assembled a cupboard before. This is no different as computers work: they execute instructions specified in some way (language), while the underlying tasks (adding, subtracting, and multiplying) have been previously wired in them, in the sense that specific underlying mechanisms will be triggered to execute the operations according to those instructions. Does this mean computers "know" how to add numbers? Not really, since what the computer is doing is applying basic pre-established rules using binary numbers to achieve the result.

When activities are composed of tasks which are unfamiliar or require precise physical operations, training takes considerably longer. Think about riding a bicycle or playing guitar; you cannot learn it from a book, or from a podcast. But robots can ride bikes and play guitar (there are plenty of example videos around). Do robots know all that? What is to know something then? *Knowing* involves comprehending in a way inferences can be made in order to adapt and apply that information in contexts that are different compared to the original context. Example: you could really practice for two or three months and learn one song on piano by furiously watching a tutorial. You could eventually get good at that one song. Does that mean you learned to play piano? If someone came and invited you for a jam session, you would be probably the worst guest ever. Computers are great at doing specific things for specific contexts, but they are incapable of adapting to other situations. The bike-riding robots will surely fall on a bumpy road, unless its control system takes that into account by design; guitar-playing robots couldn't jam either. We only know something when we can adapt that knowledge to varying contexts and still obtain satisfactory results; otherwise we are just following a recipe or executing instructions like computers and robots do. We as engineers must adapt what we know from previous experiences and projects to the new contexts we encounter in our careers. We cannot apply rigid "fetch and retrieve" rules, for they will not work. Every new design scenario is different from the previous one. There are not two equal cases.

Even though computers cannot know and understand in the human sense of the word, they can be used to process relationships, attributes, and properties of different types of entities, and find patterns, unknown contradictions, and circular loops. Computers are great at rule-checking, so their value for that is very high. It is surprising that organizations haven't adopted pervasively the practice of capturing their concepts in graphs; new insight can be discovered by using the right set of tools. Early-stage NewSpace projects, having the advantage of being young and small, should not let this opportunity pass. Establishing the practice of eliciting concept maps and knowledge graphs out of the main concepts related to what the organization is trying to accomplish, along with their relationships, and capturing all that in formats that computers can process is a great step to grow a semantic information system. This way, the organization's documentation could be semantically consistent across the board.

3.9.1 Creating Knowledge Graphs with Python

So how to create knowledge graphs then? The approach taken here is to use open-source tools and a widely known language as Python, and combined it with a proven

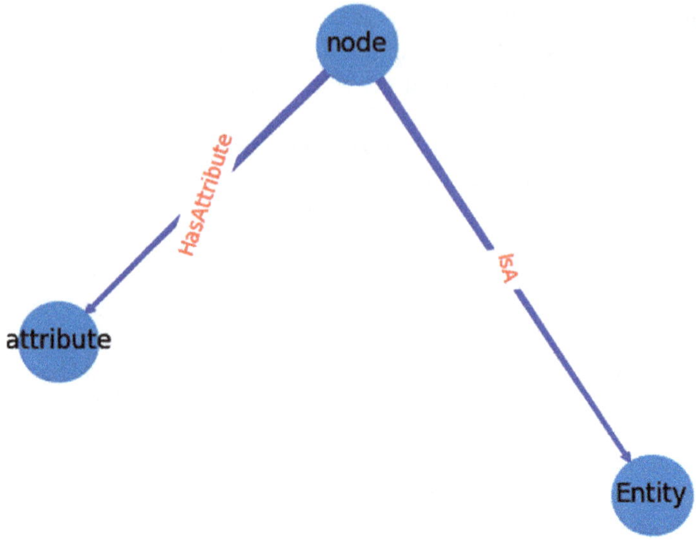

Fig. 3.16 The basic elements of a directed graph: entities, nodes, attributes, and edges

graph library like NetworkX,[7] plus the always impeccable Matplotlib,[8] and the result is a *low-cost* ontology capture tool we can play with (cost here understood not only as monetary but also learning cost). NOTE: since code autocompletion is a very nice-to-have when it comes to work with knowledge-graphs, I also recommend avoiding plain text editors and use PyDev.[9] Here, we are using Python language as the modeling language, whereas the semantics are defined by how we connect the syntactic blocks together. We start by defining four basic objects on top of which all our ontologies will be built: entity, edge, node, and attribute. An entity is just an abstract concept from where all the node objects inherit from. As depicted in Fig. 3.16:

- A node is an entity
- A node can have attributes. Nodes represent concepts.

Nodes are used to represent concepts. Concepts (nodes) are connected by means of edges.

With these very basic elements in the toolbox, we can start constructing ontologies. That's is one amazing takeaway from this exercise: Knowledge, ultimately, is a complex arrangement of very basic building blocks. Complexity is in the connections, and not in the variety of elements to construct the knowledge. The following code snippet contains the source code used for creating the KG blocks.

[7]https://networkx.github.io/.

[8]https://matplotlib.org/.

[9]https://www.pydev.org/.

```
class Edge():
        def __init__(self, label,sign=1):
                self.label = label
                self.sign = sign
                self.value = 0
                return

#Define a basic set of common relationships, they're mostly verbs

isA = Edge("IsA")
owns = Edge("Owns")
contributesTo = Edge("ContributesTo")
extends = Edge("Extends")
has = Edge("HasAttribute")
uses = Edge("Uses")
creates = Edge("Creates")

needs = Edge("Needs")
defines = Edge("Defines")
runs = Edge("Runs")
consumes = Edge("Consumes")
drives = Edge("Drives")
executes = Edge("Executes")
operates = Edge("Operates")
realizes = Edge("Realizes")
contains = Edge("Contains")

class Entity():
        def __init__(self, name):
                self.edges = []
                self.own=[]
                self.name=name
                self.values={}
                self.abstract = True
                nodes.append(self)
                return

        def addEdge(self,EdgeObject,NodeObject):
                #Arrow is always outbound
                copyNode = copy.deepcopy(NodeObject) #I keep a deep
copy the edge and the pointed object, in case I need to modify it
                copyEdge = copy.deepcopy(EdgeObject)

self.edges.append({'object':NodeObject,'name':NodeObject.name,'edg
eObject':EdgeObject,'label':EdgeObject.label,'sign':EdgeObject.sig
n})
```

```
        self.own.append({'object':copyNode,'name':copyNode.name,'edge
Object':copyEdge,'label':copyEdge.label,'sign':copyEdge.sign})
            return

    def printEdges(self):
            for edge in self.edges:
                print (self.name,"-
>",edge['label'],'(',edge['sign'],')',"->",edge['object'].name)
                #edge['object'].printEdges() #This can cause an
endless loop if there is a cycle relationship

    def getEdges(self):
            retList = []
            for edge in self.edges:
                retList.append([self.name,edge['object'].name])
            return retList



    def getEdgeLabels(self):
            retList = []
            for edge in self.edges:

    retList.append({(self.name,edge['object'].name):edge['label']
})
            return retList

class Attribute(Entity):
    def __init__(self, name,val=0.0):
            Entity.__init__(self, name) #To indicate it's abstract
            nodes.append(self)
            return

class Node(Entity):
    global nodes
    def __init__(self, parent,name,abstract=True):
            Entity.__init__(self, name) #To indicate it's abstract
            self.own = []
            self.abstract=abstract
            self._parent = parent
            self.addEdge(isA,self._parent)
            nodes.append(self)
            return
```

Snippet 3.1—Knowledge graphs with Python

You will find in this previous snippet that, when I add an edge to a node, I keep an internal copy of that edge, using copy.deepcopy. Why is that? This is related to the problem about attributes and how they are inherited. Using again the *mass* example, *flight_computer* has *hardware* as parent, but *flight_computer has* its own mass. Same for the *navigation_computer*. Then, if I modify the mass of *flight_computer*, I would be modifying the mass of any other entity using *hardware* as parent. That's why, in my particular implementation, I want to keep my own copy of it and modify it as I want. There is no silver bullet to sort this out, your implementation could be different, maybe attributes being just a number and not a formal entity. With these basic classes defined as the syntactic building blocks, then it's up to us to connect them according to the ontology we are trying to describe, create our concepts and map them according to what we want to express, for example, if I want to describe rockets and satellites:

```
from onto import *
import networkx as nx
import matplotlib.pyplot as plt
from networkx.algorithms.assortativity.correlation import
attribute_ac


if __name__ == "__main__":

     vehicle = Node(entity,"vehicle")
     rocket = Node(vehicle,"rocket",False)
     satellite = Node(vehicle,"satellite")

     hardware = Node(entity,"hardware")

     attribute = Attribute("attribute")
     mass = Attribute("mass")
     volume = Attribute("volume")

     mass.addEdge(isA, attribute)
     volume.addEdge(isA, attribute)

     hardware.addEdge(has, mass)
     vehicle.addEdge(has,mass)
     vehicle.addEdge(has,volume)

     flight_computer = Node(hardware,"flight_computer",False)
     flight_computer.values['mass']=1.0

     imu = Node(hardware,"IMU",False)

     flight_computer.addEdge(contains, imu)

     navigation_computer = Node(hardware,"nav_computer",False)
     navigation_computer.values['mass']=2.0

     rocket.addEdge(contains, flight_computer)

     rocket.addEdge(contains,navigation_computer)
```

Snippet 3.2—An ontology example

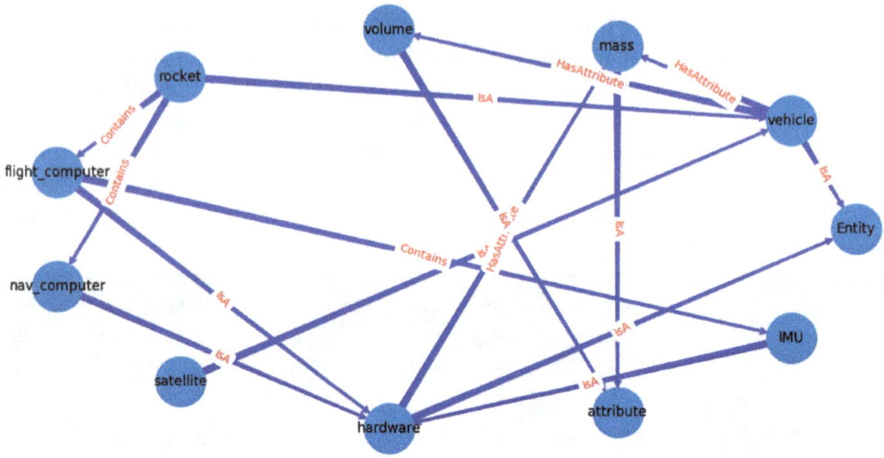

Fig. 3.17 Directed acyclic graph generated from code

Which yields the next directed graph (Fig. 3.17):

As you can see, graph layout can become tricky for a high number of nodes and edges. You can refer to NetworkX[10] docs to find the most suitable way of arranging your DAG. In order to identify loops or cycles in your DAG, you can simply run find_cycle function from NetworkX. For the previous ontology:

```
>>> nx.find_cycle(G)

networkx.exception.NetworkXNoCycle: No cycle found.
```

Snippet 3.3—Searching for loops in a directed graph

Let's create now a simple one which purposely contains a cycle or loop:

```
hates = Edge("Hates") #I create a new relationship not precisely
based in love
cat = Node(entity,"cat")
dog = Node(entity,"dog")
cat.addEdge(hates, dog)
dog.addEdge(hates, cat)
```

Snippet 3.4—A basic example of a loop

This outputs the next graph (the "hate" edges are both superimposed) (Fig. 3.18): So if we run find_cycle() it gives:

```
>>> nx.find_cycle(G)
[('cat', 'dog'), ('dog', 'cat')]
```

Snippet 3.5—Output of the find_cycle method for the graph

---

[10]Link to NetworkX repo.

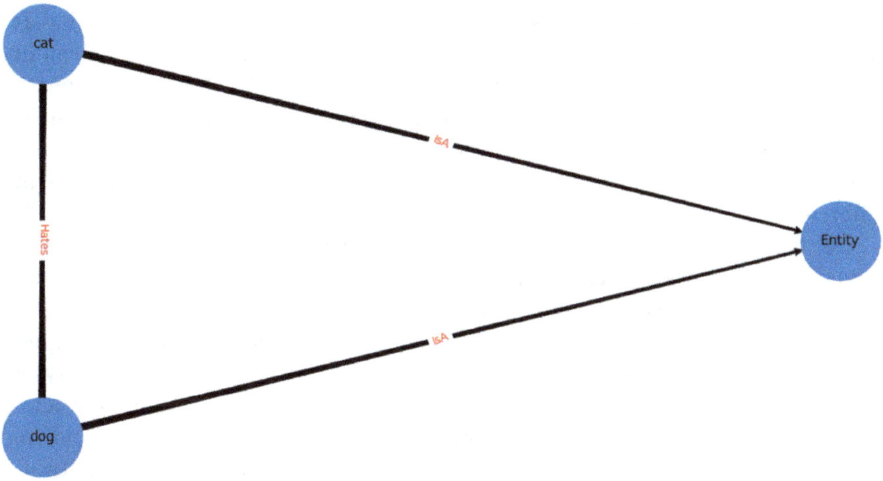

Fig. 3.18 Mutual relationship between to concepts ("dog hates cat, cat hates dog")

Note that there is nothing particularly wrong with finding cycles in your knowledge graphs; on the contrary, it often conveys useful information; for example for identifying reinforcing loops which can turn out to be problematic in some contexts. Every time you do find a cycle in a graph, it is a good exercise to stop and think if such a loop makes sense or you could be over-specifying a relationship, since sometimes it is illustrative enough to specify only one direction. This section provided a very minimalistic way of capturing computational ontologies using directed graphs by means of standard programming tools and open libraries. This could be used as the starting point of an automated way of creating concept maps for your domain, where the resulting graphs would be used for enforcing the way your hierarchies are created and your knowledge is captured and maintained.

3.10 The Brain of the Organization

While researching on the topic that gives name to this section, I came across Stafford Beer again, who I introduced before. I found he had written a book called "The Brain of the Firm".[11] I have criticized cybernetics on earlier chapters as oversimplifying by its biasing toward mechanistic metaphors (and I stand on that), so it was a bit disappointing to find a book on the topic but treated in such a way. "The Brain of the Firm" created some impact some decades ago, mostly in the 1970s, including some bizarre ramifications.[12] Despite the eccentric and intricate approach taken by Beer in his theory (as of today, I still do not fully understand big parts of it, his diagrams

[11]Beer (1995). *The Brain of the Firm*, 2nd Ed..

[12]https://en.wikipedia.org/wiki/Project_Cybersyn.

are nothing less than cryptic) it is still thought-provoking. Is there such a thing as a brain-like function in the organizations we create? Is there any analogy we can find between how our brains work and how organizations work? What is our function as individuals in a hypothetical "brain of the organization"? Beer's work subdivides the brain-like functions in layers: from very low-level actions to high-level more "intelligent" decision-making stages. Let's try to use this framework but in a slightly different way.

Our brain, as a sensemaking unit, can only unleash its full functionality if it is connected to a set of peripherals which: (a) feed it with information (sensory/sensors), and (b) allow it to execute actions to modify (to some extent) the environment according to wishes or needs (motor/actuators). This is a nervous system; or, more precisely, a sensorimotor nervous system. The main function of our nervous system is to arm us with the tools to navigate the world around us, come what may, and to provide us with an introspective assessment of ourselves. Now let's use this analogy for an organization. The organization needs to navigate through the reality it is immersed in (markets, competitors, financial context), and needs to be aware of its internal state (team dynamics, team commitment, goals, etc.). Organizations have to execute actions to alter those contexts according to wishes/desires (policy making, product lines, product variants, acquisitions, mergers, etc.). Then, it does not appear too off to say organizations can functionally be represented as a collective sensorimotor nervous system. But analogies need some care. Since we are talking about people and not exactly cells (neurons), then this analogy can sound a bit oversimplifying. Think about this for a moment: I am stating that a brain as an analogy is an oversimplification for an organization! In a brain, neurons communicate by a firing mechanism that is rather simple, at least compared to how we "fire" as members of a social network. Neurons fire according to thresholds set by chemical reactions that are triggered depending on certain stimuli. We, as "neurons" in this hypothetical organizational brain, have very complex firing mechanisms compared to a cell; we communicate in way richer manners. Our "firing thresholds" are defined by very intricate factors such as affinity, competition, insecurities, egos, and a long etcetera. Also, in our own individual nervous systems, the decision-making is more or less centralized in our brain and all information flows into it, whereas in the organizations the decision-making can be more distributed across the "network of brains"; this also means the information channels are not all routed to one single decision-maker. As in any network, a network of brains can show different topologies. This network is recursive, meaning that it is a network of networks.

It is naive to believe organizations follow only one topology; e.g. the "fully connected" topology (Fig. 3.19). In reality, a mixture of all topologies can be found; i.e. a hybrid topology. For example, organizationally, a "bus" topology represents a shared channel or forum where all actors can exchange information openly; nothing goes from node to node. Great for communication and awareness, somewhat problematic for decision-making: who calls the shots? The star topology works somewhat opposite: one actor acts as concentrator, all nodes connect to it; probably more optimal for decision-making, suboptimal for awareness and communication. The way subnetworks connect is not to be overlooked. For example, some individuals

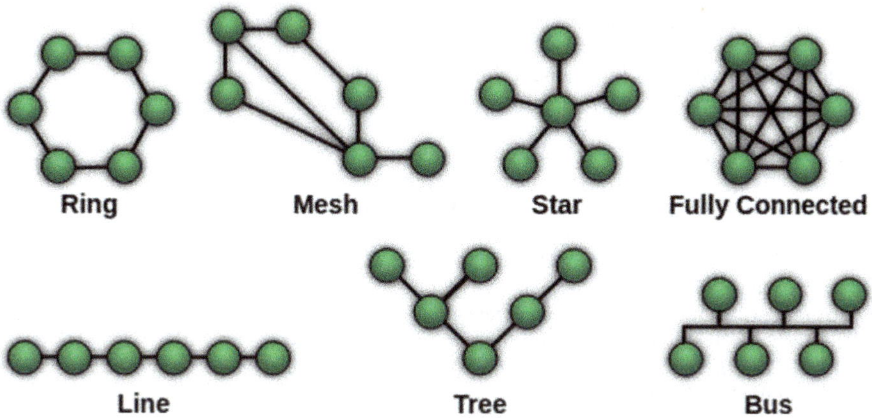

Fig. 3.19 Network topologies (credit: public domain)

can act as *gateways* between subnetworks (they connect two groups or departments for example), and some others act as "firewalls", in the sense that everything to/from a group needs to go through them and filter out information that he or she considers not relevant for the collaboration.

If the brain analogy holds, the question is: Can the organizational sensorimotor nervous system show disorders as our sensorimotor nervous systems can? Can organizations have Alzheimer for example? Or amnesia? Or PTSD?

References

Anderson, O. R. (1992). Some interrelationships between constructivist models of learning and current neurobiological theory, with implications for science education. *Journal of Research in Science Teaching, 19*(10), 1037–1058.

Colfer, L., & Baldwin, C. (2010, February 18). *The mirroring hypothesis: Theory, evidence and exceptions*. HBR—Working Knowledge. https://hbswk.hbs.edu/item/the-mirroring-hypothesis-theory-evidence-and-exceptions

Conway, M. (1968). *How do committees invent?* Mel Conway's Home Page. http://www.melconway.com/Home/Committees_Paper.html

DeMarco, T., & Lister, T. R. (1999). *Peopleware: Productive projects and teams* (2nd ed.). Dorset House Publishing Company.

Frolov, V. (2018, June 20). *NoUML*. Medium. https://medium.com/@volodymyrfrolov/nouml-afbb7f07f369

Fuchs, C., & Golenhofen, F. (2019). *Mastering disruption and innovation in product management: Connecting the dots. Springer*. https://doi.org/10.1007/978-3-319-93512-6.

Gerber, E. (2007). Improvisation principles and techniques for design. In *Conference on Human Factors in Computing Systems—Proceedings*, 1069–1072. https://doi.org/10.1145/1240624.1240786

Hayakawa, S. I. (1948). *Language in thought and action*. London: George Allen & Unwin Ltd.

Henderson, R., & Clark, K. (1990). Architectural innovation: The reconfiguration of existing product technologies and the failure of established firms. *Administrative Science Quarterly, 35*(1), 9.

Lemme, B. H. (2006). *Development in adulthood*. Boston, MA: Pearson Education Inc.

McGreal, D., & Jocham, R. (2018). *The professional product owner: Leveraging scrum as a competitive advantage*. Addison-Wesley Professional. ISBN 9780134686653

Morris, D. (2017). *Scrum: An ideal framework for agile projects. In easy steps*. Leamington Spa. ISBN 9781840787313

Novak, J. D., & Cañas, A. J. (2008). The theory underlying concept maps and how to construct and use them. *Technical Report IHMC CmapTools 2006–01 Rev 01-2008, Florida Institute for Human and Machine Cognition*. Retrieved from http://cmap.ihmc.us/Publications/ResearchPapers/TheoryUnderlyingConceptMaps.pdf

Pankin, J. (2003). *Schema theory*. MIT. http://web.mit.edu/pankin/www/Schema_Theory_and_Concept_Formation.pdf

Pasher, E., & Ronen, T. (2011). *The complete guide to knowledge management: A strategic plan to leverage your company's intellectual capital*. Wiley. ISBN: 9781118001400

Pichler, R. (2010). *Agile Product Management with Scrum Creating Products that Customers Love*. Addison-Wesley. ISBN: 0-321-60578-0

Rubin, K. (2013). *Essential scrum. A Practical Guide to the Most Popular Agile Process*. New York: Addison-Wesley. ISBN 978-0-13-704329-3.

Chapter 4
Complexity Creep

> "*Everything simple is false; everything complex is unusable*".
> —Paul Valéry.

Abstract Design never ends, if allowed to continue. Designers are *pushed* by context to take snapshots and release their designs to the intended audience when some level of maturity is reached, or as deadlines dictate. Quantifying such maturity can be highly subjective, which encourages engineers to continue adding functionalities until some sort of external force tells them to stop. Complexity creep is not just a technical problem. Complexity creeps in every single aspect of a young organization. As things get more complex, they encroach, making switching barriers higher and creating all sorts of lock-ins, which can turn architectures too rigid.

Keywords Complexity creep · Featuritis · KISS principle · Change management · Switching barriers

We discussed before that the life cycle of a system flows in continuous time and that we decide to divide such cycle in chunks, for practical reasons. We also said that those chunks are visited several times (iterations) until maturity is acceptable. This chapter is about discussing how to define when such maturity is "acceptable". When does the moment come when we can declare the design to be done so we can move ahead and produce the system? Quick answer: it never comes. Complete designs just do not exist and have never existed. I'll elaborate.

Oftentimes mistaken as a mechanistic, almost recipe-oriented activity, Engineering is, on the contrary, a highly creative endeavor. Creativity involves, not in making something out of nothing, but in integrating distinct materials and parts into a new whole. In a typical lifetime of anything engineered, there is quite some time spent by engineers creatively iterating through many different options to find the best solution possible. The process is known: a need enters on one side of the box, magic happens, and a working system or solution comes out on the other side of the box. It is a problem-solving endeavor. Now, that magic is what this chapter is trying to

I. Chechile, *NewSpace Systems Engineering*,
https://doi.org/10.1007/978-3-030-66898-3_4

describe and how we engineers sometimes overdo it. I will start with a observational postulate after years doing technology development from the depths of the trenches:

1. Provided infinite resources and isolation, an engineer will design forever.

A corollary of this, probably a bit fairer to engineers, can be:

1. Without constraints, features expand to infinity.

Overengineering is a disorder of the modern engineer. Let me explain. If you leave an engineer, no matter the discipline, with enough resources and in a total open loop (without anybody or anything to limit him/her degrees of freedom), he or she will enter an endless cycle of never being satisfied with his/her current design and continuously thinking it can be improved. It does look like we engineers are *baroque* at heart: ornamenting just needs to keep going; more is more. A software engineer will hold the release of that piece of software to profile some functions a bit more and maybe making that algorithm faster to gain some precious milliseconds. Or he will add some web interface because why not. Or add a nifty RESTful API in case someone wants to integrate this software to some other software: everything needs a RESTful API nowadays anyway. The Electrical engineer will look for this new power regulator, which has incredibly nice quiescent current and it is a 1:1 swap with the current one so why not! Or hey, this microcontroller has more RAM and more ADC channels, and it is a lower power device on the same package so, easy-piecey, more ADC channels are always welcome. Or, heck, let us redo the layout for better signal integrity on that DDR bus. The mechanical guy will seek to design the system to be easier to assemble; instead of 400 screws, he can bring it down to 350, or use a new alloy to save 2% mass. But design, as many other things in life, has to come to an end; this seems to be difficult to grasp for many of us. It wants to go on forever. Is it a trait of engineers' personality or is it the true nature of the design activity? It happens to be a bit of both. In some other activities, for example cooking, you are forced to declare something ready otherwise you will burn it and it will have to be trashed. In medical practice, a surgery is clearly defined as ready at some point otherwise risks would increase. Engineering design does not really have such a clear boundary between ready or not ready. What we can do, we can only take "snapshots" of the work and artificially flag them as ready for the sake of the project. It is a decision to freeze things or put the pencils down at some point and declare something as *done*. This can be challenging for engineers, since they need to frame their design decisions against time; a big architectural change at some late point of the development stage could be too risky for the delivery date, hence it cannot be done, or if it needs to be done, then the deadline will be missed. In tennis, there is a tendency of some players to lose matches after having a clear lead. This difficulty to close out the match is a psychological response of their minds to the fact of having a win "around the corner". Some relax too much, expecting the opponent will just give up and just close the game for them. Some others start to listen to that inner critic voice telling them "you better not to lose this one" which ends up being counterproductive, in a self-fulfilling prophecy of sorts. Engineers need to close out their design *matches* as tennis players do. Thing is, designing systems require multidisciplinary teams, so no design engineer gets to live alone and isolated. There are multiple other design engineers, from the same or different disciplines, pursuing their own design challenges, playing their own "design matches" and fighting their inner demons telling them to add more and more. Imagine multiple designers in an uncoordinated and silent featuritis.[1] A perfect disaster, and still, this basically describes every design team out there. It does appear like engineering design just does not reach steady state; like it is in a state of constant expansion and growth. The cyclic nature of the design process seems to complot against feature creep, opening the possibility of adding new things at every run.

John Gall, in his always relevant *Systemantics* (Gall 1975) states:

> Systems Tend to Expand to Fill the Known Universe

And then he adds:

> A larger system produced by expanding a smaller system does not behave like the smaller system

Why do we keep on adding features nobody requested when we are left alone with our own souls? You would imagine engineers being reluctant of complexity increase, yet reality seems to show they are more than fine with it.

Without diving into cumbersome definitions of complexity we will stick to a very straightforward definition of complexity:

1. Complexity is a direct measure of the possible *options, paths* or *states* the system contains.

Complexity creep is surely multifactor. We, as engineers, do seem to want to exceed expectations and deliver more "value" by adding unrequested, supposedly *free* functionality, while the truth is we are just making things less reliable and more difficult to operate, one small feature at a time. It is also true that not all complexity *inflation* is only engineers' fault. A product can suffer two different types of creep: internal and external. The external is coming from the customers/stakeholders, who can easily get into a feature request frenzy ("I want this, this and that"), with the expected defense barriers (Project Managers, Systems Engineers) being too porous to contain the spontaneous wishes and permitting those wishes to filter in directly to the design team. And then there is the internal creep from within the design team, radiating spontaneously from its very core outward, again with the Systems Engineering process failing to measure the complexity growth. It can be clearly seen how products are prone to complexity escalation: it comes from every direction. Systems Engineering plays a key role as firewall for *featuritis* and failing to do so greatly impacts the work. The external (from customers) creeps top–bottom toward the design engineers, while the internal creeps all the way from designers up to the customers.

It cannot go unmentioned that when the engineering design activity is immersed in a surrounding atmosphere of complexity growth like the one you can find in

[1] https://en.wikipedia.org/wiki/Feature_creep.

Fig. 4.1 Choreography (credit: "De Notenkraker" by Haags Uitburo is licensed under CC BY-NC-SA 2.0)

basically any startup environment, then such growth, seen as something desired and even expected, will be pervasive, and will influence decision making, including design. The verb "to grow" has become burdened with positive value connotations that we have forgotten its first literal dictionary definition, namely, "to spring up and develop to maturity". This view reasons that growth cannot continue perpetually, requiring maturing and stabilization. On the same note, the word "feature" happens to carry too much of a positive meaning compared to the word "complexity". *Feature growth* does not really sound that bad if you add in a board meeting slide; it can be even a great buzz phrase to impress your growth-addicted investors. "Complexity growth", on the other hand, can be a bit more difficult to sell. At the same time, the word "complexity" is a collective measure, whereas the word "feature" refers to a single thing. Hence, when we talk about features, we refer to the individual and we easily forget its impact to the collective.

But, to design is to introduce change. When you have a baseline, and you want to make it better in some way, for one reason or the other, you introduce changes to it: you alter it. Change is inevitable. Change can be good as well, but when thrown to the wild in a multidisciplinary endeavor, it should be done under some sort of agreement and by providing reasonable awareness.

Picture that you are watching a group of people dancing in a group, like in Fig. 4.1. Nothing like watching a nice choreography where everybody flows in beautiful synchronicity. Synchronicity is expected, otherwise we just cannot call it a choreography. It can be easily spotted when one of the dances screws up and loses sync with the rest. It is unpleasant to the eye, and somewhat cringy as well. Now, imagine a ballet dance where all the dancers individually decide to just add a bit of extra movement here, an extra jump there, because the audience will love it. You can see how we engineers would look if we would dance ballet as we design. Imagine an orchestra where the guy behind the cymbals decides just to ornament a move to make it cooler, on the go. A technical system, which is composed of many subsystems integrated together (a spacecraft, a UAV, an aircraft, a gas turbine, you name it) amplifies change in ways that are very difficult to predict. Integration acts as an amplification factor. Any change in an integrated system, no matter how small, creates ripple waves in all directions.

Example: An Electrical engineer in charge of a power distribution and measurement board decided to use a more efficient DCDC converter. There was just a small, tiny, insignificant (from his point of view, of course) detail that the enable pin for the power distribution chip was the inverse logic compared to its predecessor: active low instead of active high (a logical 0 was needed now to enable it). Chosen device was 1:1 pin compatible with the existing design, so it is a no brainer! Who would oppose to more efficiency, right? Now he did not exactly communicate nor document this in almost any way, he just passed it low-key as a change in the bill-of-materials, as a different part number. Changes in bill-of-materials are somewhat common: capacitors go obsolete and new ones appear, but those are passive components which usually have little impact on performance. Anyway, when the new board arrived, the integration team flashed the firmware, everything went reasonably well, and the board got integrated (let us assume here that board level verification procedure coverage was permeable enough to let this go unnoticed from tabletop verification). Now that is when the fun began: a puzzled operator during rehearsal operations saw that the subsystem hooked to the new power converter was **ON** when it was supposed to be **OFF**. It was draining current and affecting the thermal behavior of some room inside the spacecraft. But how? After many (many more that read would think!) hours of debugging, the issue was found. So, the firmware guy (me in this case…) had to go and patch the firmware for the new hardware design. A piece of source code that was reasonably robust and trusted had to be edited, tampered, increasing the chances of introducing new bugs due to the pressure of schedule. But that is not all, unfortunately. The ground software guy had to resync the telemetry databases and charts accordingly to switch the logic (now **OFF** means **ON** and **ON** means **OFF**, probably red means green and vice versa; a semantic nightmare), and the changes had to be reflected in the operations manuals, procedures and all operators need to be informed and retrained. It can be seen how a small, probably good-hearted, modification created ripple waves all over the place, well beyond the original area where the change was proposed, and this was such a small change, so picture how bigger changes can go.

The challenge with change is: every time we introduce one, no matter how small, we depart from what we know and we set sails to the unknown. Then we have to build trust again in the new version/iteration of it, which takes time and effort and needs verification. Changing means stepping into the unknown. This is not a problem per se, but when we are dealing with highly integrated mission-critical systems, well, the unknown is usually not your friend. We will never be able to completely foresee how our changes propagate. Systemic risk can build up in a system as new elements are introduced, risks that are not obvious until after something goes wrong (and sometimes not even then). Is not all this enough pressure to think twice before adding an unrequested feature? Issue is that often designers' visibility does not span wide enough to think about all the other parts of the system where their unsolicited changes can (and will) impact; they are probably impeded to see the consequences of their acts.

Complexity creep and change are two sides of the same coin. Is the solution then just to avoid change? Not at all, we cannot escape from it, and innovation blooms in change-friendly environments. But complexity creep in a system has implications; it always impacts somewhere, and that somewhere often is usability. Complexity creep kills, wherever the direction it creeps from. When resources are scarce, the last thing to look for is unnecessary functionality nobody required (internal creep). In the same way, being early stage should not mean signing off to any random complexity requested from external parties, unless this external party is a customer. Reviewers, board members, or advisors should not have direct access to request complexity to grow. At least not without some controls in place to make sure the leap of complexity is justified in some way. In any case, the question that surfaces now is: how to measure all this in order to prevent it? Unrequested features finding their way into the product can be hard to catch after all.

We started this chapter with a thesis that an engineer in "open loop" will creep complexity ad infinitum. The key part of this is the "open loop" remark, and it is the subject of the next section.

4.1 Quantitative Systems Engineering

> We can't control what we can't measure
>
> Measure twice, cut once

Systems Engineering is a closed-loop activity, like many others. Feedback is an essential part for any engineering effort to sustain its life. And here we are not referring about feedback in the technical machines we create, but in the design process itself. Designing and implementing technical systems mean creating multiple disparate outcomes, which require some level of control, or regulation; this means keeping those outcomes within desired margins or bands, aligned with an overarching strategy. For that, outcomes to be controlled are fed to regulating processes where they will be compared to a desired set point, and actions (ultimately, decisions) will be taken if deviations are perceived; the intensity of the action is expected to be proportional to the magnitude of the deviation from the goal. Feedback loops are ubiquitous in Engineering practice; loops can be balancing or reinforcing, depending on the decision strategy around the difference between outcome and set point. Reinforcing loops tend to increase outcomes indefinitely, or until saturation. Balancing loops apply corrective action in order to minimize the difference between what's measured and what's desired. Figure 4.2 depicts a typical closed-loop generic process:

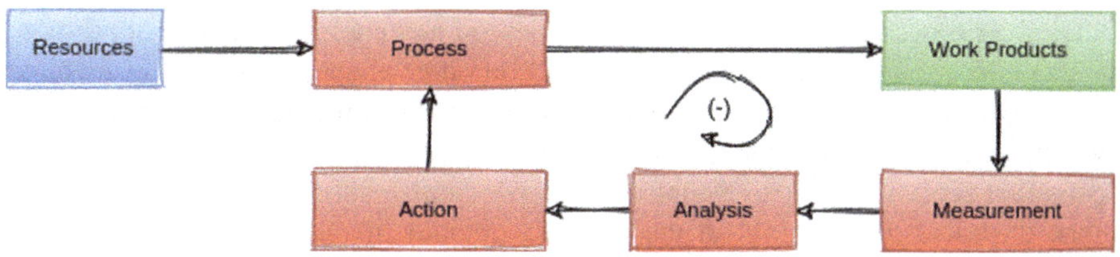

Fig. 4.2 A closed-loop process

The Systems Engineering practice is composed of a collection of processes that are executed on a System-of-Interest throughout its life cycle. Keeping the outcomes of these processes within desired margins relies inevitably on measurement. Outcomes from system design processes are of typically two types:

- Data and Information: blueprints, schematics, source code, requirements, manuals, etc.
- Realized/tangible/physical: subsystems, modules, boards, software executables, libraries, etc.

Now how do we quantify these things if we need to keep them bound to some margins? The variables we need to measure are often very hard to quantify. How do we extract a descriptive quantity from a pile of documents or operation manuals? What metrics can be mined from the process of integrating or verifying a system? How can we quantify the complexity of, say, software to gauge if it has grown outside desirable margins? Imagine I have a fever, but instead of a thermometer, I only have a report about my general condition, where most of the information is about non-numerical observations like the way my face looks. What I need is a number, a quantification (temperature, provided by a sensor) which turns out to be a measure of my health condition at that moment. What I get instead is a more indirect and abstract thing: a few pages of verbal text. Another classic example is a résumé; a sheet of paper stating a person's education and experience. How can we extract a "suitability for the job" metric which we could use for comparing and benchmarking between candidates and help us decide better? Such quantification just refuses to be known. Eventually, some sort of suitability valuation will have to take place for us to decide. But such assessment will be highly qualitative, based on our own and most likely highly biased analysis and interpretation.

This is some sort of a classic flow: when we lack metrics, we rely on our qualitative gut feelings, which are far from being objective since they are impregnated with our own beliefs and past experiences. Problems appear when qualitative methods are forced to provide metrics, and then those numbers are fed into feedback loops and used for decision-making. Qualitative methods have an inevitable coating of human factors. Our estimations are highly biased and greatly affected by our past experience, among many other cognitive factors. Many activities around Engineering and Project Management are carried using information infested with biases and cognitive offsets. One way of overcoming this issue is by means of *operationalization*, which is a process of defining the measurement of a phenomenon that is not directly measurable, though its existence is inferred by other phenomena. Project management (acknowledging its highly qualitative nature as well) has explored quantitative approaches, for example Quantitative Project Management.[2] The phrase which opens this section: "we can't control what we can't measure" mantra captures, in short, that decision-making on the grounds of poor quantitative information is a dangerous game. It could be rewritten as: we can not correctly decide upon what we cannot quantify. We still, though, decide "without numbers" every single day of our lives, from observation, interpretation, and operationalization. But we cannot easily benchmark or compare without quantification, we cannot easily visualize without quantification, nor we can properly predict, forecast, identify trends or extrapolate without some level of quantification.

Systems Engineering needs to be as quantitative as reasonably possible, and steer away from forcing qualitative methods to look quantitative, unless qualitative decision-making is fed with good heuristics (experience). The highly recursive and iterative nature of Systems Engineering needs clear metrics we can supervise to understand if the processes are working correctly, and ultimately if we are building the right system, the right way. But how do we capture those numbers? How should we hook our "sensors" around it? There are two different places to *sense*: the product itself and/or the processes.

Product or System measures: What is the group of critical numbers and figures we could gather from our system in terms of performance? What are those numbers that basically matter the most? These are usually called Technical Performance Measures (TPMs). In a nutshell, it basically refers to a set of meaningful numbers that will describe the way our product is expected to perform under certain conditions, and for the right context. These TPMs can be, for example, fed to pilot customers, through marketing or sales, to get feedback from them, to potentially adjust the design if TPMs do not seem to match a solid business or use case. TPMs are a very important player in "closing the loop" with the market, and essential for small organizations building complex systems.

Process measures: The multiple processes we need to execute upon our system-of-interest are a goldmine of information, and often we let that information go down the drain. For example, for the design phase, we could define metrics on complexity to prevent it from creeping as it usually does; it could be inferred from number of lines of source code, or number of levels of the system hierarchy, or scoring from design reviews, from bug reports, etc. For the verification process, we could monitor a figure about defects found per unit of time, which would allow us to understand the effectiveness of the process itself but also our improvement over time, or collect the time a particular test takes in order to assess learning curves and predict how long it is expected to take in the future, for better planning. Or quantify defects found on critical third-party components from different suppliers, in order to assess their comparative performance or obtain a metric about how bad our vendor lock-in is.

[2]Quantitative Project Management incorporates statistical and probabilistic methods to achieve the project's established quality and process performance objectives.

The information is there, it requires a reasonable level of creativity and data mining mindset to acquire the figures.

Engineering technical things is an information-intensive activity. Reasonable resources should be allocated to define a methodology of sensing, storing, and retrieving the critical metrics around it. This method should dictate what numbers will be mined and how, making sure metrics with proper quality are obtained. The importance of any measure is that it will most likely feed a loop, or serve as an indicator, early warning or flag that something has changed in the process, and that change may be good, detrimental, or neutral to our system development, augmenting our decision-making.

- Feedback loops are ubiquitous in Systems Engineering due to its recursive and iterative nature.
- Systems Engineering practice relies heavily on qualitative methods due to the difficulty of obtaining quantifiable numbers in many of its processes and activities.
- A more quantitative Systems Engineering approach requires identifying key metrics, and defining a consistent method of collection, storing, and retrieval of such metrics for the sake of comparison, prediction, extrapolation, and better estimation.
- A more quantitative Systems Engineering approach needs to be aware that when a measure becomes a hard target, it ceases to be a good measure, as Goodhart's law observes.[3] This can occur when individuals or organizations try to anticipate the effect of a policy or a change by taking actions that alter its outcome, in a sort of self-fulfilling fallacy.

4.2 Change Management

Iteration is at the very heart of engineering design. Things evolve, and evolution cannot take place without change. Change is unavoidable, and constant. The world around us changes continuously. In startups this is even truer; things are fluid and in a week time many things can look different. This is another effect of the learning process while young organizations grow; as things are better understood, changes are applied to correct previous partial understandings. Feedback loops require changing and adjusting to meet some goal. An early stage organization that does not change and adapts to the ever-changing context is condemned to disappear.

But change needs coordination and alignment to an overarching strategy. Uncoordinated, disparate changes are as damaging as not changing at all. At the same time, the way change is handled speaks, and loud. For example, constant changes in design decisions can spread the message that "the sky's the limit", ramping up complexity and causing project overruns. This can put the Systems Engineer or Project Manager

[3]Goodhart's law states that when a feature of the economy is picked as an indicator of the economy, then it inexorably ceases to function as that indicator because people start to game it.

reputation and competence on the line as well, since they are expected to act as "shock absorbers" or "firewalls" preventing noise filtering into the design team.

Changes can take place in the system under design or in organization, or both (big system). Organizational changes include for example altering structures, strategies, procedures, or cultures of organizations (Quattrone and Hopper 2001). The term encompasses both the process by which this happens (i.e. "how") and the content of what is being altered (i.e. "what"). By definition, change implies a shift in the organization from one state to another. This shift may be deliberate, with the aim of gaining or losing specific features of the organization to attain a defined goal, or it may be less deliberate, perhaps occurring as a consequence of developments outside the control of the organization. Moreover, during the change process, additional parts of the organization may be unintentionally affected, particularly when change is experienced as excessive (Stensaker et al. 2001). When does change become *excessive*? Situations that can result in excessive change include: (1) the organization changes when the environment or organizational contingencies do not suggest the need to change; (2) when the organization changes for change's sake, and (3) that excessive change occurs when the organization changes one element but fails to change other dimensions accordingly (Zajac et al. 2000). The number 3 is especially relevant. Changing particular elements while leaving other elements unchanged can result in imbalances. This does not imply the whole organization must change at once; it just means organizational change must be structure-aware and seek balance.

In engineering, change is no stranger. Managing change has been part of the engineering practice since the early days. The discipline that combines the management of changes and configuration of the system in a structured way is called Configuration Management. It is the Management discipline that:

- Applies technical and administrative direction and surveillance to identify and document the functional and physical characteristics of the system under design;
- Controls changes to those characteristics;
- Records and reports change processing and implementation status;

The purpose of Configuration Management is therefore to establish and maintain consistency of the system's performance, functional and physical attributes with its requirements, design and operational information throughout its life. It is typical to distinguish four different Configuration Management disciplines (Altfeld 2010).

- Configuration Identification: is the Configuration Management activity that assigns and applies unique identifiers to a system, its components, and associated documents, and that maintains document–revision relationships to the system configurations.
- Configuration Control: involves controlling the release of and changes to baselined products throughout their life cycle. It is the systematic process to manage the proposal, preparation, justification, evaluation, coordination and disposition of proposed changes, approved by an appropriate level of authority.

- Configuration Auditing: verifies that the system is built according to the requirements, standards, or contractual agreements, that the system design is accurately documented, and that all change requests have been resolved according to Configuration Management processes and procedures.
- Configuration Tracking: involves the recording and reporting of the change control process. It includes a continuously updated listing of the status of the approved configuration identification, the status of proposed changes to the configuration, as well as the implementation status of approved changes. This configuration information must be available for use by decision-makers over the entire life cycle of the aircraft. In an environment where a large number of multifunctional teams operate concurrently it is particularly important to understand how Configuration Control works.

In smaller organizations, change and configuration management is usually not enforced a very rigid way. During the proof-of-concept stages, this may be a purposely decision to keep things fluid and going. Eventually, a compromise must be reached between openness to new functionality and maturing a system or product design. This is a bit like tightrope walking if you want: too rigid change management can kill quick turnarounds and innovation; too loose approach opens the door for complexity creep. A lightweight, tool-agnostic approach to avoid feature creep could be:

- We must be aware of the impact we make when we introduce even the slightest improvements to a subsystem, which is integrated in a bigger system: think systemically. A small change can and will impact even the farthest areas.
- We must *close the loop* with other designers with periodic reviews and processes to make sure our perception of improvement is aligned with the organization grand plan and tame our own cravings to add "value" nobody requested.
- Put safeguards in place to catch uncoordinated changes.
- Do not prefer an awesome unknown over a suboptimal known: i.e. do not believe new designs will be *per se* better than old designs. A corollary: beware of making projects depend on unknown functionalities or features that are yet to be created from ground up.
- Always prioritize *value* as the criteria to decide about introducing changes and unknowns but agree on what *value* means.
- We are fallible, and fully capable of making mistakes. We can overestimate our own capacity of assessing complexity and its implications. What is more, we tend to hold overly favorable views of our own abilities in many social and intellectual domains. Our own incompetence can rob us of the ability to realize we are incompetent. (Dunning and Kruger 1999).
- Go incremental. Small steps will get us more buy-in from stakeholders and ease our way. Show ideas with proof-of-concepts and improve from there. Never try to go from zero to everything. John Gall observes that "a complex system that works is invariably found to have evolved from a simple system that works" (Gall 1975).
- Design systems that can provide their function using the least variety of options and states as reasonably possible.

- Change, when handled, fuels innovation and differentiation.

In NewSpace, time to market is essential. Failing to release a product to the market and losing an opportunity to a competitor's hands can be fatal. Sometimes, having a product in the market, even a prototype far from being perfect, might be the winning card to win a contract or to be the first and capture essential market share. Market forces dictate that *better is the enemy of good enough*, which could be counter-intuitive for our somewhat extravagant technologically *baroque* minds, but essential for product and company success.

4.3 Good Design Is Simple Design

The saying goes: "perfection is attained not when there is nothing left to add, but when there is nothing left to remove". Making things simple can be, ironically, complicated. Another way of putting it (which comes from the original phrasing of Ockham's razor):

1. Entities should not be multiplied beyond necessity

The key there is *necessity*. We can always invent necessity where there is none. It takes effort and energy to prevent complexity from increasing. An observation by Larry Tesler, one of the leaders in human–machine interaction, states that the total complexity of a system is a constant. This means, as we make the user interaction simpler, the hidden complexity behind the scenes increases. Make one part of the system simpler, said Tesler, and the rest of the system gets more complex. This principle (or postulate) is known today as "Tesler's law of the conservation of complexity". Tesler described it as a tradeoff: making things easier for the user means making it more difficult for the designer or engineer. According to this, every application or device has an inherent amount of irreducible complexity. The only question is who will have to deal with it, if the user or the engineer (Saffer 2009).

This observation can have a corollary about the relationship between abstraction and simplicity. Consider the automatic transmission in the automobile, a complex combination of mechanical gears, hydraulic fluids, electronic controls, and sensors. Abstracting the driver from having to shift the gears manually are accompanied by more complexity in the underlying machinery. Abstraction does not mean overall system simplicity. Good abstractions only cleverly disguise and hide complexity under the hood, but complexity is very much there. What Tesler observes is that what is simple on the surface can be incredibly complex inside; what is simple inside can result in an incredibly complex surface (Norman 2011, 47). The "constant complexity" observation from Tesler connects to the initial definition of complexity in this chapter: the amount of "options" or "states" the system has is an inherent property of it. How that variety of states reaches the user is a design decision. Making things simpler for the user (and here a user can be also another subsystem in the architecture), requires encapsulating that variety in some way. When the system is composed of many elements and the architecture is already synthesized, moving this variety around can create a wave of cascade change across it. A typical example: imagine two subsystems in a system, A and B. Subsystem A commands subsystem B. The list of commands of B is very extensive and granular (which indicates B has a lot of states and options to configure) and it gives A the possibility of controlling the functions of B very precisely. Due to the amount of commands needed, subsystem A (who is the *user* of subsystem B in this case) requests an architectural change to be made in order to make the operation of Subsystem B *simpler* (from its own perspective). Which approach is simpler?

- Does A care about the options B provides?
  - If true, perhaps grouping (encapsulating) all commands to B in one single command could work, but this will impose changes in B. Note that the amount of options/states of B does not change with this. Subsystem A used to have the possibility of sequentially commanding every single option B provided but if now there will be one single command then B will have to handle the sequence by itself. B complexity increases. Both A and B must change.
  - If false, then an abstraction layer can be created for A to have only a subset of B commands. If still those commands A will not use anymore are needed for B to perform, B will have to execute those itself. B complexity increases. Both A and B change.

Any path taken will require changes in both A and B. It also means that in case of failure of the new encapsulated command (for example due to checksum error), the whole chain of functions will fail with it. This approach is transferring complexity around (in the Tesler sense of the word) but not reducing it. Redesigning subsystem B to have less internal options to offer can also be an alternative, but in this case still A and B must change.

4.3.1 Complex Can Be Simple

All the systems we design have a user. The user has a key role in mingling with the complexity of what we do. Again, here "user" is a very broad term outside the typical "human customer" role. A user can be either a human or just another system connected to our system. Computer science tends to use the word *actor* when referring to a computer interacting with another computer. The word *user* better captures the essence of what we are discussing here, instead of actor. A user is an *entity* that interacts with a system in order to utilize the system's main function for some purpose. Users do not need to be human. Then, computers can be users of other computers. On spacecraft, the on-board computer is the user of the radio for example, for transferring data to ground. Same as the power subsystem is the user of solar panels; it distributes that power around the place as needed. Users couple to the systems they are trying to use, and of course this coupling is done by means of an interface. If the user is a human trying to use a computer system, it will be commonly called Human–Machine–Interface (HMI). In any case, while coupling to the system through an interface, the user is exposed to what is given to it for using the system of interest. In this process is where choices can be made for showing or purposely hiding complexity from the user. Complex things can be *perceived* as simple things.

Often, when we interact with the world around us, we realize it may be described in terms of underlying rather simple objects, concepts, and behaviors. But when we look a bit closer, we see that these simple objects are composite objects whose internal structure may be complex and have a wealth of possible behavior. That wealth of behavior can be irrelevant at the larger scale. Similarly, when we look at bigger scales, we discover that the behavior at these scales is not affected by objects and events that appear important to us. The world around us can be described by means of levels or layers of description. A level is an internally consistent picture of the behavior of interacting elements that are simple. When taken together, many such elements may or may not have a simple behavior, but the rules that give rise to their collective behavior are simple. The existence of multiple levels implies that simplicity can also be an emergent property. This means that the collective behavior of many elementary parts can behave simply on a much larger scale. The study of complex systems focuses on understanding the relationship between simplicity and complexity. This requires both an understanding of the emergence of complex behavior from simple elements and laws as well as the emergence of simplicity from simple or complex elements that allow a simple larger-scale description to exist (Bar-Yam 2018). Complexity, at the end, heavily depends on the point of view (Fig. 4.3).

Complexity is part of the world, but it should not be puzzling: we can accept it if we believe that this is the way things must be. Just as the owner of a cluttered desk sees order in its structure, we will see order and reason in complexity once we come to understand the underlying principles. But when that complexity is random and arbitrary, then we have reason to be annoyed. Modern technology can be complex, but complexity by itself is neither good nor bad: it is confusion that is bad. Forget the complaints against complexity; instead, complain about confusion. We should complain about anything that makes us feel helpless, powerless in the face of mysterious forces that take away control and understanding (Norman 2011).

4.4 Complexity Is Lucrative

If only complexity would be just a technical problem. But complexity creeps in every single aspect of an organization. As the organization evolves, everything in it gets more complex: logging into corporate online services gets more complex, door access controls get more complex, devices for video calls get more complex, coffee machines get more complex, even hand-drying equipment in toilets gets more complex. More people, more departments, more managers, more everything. More is more. As it can be seen, the Universe does not really help us engineers to tame our complexity issues: everything around us gets more complex. As more souls join the

Fig. 4.3 All the complexity of the world is just a dot in the sky for some observer (credit: NASA/JPL/Cornell/Texas A&M, CC BY 2.0 https://creativecommons.org/licenses/by/2.0/)

organization, division of labor gets extremely granular, communication complexity combinatorially skyrockets, and cross-department coordination becomes a game of chess. Departments start to feel they have difficulties correlating their information with other departments' information. There is an increasing sense of "lack of": "lack of control", "lack of tracking", "lack of awareness".

The result: an explosive ecosystem of tools and methods. What specific tools or methods will depend perhaps on what company the managers are looking up to at the moment (Google and Apple are always ranking at the top), or what idea the new senior VP has just brought from his or her previous employer: OKRs, Six Sigma, Lean, SAFe, MBSE, or worse, a hybrid combination of all those together. With more and more elements in the org hierarchy appearing, gray areas bloom. Nothing like the ambiguity of something which perfectly fits in two or more different narratives, depending on who you ask. All of a sudden, spreadsheets fall in disgrace and everybody wants to replace them for something else, and a frenzy for automating everything kicks in. The automation, tooling and methods *skirmish* can be tricky because it transversally cuts the organization. Then, tooling grows detached from particular projects or departments' realities and needs. They become "global variables" in the software sense of the word. You change one and the implications are always ugly in many places. This is a sort of paradox: tools are usually used by the practitioners, but for some strange reasons the practitioners rarely get to choose them, but managers do (they are defined top bottom). As the big system evolves, practitioners seek for tools more specialized for their roles. For example, a tool that tracks customer orders and focuses on service-level agreements (SLAs) differs considerably from one that tracks issues in a software component or one that's targeted at business analysts for modeling customer use cases and workflows. If in the past the organization could use a single tool for those different tasks, as it grows that stops being the case. The one-size-fits-them-all approach eventually stops working but in the process, a lot of energy is spent stretching tools to use cases they were not designed for. In all this confusion, a great deal of revenues is made by the tooling world. This is a moment when the organization has its pants at the lowest for vendor lock-in.

On the other hand, allowing every team or project to define its own tooling and methodologies would be impractical. Hence, as the tech organization expands, a key part of the tech leadership needs to look after cross-cutting integration, with a very strong system orientation, with the goal to harmonize tools and methods to work the best way possible well across the *multiverse* of projects and systems.

4.5 Switching Barriers

Vendor lock-in is just another *switching barrier* among many other switching barriers that organizations grow with time. Switching barriers are not only a result of greedy suppliers wanting us to stay captives of what they sell. We create our own internal, domestic switching barriers as well. Switching barriers can take many forms, but in tech organizations it frequently materializes in the system architecture and in the ecosystem of *ad hoc* homegrown components and tools, which have expanded and encroached beyond initial plans, where all of a sudden changing them becomes (or feels like) impossible, since the tool or subsystem is now affecting or influencing too many places and processes. You can feel the horror when thinking of changing those things for something else; usually a shiver goes through your spine, it feels just impossible or prohibitively costly. This brings up a sort of "paradox of the bad component" where the apparent impossibility of changing them makes organizations decide to continue extending the conflicting components only to make the problem worse under the illusion that they would eventually get good in some uncertain future provided enough tampering and refactoring is performed. By adding more complexity to poorly architected things, we just make them worse.

This is one of the main beauties of modular design (to be addressed in the next chapter): it brings back the power of choice to designers and architects. Modular architectures make switch barriers lower, which is an asset when products are in the maturation process. Modular thinking must be applied at all levels of the *big system*, and not just on the things that fly. Ideally, all architectural decisions should be made with the mindset that no element of the system is irreplaceable.

References

Altfeld, H.-H. (2010). *Commercial Aircraft projects, managing the development of highly complex products*. New York: Routledge.

Bar-Yam, Y. (2018, Sept 16). *Emergence of simplicity and complexity, New England Complex Systems Institute*. New England Complex Systems Institute. https://necsi.edu/emergence-of-simplicity-and-complexity

Dunning, D., & Kruger, J. (1999). Unskilled and unaware of it: How difficulties in recognizing one's own incompetence lead to inflated self-assessments. *Journal of Personality and Social Psychology, 77*(6), 1121–1134. https://psycnet.apa.org/doi/10.1037/0022-3514.77.6.1121

Gall, J. (1975). *Systemantics: How systems work and especially how they fail*. Quadrangle/The New York Times Book Co.

Norman, D. (2011). *Living with Complexity*. The MIT Press.

Quattrone, P., & Hopper, T. (2001). What does organizational change mean? Speculations on a taken for granted category. *Management Accounting Research, 12*(4), 403–435. https://doi.org/10.1006/mare.2001.0176.

Saffer, D. (2009). *Designing for interaction: Creating innovative applications and devices*. New Riders.

Stensaker, I., Meyer, C., Falkenberg, J., & Haueng, A.-C. (2001). Excessive change: unintended consequences of strategic change. Paper Presented at the Academy of Management Proceedings, Briarcliff Manor, NY.

Zajac, E., Kraatz, M., & Bresser, R. (2000). Modeling the dynamics of strategic fit: A normative approach to strategic change. *Strategic Management Journal, 21*(4). https://doi.org/10.1002/(sici)1097-0266(200004)21:4%3c429::aid-smj81%3e3.0.co;2-#.

Chapter 5
Modular Spacecraft Architectures

As an architect, you design for the present, with an awareness of the past, for an unknown future.
—Norman Foster

Abstract Designers will claim by default that their designs are modular. But experience shows they tend to create largely integral architectures, often under the pretense of vertical integration and in-house know-how protection, incurring in troubling non-recurring costs, creating high-switch barriers, and more importantly, delaying time to orbit (and to market). Reality is that both vertical integration and IP protection can be achieved using modular open architectures. The important bit is to identify which blocks of the architecture are worth reinventing from scratch. In NewSpace, core differentiating technologies tend to be on the payloads and sensors. Hence, the spacecraft bus architecture can be commoditized by choosing standard backplane-based form factors, high-speed serial communication, and configurable interconnect fabrics.

Keywords Modularity · Conceptual design · Switched fabrics · Interconnect fabrics · VPX · OpenVPX · SpaceVPX · CompactPCI serial space · Integrated modular avionics · IMA · IMA-SP · Switch barriers

With the previous chapter about complexity creep vigilantly watching us, let's start addressing actual spacecraft design matters.

We spoke about how life cycles and breakdown structures are arbitrary discretizations which we perform for organizing and managing better the task of synthesizing a system (or, more generally, to solve a problem); discretization is an essential part of engineering design. As we break down the system into its constituent parts, we create a network of entities whose relationships define, among other things, ownership and composition relationships. By creating such a network, we allow the global "main" function to emerge. Meaningful functions needed for the system to perform can be grouped into functional components. Functional components are, as the name indicates, function *providers*; they can provide, at least, one function. Functional components are seldom atomic; they have their own internal structures themselves.

I. Chechile, *NewSpace Systems Engineering*,
https://doi.org/10.1007/978-3-030-66898-3_5

For a functional component to be able to provide its function, an interface needs to exist. An interface in a function-coupling mechanism or artifact. An interface allows a component to convey, contribute, and/or connect its function across boundaries to other component or components, depending on the topology, equally receiving the same from other components. Component boundaries define where its functional entity begins and where it ends; an interface is a contract between functional component boundaries. This contract specifies how the function will be correctly supplied and consumed. Often, the function-coupling between components requires transferring something (information, matter, or energy), in which case the interface acts as a channel for that *something* to flow. Think about a car as an object: the door is a component with an interface which couples to the car body and contributes its main functionality of communicating the inside of the cockpit with the outside world and vice versa. The car door provides secondary functions as contributing to structural sturdiness of the body, and to the general aesthetics; i.e. to make it look nice. Car doors have interfaces for them to couple to the higher-level system, but at the same time, car doors are interfaces in their own right. Think about a bridge connecting two roads: the bridge itself is a component with a reasonable complexity, but from the road system perspective, the bridge is just an interface. It is the point of view in the system hierarchy from where we observe the bridge which makes it look like a component or an interface.

When does a component become *modular*? It is modular if its relationship to the higher-level system is not "monogamous" if such term can be used as analogy. This means that a component can provide its function to some other higher-level system (either with similar or a total dissimilar goal) without the need of introducing any change to it. Are car doors modular? Certainly not. We cannot take a door from a Hyundai and put it in a Mercedes and expect it to work, due to the fact that the *functional coupling* depends on the mechanical form factor: the door was designed to specifically fit in the Hyundai. On the other hand, plenty of car parts are modular, like tires, windshield wipers, lightbulbs, and so on. We could take a tire from a Hyundai and put it in a Mercedes, provided cars are of roughly similar type; performance might be somewhat impacted, but function would be reasonably coupled, because the interface to the axle is standardized across vendors. If a software library can be used on different applications (provided the way to interface it is understood and documented), it is modular. In fact, a software library can in turn contain other libraries, and so on. Software is a great example of how modularity can enable great ideas. Whole open-source community sits on modular thinking. Software gets well with modular approaches because the interfaces are extremely malleable, development cycles are way shorter, enabling "trial and error" loops to happen fast.

Functionality never exists in isolation: we have to connect functionality to contexts through different mechanisms. Take for example the Unix operating system. Unix comes with a rich set of shell commands such las *ls*, *cat*, *nc*, and *sort*. Let's take as an example the useful netcat (or *nc*). Netcat is a versatile tool which is used for reading from and writing to network sockets using TCP or UDP. Netcat can be piped (combined) with other commands and provide a rich set of tools to debug and probe networks. Netcat provides its functionality by means of the shell context. Typically, when we code, for example in C, we frequently find ourselves needing netcat-like functionality in our code. We very often end up creating handcrafted downscaled clones of netcat. We could just use netcat, but we cannot (at least not easily or effectively). Since our context (a process of our own) is different compared to what netcat has been designed for (a shell), we need to grow a homegrown netcat() ourselves.

Modularity is a design strategic decision, for it can impact factors such as cost, schedule, and risk. It should be said that when system integrators own the whole architecture and also own the life cycle of some of the components in the architecture, modularity can be difficult to justify, unless there is some potential for a component to be used in another system across the product line. One may think that there is no real reason why ***not*** going modular, but it might be a tricky analysis. Space industry is famously known for its non-modular choices. Most of the satellite platforms in the world are designed for specific payloads. This means, the system and the component (payload) have a very *monogamous* relationship; changing the payload for a totally different platform is nearly impossible in every single way (mechanically, electrically, etc.). Years ago, a project involving a spacecraft bus provider X and a NewSpace payload provider Y was kicked off as a joint endeavor. The bus provider was quite "classic space" minded, whereas the payload company had a concept of operations (CONOPS) which was very NewSpace. Neither the bus nor the payload had been designed in a modular way, meaning that to make their functionalities couple, a good deal of alterations at both ends were needed. Bespoke software was created for this troublesome coupling to happen. This was largely *ad hoc* software that did not get the chance of undergoing extensive testing due to the tight schedule. Payload failed after a few weeks of operations due to a software bug. The failure is quite poetic in itself: hacking your way to make two components to couple when they were not designed to couple tends to end badly. You can still put a puzzle together with your eyes closed; pieces will eventually fit if you push hard enough. But when you open your eyes it will not look nice.

Modularity can also be unattractive from a business perspective. When you go modular, you risk yourself to start losing your identity and being treated as a commodity. If a customer can quickly replace your module with a competitors' one, they might. The automotive industry has a very selective criteria on modularity, and the reason is cost but largely differentiation and brand image. Automakers go modular on things that do not impact their brand image or aesthetic integrity of their products, but do not go modular on things that do impact such integrity. Organizations can choose to design large and complex things in a modular way if it is for their own benefit; for example, automotive companies make their engines and chassis modular in the way they can use them across their product line. In the aerospace industry, turbofan engines are highly complex devices but are also modular; they are designed in such a way that they can work across different aircrafts. Modularity is highly connected to the system hierarchical breakdown. It is the decision of the architects to define the scope of the modular approach in a system design. For example, in space applications, the whole payload can be taken as a functional component. This way,

Table 5.1 A spacecraft functional breakdown

Functional component	Function it provides	How function is coupled to the context
Power management	Convert energy from Sun Store Energy; Distribute Energy; Determine electrical state by measurement	Mechanical (fixations to structure) Thermal (conduction) Electrical (power and data connectors), Software (command and data handling)
Structure	Provide physical housing for bus and payload Protect from launch loads	Mechanical Thermal
Attitude management	Determine orientation Alter orientation	Mechanical (fixations to structure), Thermal, Electrical (power and data connectors), software (command and data handling)
Thermal management	Determine thermal state Alter thermal state	Mechanical (fixations of heaters and thermostats to structure), Thermal, Electrical (power and data connectors), software (command and data handling)
Comms management	Convey data to/from spacecraft	Mechanical (fixations to structure), Thermal, Electrical (power and data connectors), Software (command and data handling)
Resources management	Store data Collect data Detect and isolate failures	Mechanical (fixations to structure), Thermal, Electrical (power and data connectors), software (command and data handling)

the coupling between the bus and the payload could be designed to be modular if needed; a multimission platform. But then, inside the bus for example, the avionics could be designed to be modular as well. So, modularity is recursive, as anything related to the recursive breakdowns we have been working on since a few chapters.

But, let's get a bit less abstract and run the exercise of trying to identify the main functional components of a typical spacecraft (Table 5.1).

Having these listed in a table does not help visualizing the hierarchical nature of all this, so let's draw accordingly.

The diagram Fig. 5.1 depicts a functional breakdown which includes the minimum amount of functions for a spacecraft to perform its main (or master) function: "Operate a Payload in Orbit". The main function flows down into subfunctions as

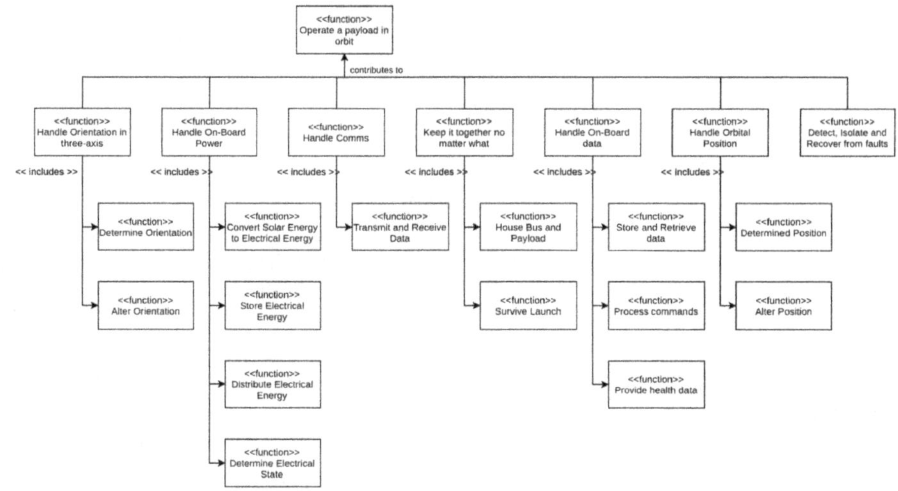

Fig. 5.1 A more graphical functional breakdown of a spacecraft

depicted in the diagram. All the subfunctions contribute to the main one. These functional blocks can also be seen as handlers; i.e. they are supposed to *take care* of something.

Now, this functional breakdown structure is totally immaterial. There is not a single mention of how the physical breakdown structure looks like. All these blocks in Fig. 5.1 are not physical entities, just abstract blocks that ideally provide a function that the system needs for its core function. Function, by itself, is weightless or intangible, which is great news for designers because it makes the functional analysis very flexible. What makes designers' life harder is the way the elements are put together to realize the function. Abstract functional architectures will not get the job done, so the design team must distribute functions across the physical architecture (i.e. allocate functions to physical blocks). Bear with me here that software can also count as "physical" in the sense that it is an entity (binary/object code) which is supposed to perform the function. Physical here means more "an entity which executes the function in the operative environment" than actually physical in the sense that you can hold it in your hand.

This is a point when it starts to be the right time for designers to evaluate whether to take the modular path (i.e. go modular) or not, hence the importance of functional analysis when it comes to modular thinking. It is from the functional breakdown structure that modularity can be evaluated. The game of connecting functions to physical/executable blocks is a many-to-many combinatorial challenge. Many functions can be allocated to one single physical/executable element, as well as one single function can be broken down to many physical/executable elements.

Let's take as an example the function: "Handle Power". This functional component is composed of the subfunctions listed in Table 5.2.

Table 5.2 Subfunctions allocation for the "Handle Power" function

Subfunction of "Handle Power"	Physical element to realize it
"Convert Solar Energy to Electrical Energy"	Solar Panels
"Store Electrical Energy"	Battery
"Distribute Electrical Energy"	PCDU (Power Control and Distribution Unit)
"Determine Overall Electrical State"	PCDU (Power Control and Distribution Unit)

Each one of the physical elements listed in Table 5.2 must couple to the system in a way they will be able to deliver their functions accordingly. Coupling will need to take place in several ways: mechanically, electrically, thermally, and software coupled (in case the element contains software, commands, telemetry, etc.). From the example above, a first challenge comes up. "Distributing energy" and "Determining Overall Electrical State" share the same physical entity: PCDU, which is, in this case a metal box with a PCB inside and connectors. We could have chosen differently, by creating a new board called "PMU" (power measurement unit) which would map 1:1 with "Determine Overall Power Status" function. But is it worth it? Probably it is not; more mass, more software, more complexity; i.e. little gain. Hence, we decide to have two functions mapped to one physical component. The system design quickly imposes a challenge: the functional architecture does not map 1:1 to the physical architecture (what is usually called a federated architecture), which kicks in the decision about centralization vs decentralization, among other similar dilemmas. The next section will present a graphical way of depicting how the functional breakdown structure maps to the physical one, by means of using concept maps as function-structure diagrams.

Architectural decisions are shaped by context and constraints. For example, cost is a common driver in NewSpace architectures, which sneaks in during the decision-making about function allocation. Mass is the other one; although launch prices have decreased in recent years, still less mass means a cheaper launch, easier transportation, etc. This is a strong incentive to concentrate functions in as few physical elements of the architecture as possible (integrated architecture). NewSpace projects understandably choose integrated approaches, as they typically allocate as many functions to as few physical blocks as possible. There are dangers associated with this approach of adding "all eggs into one basket", related to reliability and single point of failure. A reasonable failure analysis should be performed, and risks taken should be understood and communicated. New technologies are also leveraging integrated architectures, such as hypervisors and virtualization. These technologies are being quickly matured and standardized for mission-critical applications.

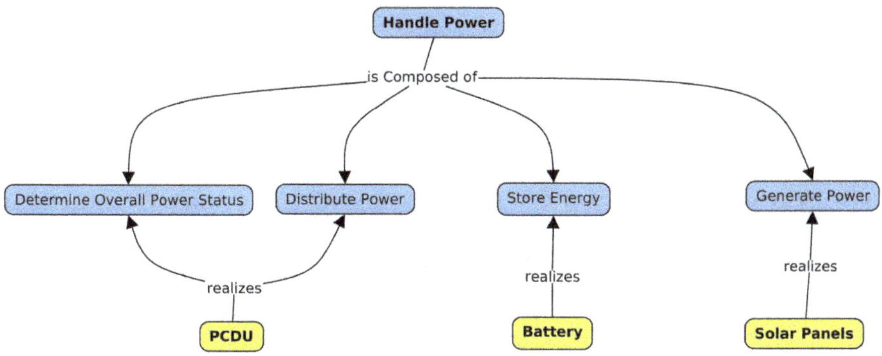

Fig. 5.2 Function-structure diagrams

5.1 Function-Structure Diagrams

Crossing the boundary between functional analysis and the physical/executable architecture is never an easy one. As said before, when the functional and physical/executable architecture depart from each other, two sorts of "universes" are spawn. It is important to keep visual clues about how those two domains (functional and physical) remain mapped to each other. By doing this, the design team can track design decisions which might be reviewed or changed later in the process. Concept maps can be used for this. In the diagram Fig. 5.2, the boxes in blue are abstract functions and subfunctions, whereas the yellow blocks are actual physical/executable elements of the architecture. Note that this type of diagrams can accompany the early-stage analysis of requirements (if there are any). Typically, requirements are early stages that point to functions and not to specific synthesis of functions (how functions are realized). This means a concept map with the initial understanding of the functions required can be transitioned to a function-structure diagram as the architectural decisions on how to map functions to blocks is performed (Fig. 5.3).

5.2 Conceptual Design

There is great bibliography about spacecraft design, such as Brown (2002). Entire books are written about this topic, so this section only aims to summarize and give a quick glimpse on how the design process kicks off, and what defines how a spacecraft ends up looking like. This section is mostly aimed for a non-technical audience who are willing to understand more about what types of tasks are performed during early stages of the design process of a spacecraft. This section is for newcomers to space. If you're a space nerd or you have put a man on the Moon, you can safely skip it before you get offended.

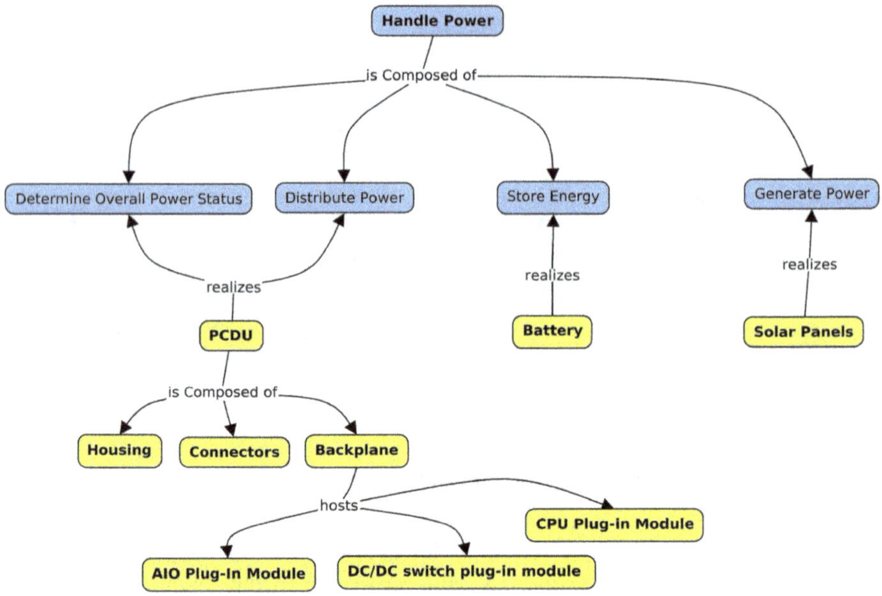

Fig. 5.3 Adding more details to function-structure diagram

In this section, it will be assumed that the project has been just kicked off, and everything starts from the scratch. That's what conceptual design is about: starting off from a few scribbles on a whiteboard, but not much else. This is a stage of very high uncertainty, and as such it requires a lot of analysis and firm decision-making. Typically, during conceptual design there might be (as in, there will be) conflicting opinions about some topic or idea, and it is the systems engineer's task to resolve those by analyzing the alternatives carefully and decidedly choosing one and allow the team to move ahead. It is often better to move ahead with an idea which might still need refining than stopping everything until the last detail is sorted out. Frequently, during conceptual design stages teams get themselves in *analysis-paralysis* swamps where discussions can take days, and nothing is decided. The design activity freezes in favor of pointless deliberations which someone needs to break. At this stage the role of an experienced systems engineer is essential, otherwise a great deal of time, energy, and momentum is lost. Eternal deliberations keep the whole organization away from the market. While teams stop to argue on how to split the atom, a competitor is getting closer to a win. Conceptual design must move ahead.

Conceptual design relies greatly on sketching; this should be at this point no surprise, we emphasized enough, a few chapters ago, how the diagram-driven engineering is. Sketches give form to abstract ideas and concepts, and they become the first two-dimensional shapes a project can offer to the people involved. Sketching is a form of communication, and it should be not only encouraged but actively trained across the organization. Unfortunately, it seems sketching does not get as much attention in engineering schools as it should. There should be more courses offered in sketching and drawing in support of design projects. Teaching basic techniques in freehand sketching would help students generate quicker and more effective visualizations ideas during their professional careers.

Note: It is very useful to take photos and archive sketches drawn in whiteboards and/or notebooks, for they become the first visual clues of the project and an important archive of design decisions as the organization matures.

Now, what should we expect from conceptual design as an activity? The main outcome of this stage should be (as a minimum):

- A rough physical configuration (mainly the bus form factor, solar panels tentative location and area, payload location).
- An understanding on the power margins (i.e. power generation vs power consumption ratio), along with power storage (i.e. battery sizing).
- An understanding on what sensors and actuators the spacecraft will use to manage its orientation in space and its position in orbit.
- A preliminary understanding on how to keep things in safe thermal margins.
- A rough understanding about the mass.
- A block diagram.

None of the bullets previously listed are totally isolated from the rest. They are all interrelated, and at this stage they are all very preliminary and rough. Many things are very fluid to be taken very seriously, and the idea is to refine those as the process goes. As we said, we don't go through the conceptual stage only once. We run through it several times: we propose something, things get analyzed and reviewed; if changes are needed, changes are done, and process repeats. The design evolves and often ends up looking quite differently compared to how it was conceptualized. It is always an interesting and somewhat amusing exercise, once a project has reached production, to go back to the conceptual stage notes and diagrams and see how different things turned out to be at the end. Conceptual design does not rigidly specify anything but suggests paths or directions which will or will not be taken, depending on many factors.

At the beginning, spacecraft engineering does not look very fancy; it is more or less a bunch of spreadsheets, some loose code, and analysis logs. Mostly numbers. Good thing is to get started with conceptual design, not many things are needed: a rough idea of the payload concept of operations (CONOPS) and some ideas about the target orbit.

The payload's concept of operations will provide a good indication of what type of attitude profile the spacecraft will have to maintain throughout its life. For example, a synthetic-aperture-radar (SAR) payload will need to operate by side-looking, meaning that frequent movements (i.e. slewing) to achieve different look angles will be needed. Or, if the radar will always look at one particular angle, we could design it in a way that side-looking attitude would be the "default" attitude.[1] Other types of payloads, for comms missions, typically stay with one axis pointing to earth so that their antennas can continuously cover the desired areas. Some other

[1] See, for example, SAOCOM mission.

more advanced payload could be gimballed and do the tracking on their own, leaving the satellite with the freedom to point wherever. Payloads greatly define how accurate and precise the orientation determination and control need to be, from coarse pointing to very precise. For example, a laser comm terminal, for very high-speed data transfer, or a telescope, will impose stricter requirements than, say, a comms relay spacecraft with omnidirectional antennas on-board.

The other important input for the conceptual design process is the orbit chosen for the mission, so let's quickly discuss how orbits are defined. Orbits are geometrically described by orbital (Keplerian) elements, graphically depicted in Figs. 5.4 and 5.5. Orbital elements are a set of parameters that analytically define the shape and size of the ellipse, along with its orientation in space. There are different ways of mathematically describing an orbit, being orbit elements the most common:

- Eccentricity (e)—shape of the ellipse, describing how much it is elongated compared to a circle. Graphically, it is the ratio of the distance from the center to the foci (F) and the distance from the center to the vertices. As the distance between the center and the foci approaches zero, the ratio approaches zero and the shape approaches a circle.
- Semimajor axis (a)—the sum of the periapsis and apoapsis distances divided by two. For classic two-body orbits, the semimajor axis is the distance between the centers of the bodies, not the distance of the bodies from the center of mass.

Two elements define the orientation of the orbital plane in which the ellipse is embedded:

- Inclination (i)—vertical tilt of the ellipse with respect to the reference plane, measured at the ascending node (where the orbit passes upward through the reference plane, the green angle i in the diagram). Tilt angle is measured perpendicular to line of intersection between orbital plane and reference plane. Reference plane for Earth orbiting crafts is the Equator. Any three points on an ellipse will define the

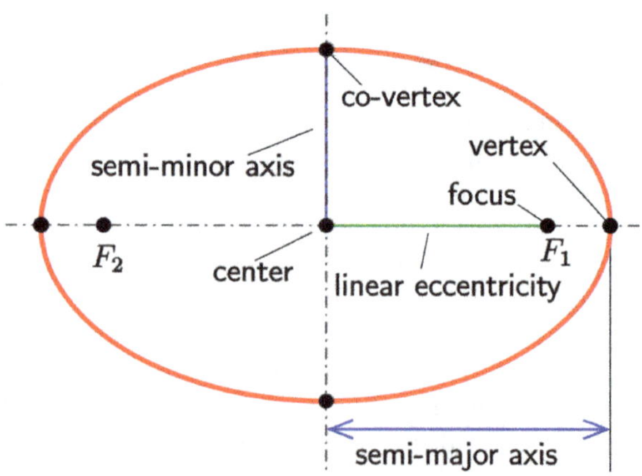

Fig. 5.4 Elements of the ellipse (By Ag2gaeh—Own work, CC BY-SA 4.0, https://commons.wikimedia.org/w/index.php?curid=57497218)

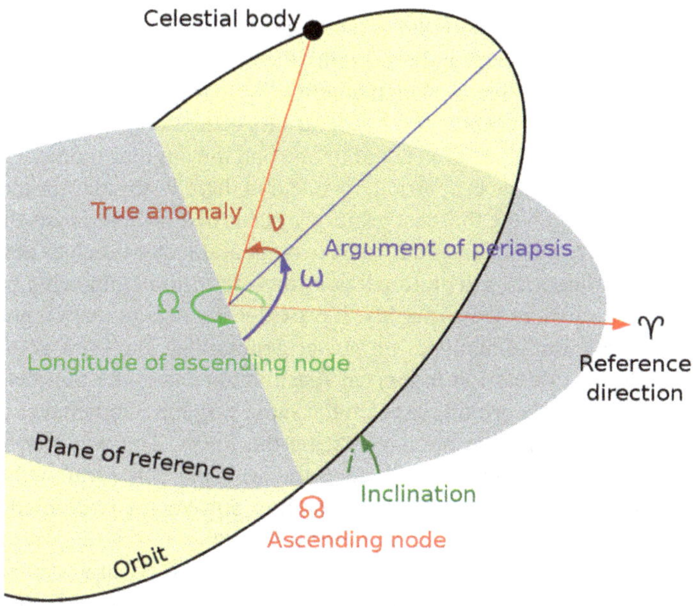

Fig. 5.5 Orbit elements (*credit* Lasunncty at the English Wikipedia/CC BY-SA (http://creativecommons.org/licenses/by-sa/3.0/))

ellipse orbital plane. The plane and the ellipse are both two-dimensional objects defined in three-dimensional space.

- Longitude of the ascending node (Ω)—horizontally orients the ascending node of the ellipse (where the orbit passes upward through the reference plane, symbolized by ☊) with respect to the reference frame's vernal point (symbolized by the Aries Ram symbol). This is measured in the reference plane and is shown as the green angle Ω in the diagram.

The remaining two elements are as follows:

- Argument of periapsis (ω) defines the orientation of the ellipse in the orbital plane as an angle measured from the ascending node to the periapsis (the closest point the satellite object comes to the primary object around which it orbits; the blue angle ω in the diagram).
- True anomaly (ν, θ, or f) at epoch (t_0) defines the position of the orbiting body along the ellipse at a specific time (the "epoch").

The mean anomaly M is a mathematically convenient fictitious "angle" which varies linearly with time, but which does not correspond to a real geometric angle. It can be converted into the true anomaly ν, which does represent the real geometric angle in the plane of the ellipse, between periapsis (closest approach to the central body) and the position of the orbiting object at any given time. Thus, the true anomaly is shown as the red angle ν in the diagram, and the mean anomaly is not shown. The angles of inclination, longitude of the ascending node, and argument of periapsis can also be described as the Euler angles defining the orientation of the orbit relative to the reference coordinate system. Note that non-elliptic trajectories also exist, but are not closed, and are thus not orbits. If the eccentricity is greater than one, the trajectory is a hyperbola. If the eccentricity is equal to one and the angular momentum is zero, the trajectory is radial. If the eccentricity is one and there is angular momentum, the trajectory is a parabola. Note that real orbits are perturbed, which means their orbital elements change over time. This is due to environmental factors such as aerodynamic drag, Earth non-oblateness, gravitational contribution from neighboring bodies, etc.

If the orbit is known, at least roughly, it provides the design team a good deal of information to drive initial analyses on power generation. The orbit geometry, and more specifically its orientation in inertial space defines how the Sun behaves with respect to the spacecraft's orbital plane; in the same way the Sun behaves differently in the sky depending on what point we sit on the Earth. The way the Sun vector projects into the orbit plane will influence the way the spacecraft will scavenge energy from sunlight. Let's quickly see how this Sun vector is defined. We shall start by describing the geometry of a celestial inertial coordinate system (with X-axis pointing to vernal equinox,[2] Z-axis aligned with the Earth's axis of rotation, and Y-axis completing the triad following the right-hand rule) and identifying some important planes: ecliptic plane (the plane Earth orbits around the Sun) and equatorial plane, which is an imaginary extension in space from Earth Equator (Fig. 5.6).

We define the solar vector, $\hat{s}$, as a unit vector in this coordinate system, which points toward the Sun (Fig. 5.7).

The apparent motion of the Sun is constrained to the ecliptic plane and is governed by two parameters: Γ and ε (Fig. 5.8).

ε is the obliquity of the ecliptic, and for Earth, its value is 23.45°; Γ is the ecliptic true solar longitude and changes with date. Γ is 0° when the Sun is at the vernal equinox. In order to find the position of the Sun starting from the inertial frame of reference, we must devise a frame of reference which is aligned with the ecliptic plane. Let's create a basis of orthogonal unit vectors at the center of this inertial system and define a unit vector $\hat{\mathrm{s}}$ along the X-axis of the test system (Fig. 5.9a). What we need to do in order to point $\hat{s}$ to the Sun is to tilt the test system an angle ε around its X-axis in order to align its xy plane with the ecliptic plane (Fig. 5.9b). Next, we just need a rotation of an angle Γ about the z-axis of the rotated *test* system; $\hat{s}$ points now in the direction of the Sun (Fig. 5.9c).

The vector $\hat{s}$ now revolves around the ecliptic plane. To evaluate better the behavior of the Sun with respect to a specific orbit, we need to geometrically define the orbit orientation. For this, we must find a vector normal to the plane described by our orbit, so let's define the geometry accordingly:

From Fig. 5.10, i is the test orbit inclination, i.e. the orbit's angular tilt from the equatorial plane; Ω is the right ascension of the ascending node, which is an angular

[2]The Vernal Equinox occurs about when the Sun appears to cross the celestial Equator northward. In the Northern Hemisphere, the term *vernal point* is used for the time of this occurrence and for the precise direction in space where the Sun exists at that time.

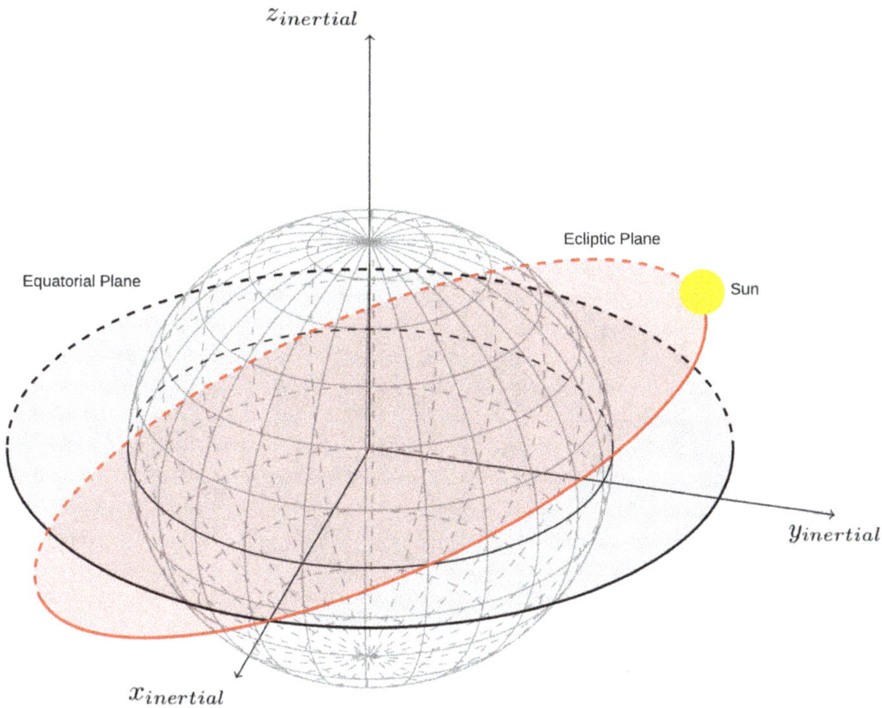

Fig. 5.6 Equatorial plane, ecliptic plane, and Sun position

measure between the inertial x-axis at the point where the orbit crosses the equatorial plane going from south to north (Fig. 5.11).

The beta (β) angle is the angle between the projection of vector s in the orbital plane and s itself. The easiest way to calculate the angle between a vector and a plane is to determine the angle between the vector and a vector normal to the plane, denoted in Fig. 5.12 by (denoting now *o* and *s* vectors with "hats" since they are unit vectors):

$$\cos\phi = \widehat{o}.\widehat{s} = \cos(\Gamma)\sin(\Omega)\sin(i) - \sin(\Gamma)\cos(\varepsilon)\cos(\Omega)\sin(i) + \sin(\Gamma)\sin(\varepsilon)\cos(i)$$

But:

$$\beta = \phi - \pi/2 = \sin^{-1}\{\cos(\Gamma)\sin(\Omega)\sin(i) - \sin(\Gamma)\cos(\varepsilon)\cos(\Omega)\sin(i) + \sin(\Gamma)\sin(\varepsilon)\cos(i)\}$$

We observe that β is limited by:

$$\beta = \pm(\varepsilon + |i|)$$

The beta angle is not static and varies constantly. Two factors that affect beta variation the most are:

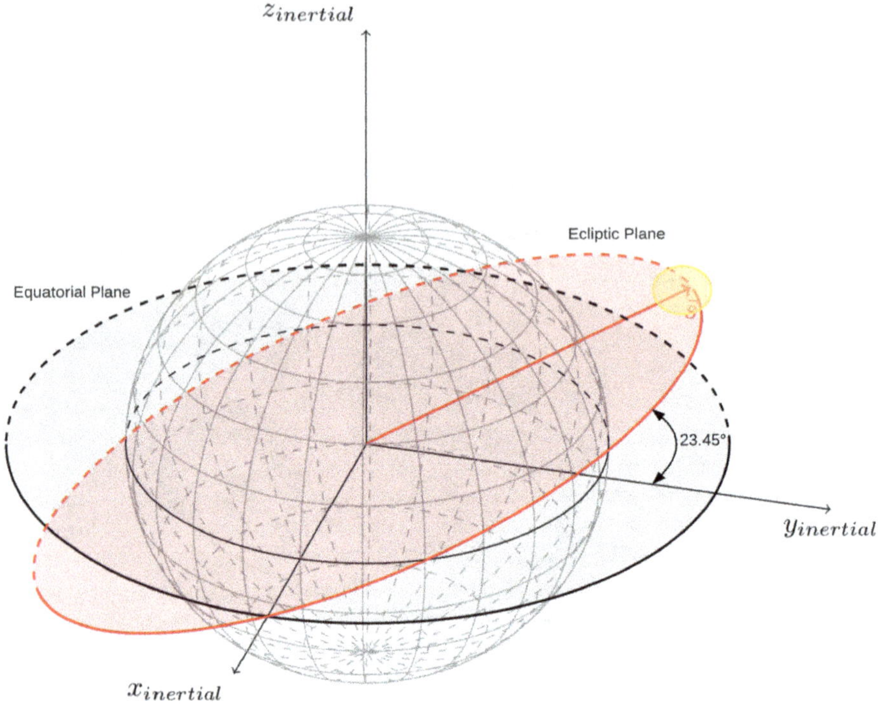

Fig. 5.7 Vector s points to the Sun

1. The change of seasons (variation in Γ)
2. Perturbation of the orbit due to the oblateness of the planet (variation in Ω, and rotation of apsides).

The variation that concerns the most design engineers is the variation due to precession of the orbit. Earth is not a perfect sphere; hence the equatorial bulge produces a torque on the orbit. The effect is a precession of the orbit ascending node (meaning a change of orientation in space of the orbital plane). Precession is a function of orbit altitude and inclination (Fig. 5.13).

This variation is called the ascending node angular rate or precession rate, ω_p, and is given by:

$$\omega_p = -\frac{3}{2}\cdot\frac{R_E^2}{\left(a\left(1-e^2\right)\right)^2}J_2\omega\cos i$$

where

- ω_p is the precession rate (in rad/s)
- R_E is the body's equatorial radius (6,378,137 m for Earth)
- a is the semimajor axis of the satellite's orbit

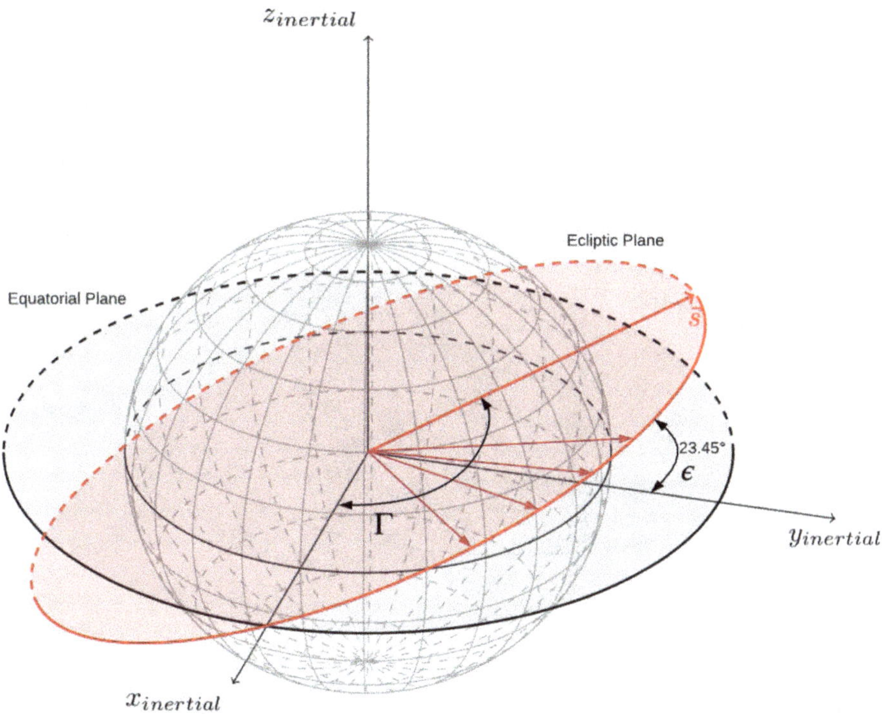

Fig. 5.8 Sun revolves around the ecliptic plane

- e is the eccentricity of the satellite's orbit
- ω is the angular velocity of the satellite's motion (2π radians divided by its period in seconds)
- i is its inclination (in degrees)
- J2 is the body's "second dynamic form factor" $= 1.08262668 \times 10^{-3}$ for Earth).

Rotation of the apsides is the other perturbation to consider. It is an orbit perturbation due to the Earth's bulge and is similar to regression of nodes. It is caused by a greater than normal acceleration near the Equator and subsequent overshoot at periapsis. A rotation of periapsis results. This motion occurs only in elliptical orbits (Brown 2002).

As beta angle changes, there are two consequences of interest to thermal engineers:

(1) The time spent in eclipse (i.e. planet shadow) varies.
(2) The intensity and direction of heating incident on spacecraft surfaces changes.

Orbit orientation will influence the duration of the daytime and nighttime on board (i.e. sunlight and eclipse times). This factor will impact not only power generation and storage but also thermal behavior, since things can get quite cold during eclipses. Physical configuration and subsystem design are greatly impacted by the beta angle behavior with respect to the body reference of reference. It does not affect

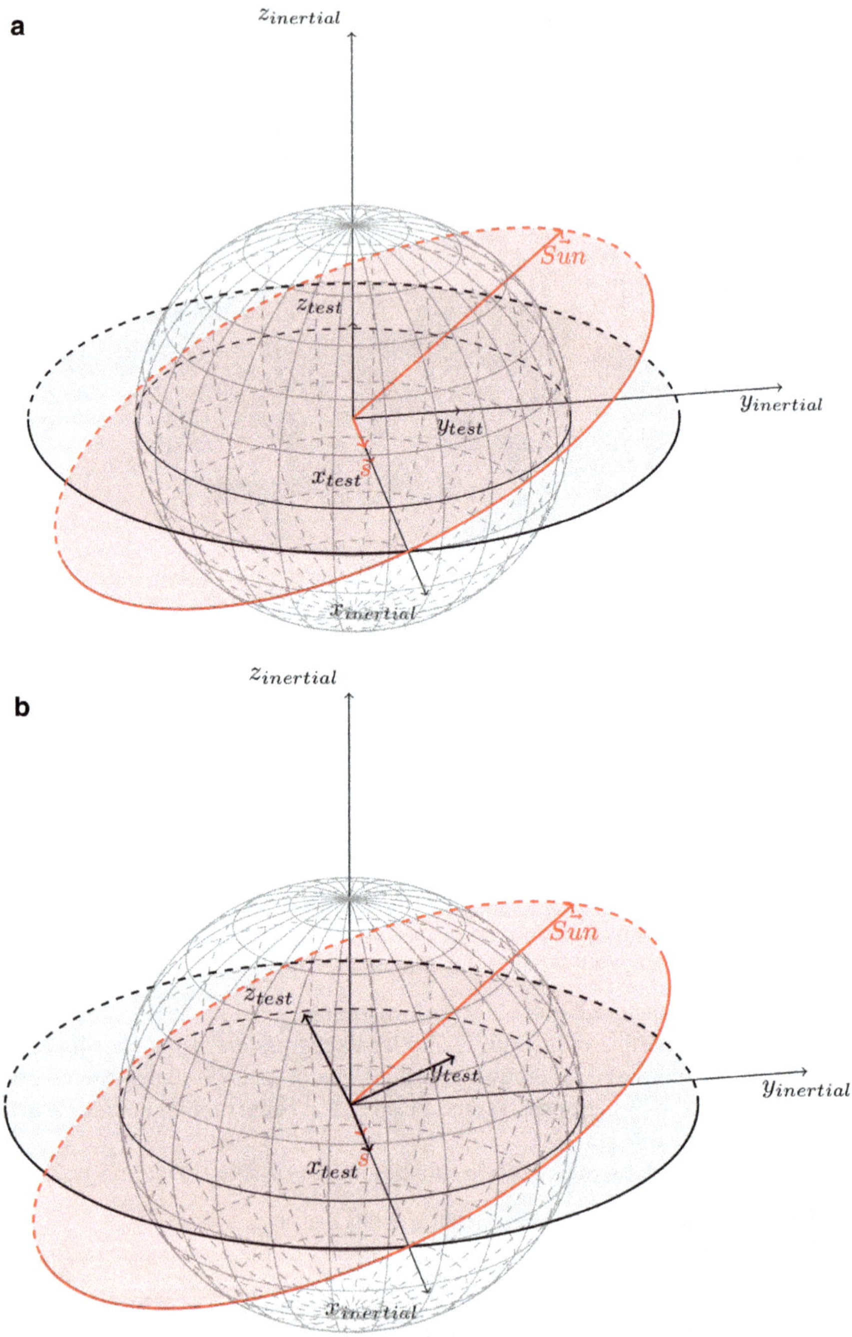

Fig. 5.9 (a, b, c) Obtaining a vector s to point to the Sun

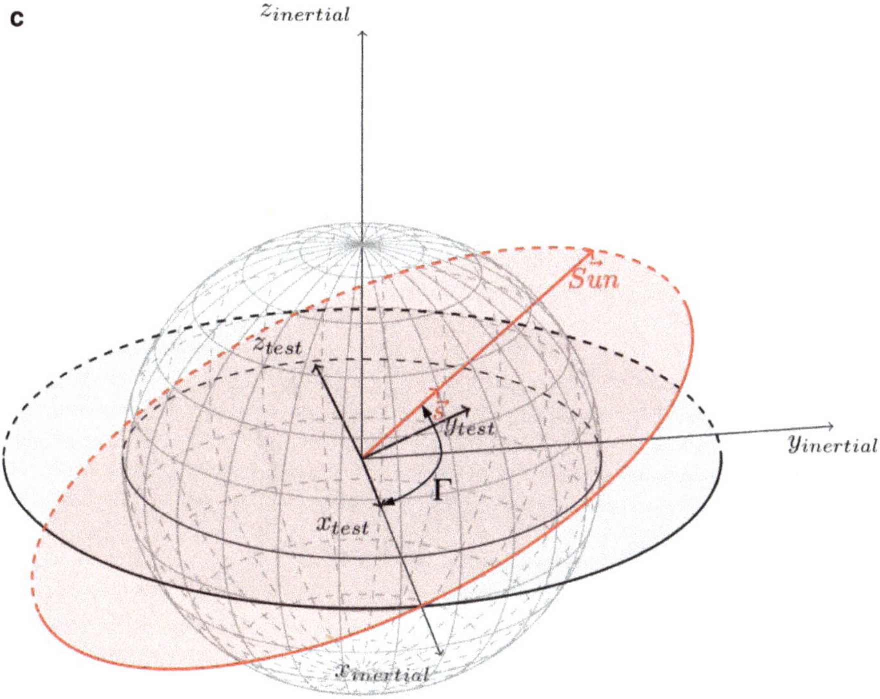

Fig. 5.9 (continued)

the dimensioning of the power-generating surfaces but defines the location of sensors that can be perturbed by the Sun. This can directly affect orbit capacity and payload availability: a suite of sensors being frequently affected by sunlight can reduce the readiness for the spacecraft to perform its mission.

5.2.1 Ground Track

Our planet remains a very popular science subject. Despite being our home for quite a while already, we keep on observing it, and we continue learning from our observations. It is a big and complex planet, for sure, and the reasons behind our need to observe it are very reasonable and disparate. We observe it to understand its physical processes (atmosphere, weather), to monitor how things on its surface are changing over time (e.g. surveillance or defense), or to understand anomalous or catastrophic events (flooding, fires, deforestation). In any case, we place satellites in orbit with a clear idea of what we want to observe beforehand; this means, what areas we want the spacecraft to sweep while it revolves around the Earth, in order to

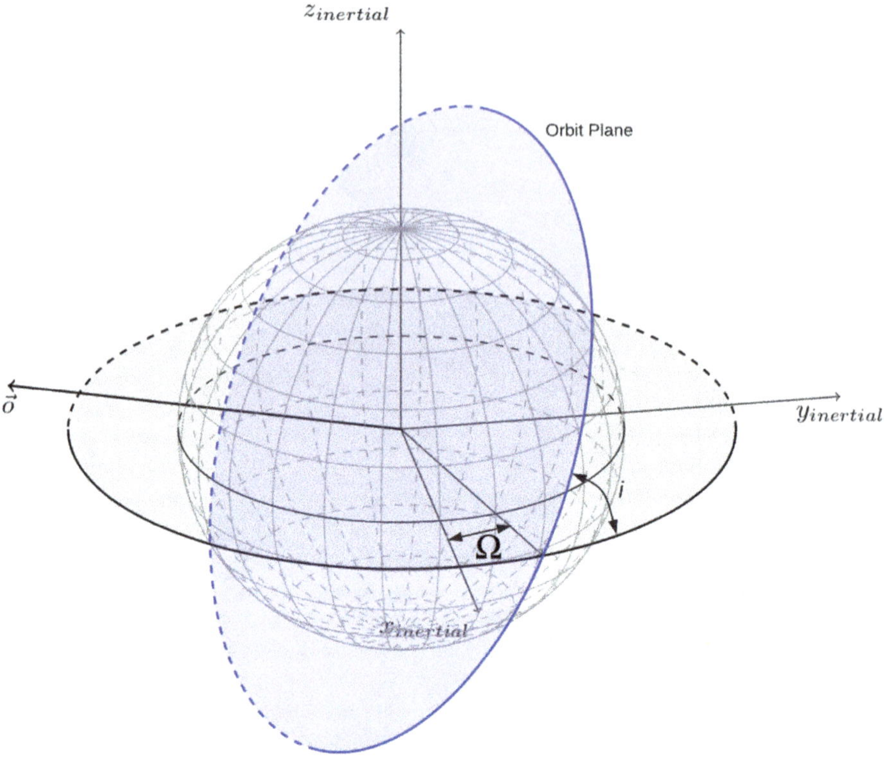

Fig. 5.10 Orbit plane geometry with respect to inertial frame of reference and vector normal to the orbit plane

fulfill its mission. The flight path of the spacecraft defines a ground track, and the ground track is a result of few things:

- The motion of the spacecraft
- The rotation of the central body
- Orbit perturbations.

The geographic latitudes covered by the ground track will range from –*i* to *i*, where *i* is the orbital inclination. In other words, the greater the inclination of a satellite's orbit, the further north and south its ground track will pass. A satellite with an inclination of exactly 90° is said to be in a polar orbit, meaning it passes over the Earth's north and south poles. Launch sites at lower latitudes are often preferred partly for the flexibility they allow in orbital inclination; the initial inclination of an orbit is constrained to be greater than or equal to the launch latitude. At the extremes, a launch site located on the Equator can launch directly into any desired inclination, while a hypothetical launch site at the north or south pole would only be able to launch into polar orbits (while it is possible to perform an orbital inclination change maneuver once on orbit, such maneuvers are typically among the costliest, in terms

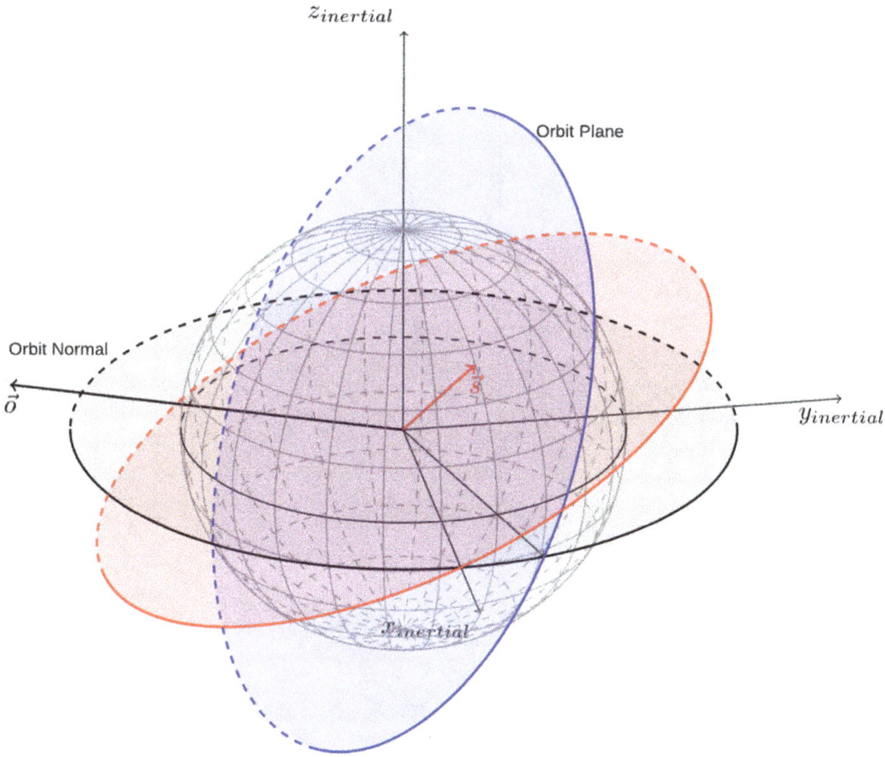

Fig. 5.11 Orbit plane, orbit normal, and ecliptic plane

of fuel, of all orbital maneuvers, and are typically avoided or minimized to the extent possible.).

In addition to providing for a wider range of initial orbit inclinations, low-latitude launch sites offer the benefit of requiring less energy to make orbit (at least for prograde orbits, which comprise the vast majority of launches), due to the initial velocity provided by the Earth's rotation.

Another factor defining ground track is the argument of perigee, or ω. If the argument of perigee of the orbit is zero, it means that perigee and apogee lie in the equatorial plane, then the ground track of the satellite will appear the same above and below the Equator (i.e. it will exhibit 180° rotational symmetry about the orbital nodes.). If the argument of perigee is non-zero, however, the satellite will behave differently in the Northern and Southern hemispheres.

As often orbital operations are required to monitor a specific location on earth, orbits that cover the same ground track periodically are often used. On Earth, these orbits are commonly referred to as Earth-repeat orbits. These orbits use the nodal precession effect to shift the orbit, so the ground track coincides with that of a previous revolution; thus this essentially balances out the offset in the revolution of the orbited body. Ground track considerations are important for Earth observation

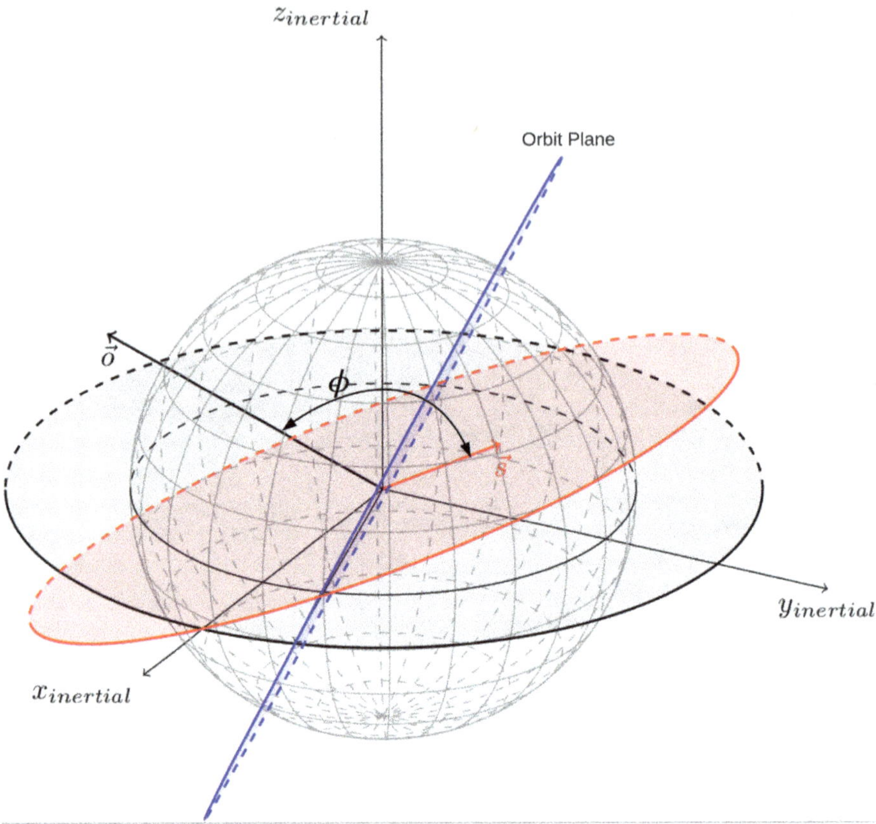

Fig. 5.12 Phi (Φ) is the angle between the Sun vector and the normal to the orbit plane

mission design since it defines the geometry on how the spacecraft will revisit a specific area. Also, for connectivity missions, antenna footprints will still depend on the way the craft will fly on specific regions.

5.2.2 Physical Configuration and Power Subsystem

A quick googling for spacecraft images gives a clear idea that basically they all look different. If you do the same with cars or aircraft, you will realize they look pretty much the same, regardless of the manufacturer and the aesthetic slight differences. An A320 and a Boeing 737 might have different "faces" (nose, windshields), and some differentiating details here and there, but mostly everything else is pretty much the same: windows, turbines, wings, ailerons, empennage, etc. The farther you observe them from, the more similar they look, at least for the untrained eye. Aircrafts, more or less, look all the same across manufacturers. Think about bulldozers, or digging

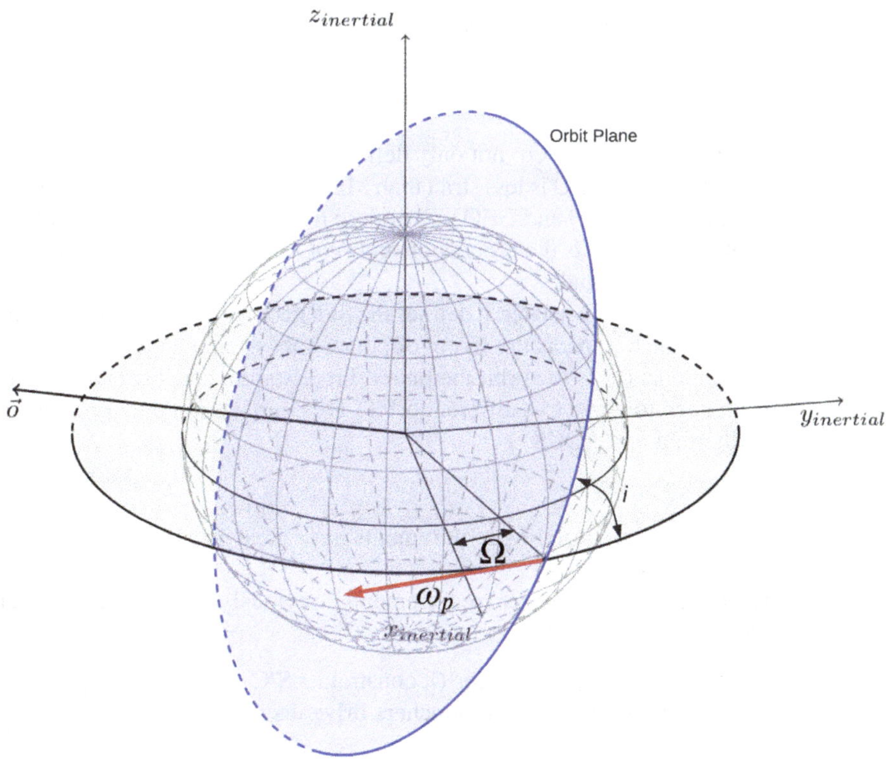

Fig. 5.13 Orbit precession

machines, same applies; you cannot really tell from a distance (provided colors and legends are not perceived as clues) which brand or model they are. Space is somewhat different. Two spacecrafts from different vendors and for a similar mission might look totally different. They will have different body shapes, different antennae, different solar panels. They will most likely be visually very different. A probable exception is GEO spacecraft which sort of look similar to each other (basically a box with two long and distinctive solar panel wings, some dish antennas and feeders). For LEO, the shape variety is way more noticeable. GEO space industry is a very market-oriented sector, similar to commercial aerospace and digging machines. A wild theory could say that companies tend to imitate more physical configurations that seem to do the work effectively, when physical configuration is not a branding factor (like in cars). If this holds true, spacecraft in LEO (for the same type of application) should start to look more alike in some future.

During conceptual design, the rough physical configuration (i.e. shape, size, or volume) usually comes as the first thing to tackle. The process of defining the spacecraft's physical configuration is multidisciplinary and highly inter-coupled and, of course, dependent on mission needs; a highly iterative one. The whole conceptual design stage is highly iterative. Initially, an IPT can be made responsible to carry out this task (in startups typically the stem/core IPT). The early factors that shape the configuration are:

- Type of orbit:
  - Altitude (LEO, MEO, GEO): not only defines the reliability approach (radiation environment for LEO is less strict than MEO) but also defines the architecture. For example, MEO and GEO will not use torquers but thrusters to unload the wheels. This impacts the propulsion and AOCS directly, and indirectly the mass budget and thermal.
  - Inclination: as we will see later, it defines the Sun geometry from the spacecraft perspective, which impacts the sizing of power subsystem elements.
  - Orbital maintenance: Is the orbit the final orbit ± station keeping? Or will the spacecraft need to move to a different orbit? This impacts not only propulsion but also thermal and AOCS.
- Payload:
  - Pointing: the payload pointing requirements have great impact on the attitude control subsystem.
  - Power: the payload's power requirements obviously impact the power subsystem sizing.
- Launch vehicle: fairing and interfaces (it constraints S/C volume envelope, mass requirement). Vibration profiles of launchers drive decisions about primary and secondary structures.

5.2.2.1 Structure and Mechanisms

The spacecraft structure is a key subsystem in a spacecraft since it is the "house" where all the subsystems are accommodated. One of the principal functions of the structure is to keep the spacecraft together during the launch environment and protect the equipment from vibrations and loads. The structure defines spacecraft's key components locations, as well as field of view for optical sensors, orientations for antennae, etc. Space mechanisms are in charge of deploying surfaces such as antennas and sensors with enough stiffness to keep them stable. The overall structural vibration of the spacecraft must not interfere with the launch vehicle's control system. Similarly, the structural vibration, once "alone in space", must not interfere with its own control system.

Structures are usually categorized into primary, secondary, and tertiary. The primary structure is the backbone or skeleton of the spacecraft, and is the major load path between the spacecraft's components and the launch vehicle; it withstands the main loads. The primary structure also provides the mechanical interface with mechanical ground support equipment (MGSE). The design of the primary structure is mandated by the external physical configuration and the launch loads.

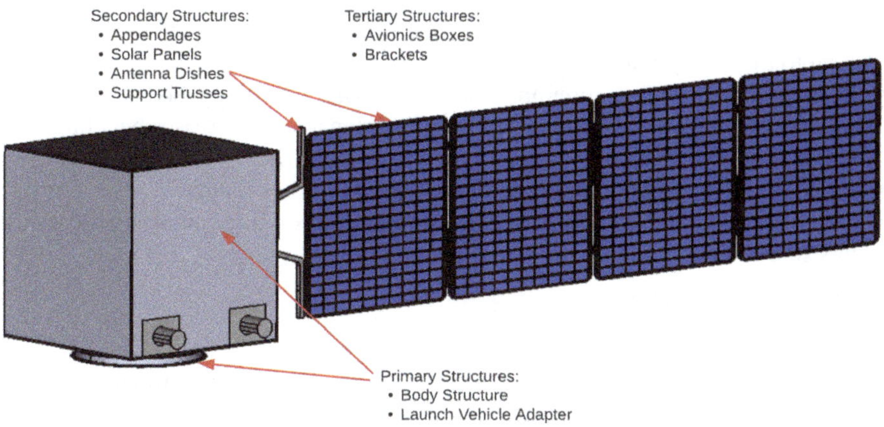

Fig. 5.14 Primary, secondary, and tertiary structures

Secondary structures include support beams, booms, trusses, and solar panels. Smaller structures, such as avionics boxes and brackets, which support harnesses are called tertiary structures.

The primary structure is designed for stiffness (or natural frequency) and to survive steady-state accelerations and transient loading during launch. Secondary structures are designed for stiffness as well, but other factors such as orbital thermoelastic stresses, acoustic pressure, and high-frequency vibrations are also considered (Fig. 5.14).

For the primary structure, typically the options are.

5.2.2.2 Trussed Structure

- A truss is a structure that can withstand loads applied to its joints (nodes) with its members loaded only axially. The arrangement forms triangles.
- The shear and torsion acting on a structure can transfer through axial loads in diagonal members, shear in individual members (in frames without skin) and shear in attached skin.
- Frames without skin are efficient only when shear transfers over a short distance.
- Buckling is typically the critical failure mode for trusses and structures with skin.
- It is a good solution since trusses are weight-efficient and spacecraft components can be mounted internally with good access.

Skin-frame structures:

- A skin-frame structure has a metal skin (sheets or panels) surrounding a skeletal framework made of stringers (members oriented in the vehicle's axial direction) and lateral frames which introduce shear into the skin.

- Intermediate frames are used to increase the buckling strength of the skin and are used to mount equipment.
- Skins in spacecraft structures are usually made of sandwich panels since they help to carry compression loads and this reduces stringer size, increases buckling load, stiff enough for unit mounting, and are weight-efficient.
- It is one of the most common spacecraft structure types found.
- Sheet skin is also used when no units are mounted on it: airplane structures use this solution for fuselage.

  Monocoque structures:

- Monocoque structure is a shell without stiffeners or ring frames.
- It is based on sandwich construction which results in a light structure; *isogrid* shells can also be made at relatively low weight.
- This solution provides a stiff structure and panels can carry all kinds of loads.
- Mounting equipment is easy and when loads are too high, local reinforcement can be added without impacting in mass to the complete structure.
- Monocoque structure is a good solution for cylinder construction. Cylinders provide the most efficient way to carry the launch loads and transfer to the launch vehicle interface.

Typical required characteristics of mechanical structures are:

- Strength: The amount of load (static) a structure can carry without rupturing, collapsing, or deforming enough to jeopardize the mission.
- Structural response: Magnitude and duration of vibration in response to external loads.
- Natural frequency: The frequency the structure will vibrate at when excited by a transient load. This characteristic depends on mass properties and stiffness. Each structure has an infinite number of natural frequencies corresponding to different mode shapes of vibration.
- Stiffness: A measure of the load required to cause a unit deflection.
- Damping: The dissipation of energy during vibration. It is a structural characteristic that limits the magnitude and duration of vibrations.
- Mass properties: Density, center of gravity, and moments of inertia.

Factors impacting structural design are:

- Launch vehicle selection: Structural impacts of launch vehicle specs, fairing sizes, and launch environments.
- Spacecraft configuration: Locates and arranges the payload and subsystems into stowed and deployed configuration.
  - Establishes operating concepts for deployable payloads, solar arrays, and antennas.
  - Establishes general load paths for the primary and secondary structures.
  - Estimatees mass properties.

- Confirms that the spacecraft-stowed envelope, mass, and center of gravity satisfy launch vehicle requirements.

Primary structure:

- Idealizes the spacecraft as an "equivalent beam" with concentrated masses to represent key components and derive design loads and required stiffness. Compares different materials, type of structure, and methods of attachment.
- Arranges and sizes the structural members for strength and stiffness.

Subsystems:

- Demonstrates that payloads, solar arrays, and antennas have their required field of view.
- Derives stiffness requirement for secondary structures to ensure adequate dynamic decoupling.
- Derives design loads.
- Derives requirements for mechanisms and develop conceptual designs.

In terms of structural loading, mechanical loads can be static or dynamic. Static loads are constant and dynamic loads vary with time. Examples:

- Static: the weight of units (mass loading when applied steady acceleration)
- Dynamic: launch vehicle engine thrust, sound pressure, and gusts of wind during launch.

Launch generates the highest loads for most spacecraft structures. It starts when the booster's engines ignite (lift-off) and end with the spacecraft separation. During flight, the spacecraft is subjected to both static and dynamic loads. Such excitations may be of aerodynamic origin (e.g. wind, gusts, or buffeting at transonic velocity) or due to the propulsion systems (e.g. longitudinal acceleration, thrust buildup or tail-off transients, or structure-propulsion coupling, etc.). The highest longitudinal acceleration occurs at the end of the solid rocket boost phase and, for example for Ariane 5 rocket, does not exceed 4.55 g (Arianespace 2016). The highest lateral static acceleration may be up to 0.25 g. A load factor is a dimensionless multiple of grams that represents the inertia force acting on the spacecraft or unit. Accelerations are directly related to forces/stresses and are "easy" to estimate and measure. There are many loads during launch in lateral and axial directions. Some loads are predicted as a function of time while some others can only be estimated statistically (random loads).

Thermal effects on structures:

Materials expand when heated and contract when cooled. In space, solar energy causes spacecraft temperatures to be neither uniform nor constant. As a result, structures distort. The different materials that make up a spacecraft expand or contract in different amounts as temperatures change. Thus, they push and pull on each other, resulting in stresses that can cause them to yield or rupture. Space structure design requires precise predictions of thermal deformations to verify pointing and alignment requirement for sensors and communication antennas. Thermoelastic forces are usually the ones driving the design of joints in structure with dissimilar materials regarding coefficient of thermal expansion (CTE), thus causing high shear loads in fasteners which join those materials. In order to minimize the effects of thermal deformation and stresses, it is important to enclose critical equipment and assemblies in MLI to keep temperature changes and gradients as low as possible. When possible, it is important to design structural parts from materials with low CTE to minimize thermal deformations. For structures with large temperature excursions, it is recommended to use materials with similar coefficients of expansion. And when attaching structural components of different materials, it is important to design the joints to withstand the expected differences in thermal expansion.

To assess structural requirements, typically a set of tests are performed (ECSS 2002):

- Design and Development Test:
  - Purpose: To demonstrate design concepts and acquire necessary information for further design.
  - Requires producing low-cost dev elements/hardware.
  - Test is performed on non-flight hardware.
- Qualification Test:
  - Purpose: Qualification tests are conducted on flight-quality components, subsystems, and systems to demonstrate that structural design requirements have been achieved. In these tests, critical portions of the design loadings are simulated, and the performance of the hardware is then compared with previously established accept/reject criteria based on mission requirements. The test loadings and durations are designed to give a high level of confidence that, if test specimens perform acceptably, similar items manufactured under similar conditions can survive the expected service environments. These loads and durations usually exceed expected flight loads and duration by a factor of safety which assures that, even with the worst combination of test tolerances, the flight levels shall not exceed the qualification test levels.
- Acceptance test:
  - Purpose: The purpose of acceptance testing is to demonstrate conformance to specification and to act as quality control screens to detect manufacturing defects, workmanship errors, the start of failures, and other performance anomalies, which are not readily detectable by normal inspection techniques.
  - Usually not exceeding flight-limit loads.
  - The acceptance tests shall be conducted on all the flight products (including spares).
- Protoflight testing:

- In classic space, structural elements that were subjected to qualification tests are not eligible for flight, since there is no demonstration of remaining life of the product.
- In NewSpace, elements which have been subject of qualification tests can be eligible for flight, provided a strategy to minimize the risk can be applied by enhancing, for example, development testing, by increasing the design factors of safety, and by implementing an adequate spare policy.

Type of environmental mechanical tests used to verify mechanical requirements:

- Random vibration:
  - Purpose: Verify strength and structural life by introducing random vibration through the mechanical interface.
  - Usually applied to electrical equipment and small spacecraft.
- Acoustic:
  - Purpose: Verify structural strength by introducing random vibration through acoustic pressure.
  - Usually done to lightweight structures with large surface areas, for example solar arrays with low mass/area ratio, reflectors, etc.
- Sine Test:
  - Purpose: used to verify natural frequencies and to achieve quasi-static loads at spacecraft level.
  - Usually applied to medium-size spacecraft that can perform a sine test to identify frequencies and mode shape and to achieve quasi-static loads.
- Shock Test:
  - Purpose: to verify resistance to high-frequency shock waves caused by separation devices.
  - Performed to electrical equipment and full integrated spacecraft.
  - Typically up to 10 kHz.
- Static Test:
  - Purpose: to verify primary structure overall strength and displacements
  - Applied to primary structure and large spacecraft structures.

The physical config IPT also needs to kick off different budgets which will be picked up by subsystems. Budgets are kept tracking the evolution of some particular factors or design metrics about a subsystem which impact the overall system design. Subsystem IPTs are responsible for keeping budgets up to date as the architecture of the subsystems evolves.

Once there is some preliminary idea of the functional architecture and the different subsystems which will be part of the spacecraft, these subsystems IPTs can be kicked off. They will:

- Elaborate the resources and equipment list (REL) which will feed the general bill of materials (BOM) of the spacecraft.
- Start keeping track of their mass + contingency margins.
- Kick off and keep track of their power profiles, plus contingency. Subsystem power directly contributes to solar arrays size, power management unit specs, and battery dimensioning.
- The AOCS IPT works closely with the physical config IPT to iterate on propellant budget and thruster definition (orientation, technology, and number of thrusters). AOCS iterates on star tracker location (including blinding analysis for all payload use cases), sizing of the wheels, micro-vibration requirements, etc.

Frequent and fluid communication between the physical config IPT lead and the subsystems IPT leads is needed to closely follow the progress and manage the evolution. Every team wants the best from their own perspective, so the pragmatic (as in, neutral) perspective of the physical config IPT leader is key in order to converge into a good solution. The physical config IPT can be "promoted" to the system IPT once the preliminary design milestones are met and the project moves forward. It is important to note that many projects never make the transition from a concept into anything else. This can happen due to funding, cancellations due to political reasons (in governmental projects), or other factors. Once a system IPT is formed, the concept stage gives way to a more organized and thorough phase; the project has been approved for the next stage, which gets closer to flight so that all the fluidity of the conceptual stage needs to start to solidify.

5.2.2.3 Power

Spacecrafts must generate more power than they consume for sustained operations. Sounds obvious, but it is not such a trivial task to analyze all the potential scenarios during the mission to guarantee a consistent positive margin. Estimating margins at very early stages can be challenging, since many things such as on-board equipment have not been defined. For generating power, typically spacecrafts are equipped with solar panels (some deep space probes[3] use radioisotope thermal generators, or RTGs, but this technology is outside of the scope of this book). It is advisable to choose very high margins during early stages of design (Brown 2002, 325). The need to define as precise as possible margins at the early stages is due to the fact that power margins impact solar panel size, physical configuration, mass, and reliability. If later in the process the team discovers the power margins were insufficient, the overall impact on the project is major, since it will require a general physical redesign. Better to overestimate in the beginning and be less sorry later.

For maximum energy generation, solar panels need to present their surface as perpendicular to the Sun as possible, and this is a factor that can impact the overall

[3]Voyager I and II are powered by RTGs. They have been working since 1977, even though power output has decreased considerably.

physical configuration. Solar panels are very ineffective devices, converting to electrical energy approximately 15–20% of the incident energy. The other remainder of the solar energy that is not converted to electrical power is either reflected or converted into heat. The heat in turn is reradiated into space (if properly thermally isolated from the rest of the bus); the net effect is to raise the solar-panel temperature. Considering this, making the solar panels to point as straight to the Sun is of paramount importance. Different approaches can be used as we shall see.

And here is where the true multidisciplinary nature of spacecraft design kicks in. The first of surely many conflicting choices designers will have to make will require a trade-off analysis (designing spacecraft is plagued by trade-offs). If the payload requires constant pointing to Earth (e.g. for comms applications, or surveillance), then mechanically moving (slewing) the whole platform to track the Sun (assuming panels are fixed to the bus) is a no-go, otherwise you will stop pointing to where the payload needs to point. This requires some decision-making: either you acquiesce to the fact that you cannot make your panels look straight to the Sun and therefore you dimension the power system to deal with that power generation downside (by choosing low-power subsystems, decreasing redundancy, etc.), or you try to find some other way of looking to the Sun with the best angle possible while keeping your payload happily looking to the ground. It can happen since your on-board power consumption will not allow you to accept panels not perpendicular to the Sun, and you will be invited (i.e. forced) to choose an option which will guarantee enough power generation, no-matter-what. There are some alternatives. One variable to consider when it comes to power generation is the total area your solar panels will expose to the Sun. The number of solar cells and how you arrange and connect them together will impact their power output. Individual solar cells generate small power, voltage, and current levels on their own. A solar array connects cells in series to increase voltages to required levels, which are called strings. Strings, then, are connected in parallel to produce required current levels. The series-parallel arrangement of cells and strings is also designed to provide redundancy or string failure protection.

But the challenge is, how to distribute the strings of solar cells in the spacecraft body. Assuming the bus is a square box of, say, 1 m each face, we cannot count on being able to completely cover it with solar cells all around. Usually only some sides only from the theoretical total of six a box has can be counted. The other faces will be taken by different objects, such as sensors, actuators, data and telecommands antennae, payload, separation mechanism/ring, other deployables, and propulsion (Fig. 5.15). Also, since the faces of the square will present different angles to the solar beta angle, this can be suboptimal for power generation reasons. Often, it can be more convenient to have all the panels in a coplanar arrangement.

To present all the panel area surfaces in a coplanar way, the option is to mechanically fold out the panels, as shown in Fig. 5.16. The layout shown still depends on the panels being able to be folded in in their "home" positions. If this is not a possibility, another layout approach is to stack the panels on one face and fold them out accordingly, as shown in Fig. 5.17.

Solar panels that fold out impact system complexity and reliability. The key factor in these mechanisms is to keep the panels secured during launch, and this is usually

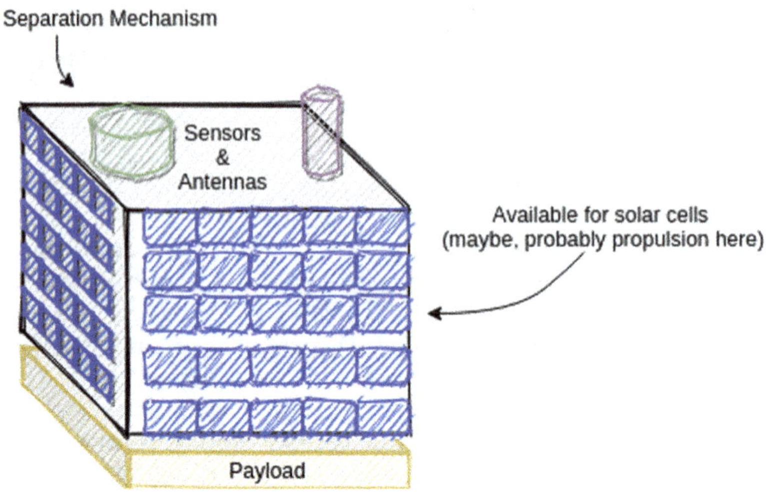

Fig. 5.15 Solar cells available area for a box-shaped spacecraft. Angles to the Sun can be suboptimal for power generation in this geometry

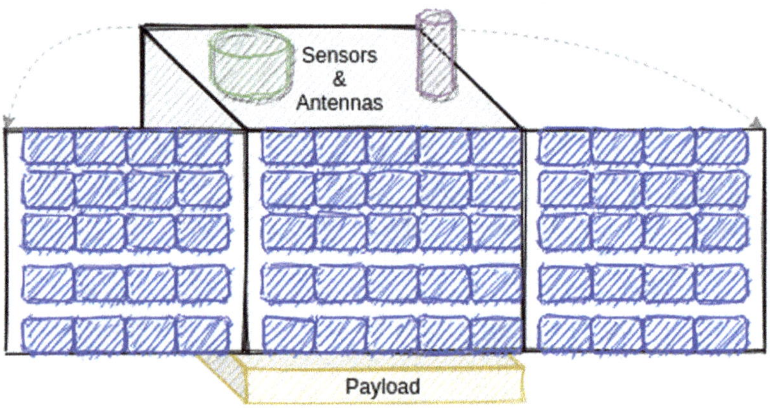

Fig. 5.16 Solar panels folding out in a coplanar manner for better Sun angle geometry

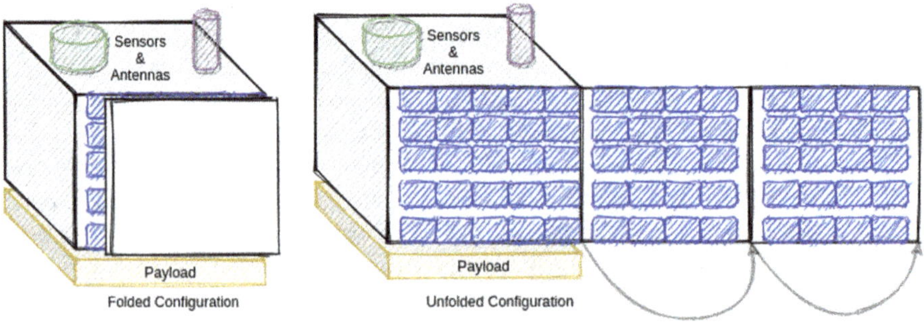

Fig. 5.17 Stacked configuration for coplanar folded solar panels

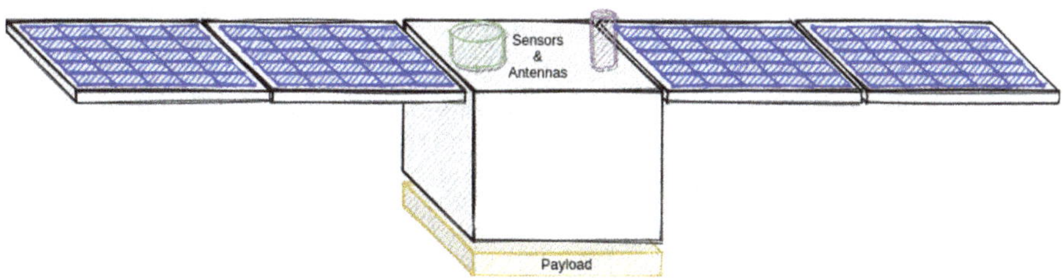

Fig. 5.18 Solar panels for Sun angles along the orbital plane

done by means of hold down release mechanism (HDRMs) which may include either release nuts (also known as separation nuts), explosive bolts, pyrotechnic cutters, or similar. In any case, the mission profile greatly shapes what path to take. For Earth pointing missions, the options shown in the previous figures are great for polar missions and Sun-synchronous orbits, where the beta angle appears consistently on one side of the orbit plane. For other orbits such as low-inclination missions, the Sun traverses high and at times along the orbital plane, hence the optimal solar panel layout is one where the panels point upwards (Fig. 5.18).

In some particular orbits, for example equatorial (0° inclination) orbits, the Sun will lay almost in the orbital plane. This means the Sun will describe a trajectory from very low angles to very high angles, which no "fixed" configuration can fulfill. An alternative for that is to use a mechanism which will tilt the solar arrays as the Sun traverses during the orbit. These devices are typically called solar array driver motors (SADM, even though naming varies depending on manufacturer), which will make the panels spin on one axis as the Sun moves (Fig. 5.19). This adds mass, complexity, and a source of mechanical vibrations on the platform, which can be problematic for applications which require very low mechanical jitter. Again, payload will define what's the best course of action. Always seeking simplicity, it is a good call to always choose the option which will minimize complexity. Regardless of the configuration chosen to create enough solar panel area for power generation, every decision must be communicated to the rest of the design team. Attitude control, for example, must be aware of how the mass distribution of the different options (fixed vs deployable) will look like, for control reasons. The mechanical teams need to understand how the physical envelope of the spacecraft will be for launcher-interfacing reasons. This

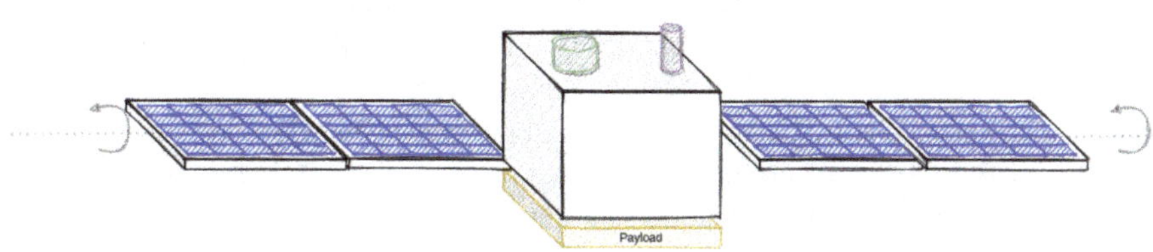

Fig. 5.19 Spacecraft sketch with rotating solar panels option

is multidisciplinary design in a nutshell: move a screw and we are surely screwing something (and someone) up.

The process of finding the right physical configuration which fulfills power generation and other subsystems' requirements is highly iterative and analysis based. But these iterations can also be eventually automated if there is a solid understanding of the most relevant input parameters and the way those parameters impact the architecture. For example, it would not be impossible to come up with a programmatic or computational way of defining the optimal solar panel locations vs beta angle, or star tracker orientation considering beta angle and payload attitude profiles. For automated solar panel and battery sizing, factors such as avionics and payload power consumption would be must-haves.

Power subsystem design requires understanding of the energy balance of the spacecraft. Both the solar panel specifications and the battery capacity requirement can be determined from such analysis. Diving a bit in more details, the energy to be supplied from the solar panels consists of four parts:

1. Energy required to supply the daytime loads.
2. Energy required to charge the battery for the nighttime loads.
3. All the energy losses involved in the system including

    a. power losses of solar panels to the daytime loads.
    b. power losses of solar panels to the battery.
    c. battery charging losses.
    d. power losses of battery to the nighttime loads.

4. All energy reserves, day and night.

Aging and degradation are factors that must take part in spacecraft design. Solar panels performance degrades with time. Batteries degrade with time. The source and rate of degradation depends on the orbit chosen and the mission type. Missions with very high altitudes like medium Earth orbits (MEO) will present accelerated degradation scenarios compared to low Earth orbits (LEO). The power system must be designed in a way its end-of-life (EOL) performance is adequate for the mission. Degradation in solar cells comes from radiation and it is caused by high-energy protons from solar flares and from trapped electrons in the Van Alien belt (Brown 2002). Power loss due to degradation is counted as approximately 25% for spacecraft in geosynchronous orbit for seven years and 30% for 10 years (Agrawal 1986).

The battery must be dimensioned accordingly. Batteries are required during eclipse periods, peak load periods, and during launch operations. There are different technologies of batteries available, Li-ion being the most popular. Battery market for space is a very lucrative market facing positive perspectives, taking into account the multiple constellations that are planning to get to orbit in the next two to five years. Some studies expect the space battery market to grow to reach US$51.3 billion in 2020 from US$ 44.8 billion in 2019. The global lithium-ion power battery market is expected to continue growing steadily during the period of 2018–2025. It is estimated that by 2025, the global lithium-ion power battery market size will exceed US$ 100 billion. Ease of availability, high energy density, low discharge rate, and long life cycle are some of the key features that make lithium-ion power batteries superior to similar products, and are expected to boost global market revenue (QY Research 2020).

The traditional definition of battery capacity is the current that can be supplied by the battery multiplied by the time from fully charged to depletion in ampere-hours (Ah). Battery ampere-hour capacity and energy capacity are proportional. It is convenient to define battery energy capacity from the energy balance and convert energy capacity to ampere-hours. The battery energy capacity is the average night-time power multiplied by the maximum eclipse time divided by the transmission efficiency battery to loads. Batteries are thermally sensitive equipment, imposing strict requirements to the thermal management subsystem.

As with any other subsystem on-board, the power subsystem design cannot be done in an isolated manner. Many decisions taken in this subsystem impact other areas, such as mechanics, attitude control, software, payload, thermal, and the like. The power control IPT (perhaps in the beginning with just one engineer) need to closely collaborate with all other subsystems to make sure the power dimensioning is not overlooking any relevant design factor.

5.2.2.4 Outcome

The physical configuration conceptual design provides important documents and analyses. Notably, it is a preliminary physical depiction of the spacecraft; i.e. a CAD design, with a good description on what type of structure will be used and preliminary location of equipment. This design should generally depict the body configuration, general dimensions, and a good grasp of deployables and mechanisms. Theoretical measures of moments of inertia will be useful for AOCS analyses. At this stage, a document showing power dimensioning rationale along with a power budget shall be generated. This includes dimensioning the solar array and battery and having a grasp on how the power will be distributed across the bus. The power budget must present how power generation will be guaranteed for worst-case scenarios. Due to the early stage of the project, it may very well be the case where power consumptions are not entirely known, so in that case by using information from quotations and adding enough margins should be enough at this stage. A block diagram of the power subsystem shall be created.

5.2.3 Thermal Management

We are totally familiar with using electronics in our everyday lives. Laptops, smart-phones, TVs are all around us. We pay less attention to the fact that electronics are designed to work at limited thermal ranges. For example, Helsinki winters remind me of this every year when my phone (a phone that is not from the highest end but also not the lowest) starts rebooting or turns off when I am outside walking with my dog. Electronics need a specific thermal environment for it to work as intended. Spacecraft electronics is no exception. It is the role of the thermal control subsystem (TCS from now on) to make sure all subsystems operate within their allowable flight temperatures.

The thermal design engineer needs to consider all heat inputs the spacecraft will have, typically the Sun, the Earth, and the on-board electronics and subsystems. All these inputs are not steady but vary with the position in the orbit or seasonally. Thermal control oversees by thermally isolating the spacecraft from the environment, except in specific parts where radiators are placed.

Thermal control engineering activity heavily relies on software tools to be performed; these tools provide simulations and numerical computations to have a good understanding of the thermal scenarios the satellite will experience during its mission way before the spacecraft is subject of environmental testing to verify the thermal design. Thermal control engineering runs for quite long times in projects solely on numerical models and theoretical calculations until the chance to verify them in some controlled environment appears. As expected, thermal subsystem interacts with many other subsystems, particularly with power subsystem. This results from the need of accounting for all dissipated electrical energy and transferring this energy to a radiator for rejection to space. Also, batteries generally have a narrow temperature operating range and often require special attention from the thermal control engineers. The thermal subsystem also interacts with on-board software, since often the software to automatically measure and control zones and rooms is done in the main computer, as well as with mechanical and structure since fixations of heaters, thermostats, and insulation blankets must be agreed with mechanical team, and it also impacts the mass budget. Multidisciplinary engineering is at its best.

Heat in space does not transfer by convection but only by conduction and radiation. This means heat produced by on-board electronics needs to be guided internally (by means of conduction) toward radiators so that it can be radiated away to space. In the same way, spacecraft can (and will) absorb radiation from the Sun and from the Earth. This radiation can be either absorbed for practical purposes (to heat things up) or reflected, to avoid overheating some critical components. Thermal control design internally divides the spacecraft in zones or rooms and makes sure all the equipment inside those rooms will stay in allowable flight temperature (AFT) margins. Heaters are located in places where the thermal balance changes over the mission lifetime. The causes of such changes are:

- Units dissipation changes. For example, important heat load being turned on or off varies greatly over time.
- External heat fluxes changes: spacecraft attitude is changed, eclipses.
- Radiator efficiency changes: changes of optical properties of radiators.

Thermal control techniques can be:

- Passive: fixed area radiators, thermal blankets, etc.
- Active: heaters (controlled by thermostats and/or software).

Thermal control uses the next devices to accomplish its task:

- Electrical heaters.
- Thermistors
- Bimetallic thermostats
- Radiator surfaces
- Thermal blankets (MLI)
- Insulation materials
- Thermal fillers
- Paints
- Heat pipes.

Again, orbits and payloads dictate which approach (passive, active) or what type of hardware and devices to use to thermally control the spacecraft. For active control, the routines to keep the zones under safe margins are executed in software running in either a dedicated CPU for thermal control or in the on-board computer. A heavyweight, active, and redundant control system will greatly impact the power consumption of the bus.

5.2.3.1 Outcome

A general notion on the thermal control strategy shall be produced at the end of the conceptual design. A preliminary location of radiators, heaters, and measurement points shall be defined, following the physical configuration. A preliminary idea of the thermal control logic, redundancies, and software considerations (e.g. a preliminary list of software modes or software requirements) shall be specified at this stage. A block diagram of the subsystem shall be generated.

5.2.4 Attitude and Orbit Control Subsystem

A spacecraft randomly tumbling in space is of very little use. A great deal of why satellites are so useful and ubiquitous is not only because they can sweep and revisit spots on the Earth very quickly but also because when they do, they can point their on-board resources in a very precise manner to perform different types of tasks. This can be a camera, a radar, a directional antenna, etc. Precise orientation is not only related to Earth-pointing needs. A space telescope or a data relay system are used in cases where very precise pointing is needed but the target might not be the Earth. The attitude and orbit control subsystem (AOCS, even though it can also be found as ADCS when orbit control is not part of it or not present) is probably the most complex subsystem in a spacecraft, and perhaps the most exotic; it easily captures a lot of attention from all directions. It is because of that complexity and its highly specialized purpose and functionality that it is the hardest subsystem to grasp in NewSpace companies. There is not an incredibly high amount of people with experience in attitude control and its extremely multidisciplinary nature of electronics, control theory, physics, algebra, math, and, of course, software does not make it any simpler. But, more importantly, AOCS subsystem interfaces literally with every single other subsystem on-board.

A very quick introduction to the general principles of attitude control will be introduced next. There are great references on this topic; for example, Wertz (1990) and Montenbruck and Gill (2001).

The motion of a rigid spacecraft is specified by its position, velocity, attitude, and the way the attitude changes over time. The first two quantities (position and velocity) describe the way the center of mass of the spacecraft translates in three-dimensional space and are the subject of celestial mechanics, orbit determination, or space navigation, depending on the aspect of the problem that is emphasized. The latter two quantities (attitude and its time evolution) describe the rotational motion of the body of the spacecraft about its center of mass. In general, orbit and attitude are coupled with each other. For example, in a low altitude Earth orbit (LEO), the attitude will impact on the atmospheric drag which will affect the orbit. On the other hand, the orbit determines both the atmospheric density and the magnetic field strength at that location, which will, in turn, affect the attitude. However, the dynamical coupling between orbit and attitude will be normally ignored and it will be assumed that the time history of the spacecraft position is known and has been supplied by some process external to the attitude determination and control system. Attitude management (or analysis) may be divided into determination, prediction, and control functionalities.

- **Attitude determination** is the process of computing the orientation in three axes of the spacecraft with respect to either an inertial reference or some object of interest, such as the Earth. Attitude determination typically involves several types of sensors on each spacecraft and sophisticated data-processing procedures. The accuracy limit is usually determined by a combination of processing activities and on-board spacecraft hardware. Typical sensors used for attitude determination are: star trackers, magnetometers, Earth sensors, inertial measurement units, gyros, and Sun sensors. Many of these sensors are complex and contain computing resources on their own. This means they are subject to any issues found in any computer-based systems operating in space such as bit-flips, resets, and, of course, bugs. The on-board fault-handling capabilities must deal with this accordingly and prevent faulty sensors affecting the overall mission reliability by isolating and correcting (is possible) the fault.
- **Attitude estimation** is the process of forecasting the future orientation of the spacecraft by using dynamical models to extrapolate the attitude history. Here the limiting features are the knowledge of the applied and environmental torques and the accuracy of the mathematical model of spacecraft dynamics and hardware.
- **Attitude control** is the process of orienting the spacecraft in a specified, predetermined direction. It consists of two areas—attitude stabilization, which is the process of maintaining an existing orientation, and attitude maneuver control, which is the process of controlling the reorientation of the spacecraft from one attitude to another. The two areas are not totally distinct, however. For example,

we speak of stabilizing a spacecraft with one axis toward the Earth, which implies a continuous change in its inertial orientation. The limiting factor for attitude control is typically the performance of the maneuver hardware and the control electronics, although with autonomous control systems, it may be the accuracy of orbit or attitude information. Some form of attitude determination and control is required for nearly all spacecraft. For engineering or flight-related functions, attitude determination is required only to provide a reference for control. Attitude control is required to avoid solar or atmospheric damage to sensitive components, to control heat dissipation, to point directional antennas and solar panels (for power generation), and to orient rockets used for orbit maneuvers. Typically, the attitude control accuracy necessary for engineering functions is on the order of fractions of degrees. Attitude requirements for the spacecraft payload are more varied and often more stringent than the engineering requirements. Payload requirements, such as telescope or antenna orientations, may involve attitude determination, attitude control, or both. Attitude constraints are most severe when they are the limiting factor in experimental accuracy or when it is desired to reduce the attitude uncertainty to a level such that it is not a factor in payload operation (Wertz 1990). Typical sensors used for attitude control are reaction wheels, magnetorquers, thrusters, control-moment gyroscopes, among others.

The attitude control functionalities described above are realized by a combination of hardware and software. The hardware is composed of sensors and actuators described in the previous paragraphs. The software is in charge of reading the data from the sensors suite and running the determination, estimation, and control routines which will compute the torque needed to orient the spacecraft according to a desired set point, and applying the computed torque by means of the actuators suite, all in a stepwise manner. This is a digital control system, or a cyber-physical system, which we will cover more in detail further ahead.

5.2.4.1 Outcome

An AOCS conceptual design would in general produce a document, few spreadsheets, and some high-level models in Octave, MATLAB, or some C code, or a mixture of all. Provided the physical configuration converges to some mass, deployables, and inertia values, the AOCS conceptual design will include a preliminary analysis on perturbation torques for the orbit chosen, which will give a first iteration on actuator dimensioning. A first understanding of AOCS modes (typically sun pointing, fine reference pointing, safe hold, etc.), tip-off rates after separation, etc. AOCS modes will specify a preliminary combination of sensors and actuators used per mode, which will be of good value for the power design. Some preliminary selection of sensors, actuators, and computers will give a good mass indication for the AOCS subsystem. A block diagram of the subsystem shall be expected at this stage.

5.2.5 Propulsion

Even though propulsion can be considered as a subsystem on its own, it is always tightly coupled to the AOCS subsystem; AOCS is the main *user* of propulsion subsystem. Propulsion provides torques and forces at the service of the orientation and orbital needs of the mission. Propulsion technologies have been diversifying throughout the years.

5.2.5.1 Electric Propulsion

Electric propulsion is a technology aimed at achieving thrust with high exhaust velocities, which results in a reduction in the amount of propellant required for a given space mission or application compared to other conventional propulsion methods. Reduced propellant mass can significantly decrease the launch mass of a spacecraft or satellite, leading to lower costs from the use of smaller launch vehicles to deliver a desired mass into a given orbit or to a deep-space target. In general, electric propulsion (EP) encompasses any propulsion technology in which electricity is used to increase the propellant exhaust velocity. There are many figures of merit for electric thrusters, but mission and application planners are primarily interested in thrust, specific impulse, and total efficiency in relating the performance of the thruster to the delivered mass and change in the spacecraft velocity during thrust periods. Ion and Hall thrusters have emerged as leading electric propulsion technologies in terms of performance (thrust, Isp, and efficiency) and use in space applications. These thrusters operate in the power range of hundreds of watts up to tens of kilowatts with an Isp of thousands of seconds to tens of thousands of seconds, and they produce thrust levels typically of some fraction of a Newton. Ion and Hall thrusters generally use heavy inert gases such as xenon as the propellant. Other propellant materials, such as cesium and mercury, have been investigated in the past, but xenon is generally preferable because it is not hazardous to handle and process, it does not condense on spacecraft components that are above cryogenic temperatures, its large mass compared to other inert gases generates higher thrust for a given input power, and it is easily stored at high densities and low tank mass fractions. In the past 20 years, electric propulsion use in spacecraft has grown steadily worldwide, and advanced electric thrusters have emerged over that time in several scientific missions and as an attractive alternative to chemical thrusters for station-keeping applications in geosynchronous communication satellites. Rapid growth has occurred in the last 10 years in the use of ion and Hall thrusters in communications satellites to reduce the propellant mass for station-keeping and orbit insertion. The use of these technologies for primary propulsion in deep-space scientific applications has also been increasing over the past 10 years. There are many planned launches of new communications satellites and scientific missions that use ion and Hall thrusters in the coming years as the acceptance of the reliability, and the cost-benefits of these systems grow. On the disadvantages of using electrical propulsion, spacecraft charging can be dangerous for on-board electronics if proper care is not taken in terms of avoiding high electrostatic potentials from building up across the structure. Performance fatigue of the neutralizer and/or the electron leakage to the high-voltage solar array can cause charging (Kuninaka and Molina-Morales 2004). Spacecraft grounding strategy and a very careful operation of neutralizers is of great importance.

Electric thrusters are generally described in terms of the acceleration method used to produce the thrust. These methods can be easily separated into three categories: electrothermal, electrostatic, and electromagnetic. Common EP thruster types are described next (from Goebel and Katz 2008, reproduced with permission):

- Resistojet
  - Resistojets are electrothermal devices in which the propellant is heated by passing through a resistively heated chamber or over a resistively heated element before entering a downstream nozzle. The increase in exhaust velocity is due to the thermal heating of the propellant, which limits the Isp to low levels.
- Arcjet
  - An arcjet is also an electrothermal thruster that heats the propellant by passing it through a high current arc in line with the nozzle feed system. While there is an electric discharge involved in the propellant path, plasma effects are insignificant in the exhaust velocity because the propellant is weakly ionized. The specific impulse is limited by the thermal heating.
- Ion Thruster
  - Ion thrusters employ a variety of plasma generation techniques to ionize a large fraction of the propellant. These thrusters then utilize biased grids to electrostatically extract ions from the plasma and accelerate them to high velocity at voltages up to and exceeding 10 kV. Ion thrusters feature the highest efficiency (from 60 to > 80%) and very high specific impulse (from 2000 to over 10,000 s) compared to other thruster types.
- Hall Thruster
  - This type of electrostatic thruster utilizes a cross-field discharge described by the Hall effect to generate the plasma. An electric field established perpendicular to an applied magnetic field electrostatically accelerates ions to high exhaust velocities, while the transverse magnetic field inhibits electron motion that would tend to short out the electric field. Hall thruster efficiency and specific impulse is somewhat less than that achievable in ion thrusters, but the thrust at a given power is higher and the device is much simpler and requires fewer power supplies to operate.
- Electrospray/field emission electric propulsion thruster
  - These are two types of electrostatic electric propulsion devices that generate very low thrust (< 1 mN). Electrospray thrusters extract ions or charged droplets

from conductive liquids fed through small needles and accelerate them electrostatically with biased, aligned apertures to high energy. Field emission electric propulsion (FEEP) thrusters wick or transport liquid metals (typically indium or cesium) along needles, extracting ions from the sharp tip by field emission processes. Due to their very low thrust, these devices will be used for precision control of spacecraft position or attitude in space.

- Pulsed plasma thruster
  - A pulsed plasma thruster (PPT) is an electromagnetic thruster that utilizes a pulsed discharge to ionize a fraction of a solid propellant ablated into a plasma arc, and electromagnetic effects in the pulse to accelerate the ions to high exit velocity. The pulse repetition rate is used to determine the thrust level.
- Magnetoplasmadynamic thruster
  - Magnetoplasmadynamic (MPD) thrusters are electromagnetic devices that use a very high current arc to ionize a significant fraction of the propellant, and then electromagnetic forces (Lorentz forces) in the plasma discharge to accelerate the charged propellant. Since both the current and the magnetic field are usually generated by the plasma discharge, MPD thrusters tend to operate at very high powers in order to generate sufficient force for high specific impulse operation, and thereby also generate high thrust compared to the other technologies described above.

5.2.5.2 Chemical Propulsion

Chemical propulsion subsystems are typically: cold-gas systems, monopropellant systems, and bipropellant systems (Brown 1996).

- Cold gas systems:
  - Almost all spacecraft of the 1960s used a cold-gas system. It is the simplest choice and the least expensive. Cold-gas systems can provide multiple restarts and pulsing. The major disadvantages of the system are low specific impulse and low thrust levels, with resultant high weight for all but the low total impulse missions.
- Monopropellant systems:
  - A monopropellant system generates hot, high-velocity gas by decomposing a single chemical—a monopropellant. The monopropellant is injected into a catalyst bed, where it decomposes; the resulting hot gases are expelled through a converging/diverging nozzle generating thrust. A monopropellant must be a slightly unstable chemical that decomposes exothermically to produce a hot gas. Typical chemicals are hydrazine and hydrogen peroxide.

- Bipropellant systems:
  - In bipropellant systems, an oxidizer and fuel are fed as liquids through the injector at the head end of the chamber. Rapid combustion takes place as the liquid streams mix; the resultant gas flows through a converging/diverging nozzle. Bipropellant systems offer the most performance and the most versatility (pulsing, restart, variable thrust). They also offer the most failure modes and the highest price tags.

5.2.5.3 Outcome

Propulsion conceptual design outcome will consist of an educated choice of the propulsion technology to be used (electrical, chemical), which will shape power and mass estimated needs. A preliminary location of thrusters in the physical configuration should be expected. A propulsion budget shall be produced; such budget will differ whether propulsion will only be used for orbital trim or station-keeping or also for attitude control. In any case, a good understanding of the amount of propellant and a block diagram of the subsystem shall be produced. A shortlist of potential subsystem providers will be gathered at this stage.

5.2.6 Data Links

Space missions require sending and receiving data from the ground segment. These data links functionally couple the spacecraft to the ground in order to transfer the commands (what the spacecraft is ought to do) and receive telemetry (health status of the satellite). At the same time, data links are used to transfer (i.e. downlink) the data the payload generates. Payload data is, of course, very mission-dependent. For an optical Earth observation spacecraft, payload data is in general raw pixels as captured by the camera sensor. For a radar mission, the raw data will be composed of digital samples of the echoes received while sweeping an area of the ground; these echoes will then be processed in order to obtain a visual representation of the area illuminated. Whether for telecommand and commanding (TTC), or for payload data transmission, radio links must be analyzed and designed according to the on-board and ground capabilities.

An important characteristic value to compute at the conceptual stage is an estimate of the signal-to-noise ratio (SNR) at both uplink and downlink, in order to understand how reliable the transmission and reception of data will be amid noise. SNR is computed considering output power, antenna gain and losses (also known as equivalent isotropic radiated power, or EIRP), atmospheric losses, free-space loss (FSL), and receiver sensitivity. After computing SNR and considering channel coding and symbol and bit rates, other important derivative parameters are obtained such as energy per symbol, energy per coded bit, and energy per bit to noise energy (Eb/No). In digital communications, the quality of signals is evaluated by the bit-error-rate

(BER). Required Eb/N0 in decibels for a BER range is usually specified by the manufacturer based on the used modulation type (Ghasemi et al. 2013).

Historically, space links bandwidths and data rates have been limited, hence information has been carefully packed between space segment and ground segment, minimizing overhead and ensuring that literally every bit transmitted and received had an explicit meaning, usually by means of defining lightweight ad hoc bit stream protocols. With higher data rates, wider bandwidths, and richer on-board computing resources, with embedded operating systems capable of running complex networking tasks, the use of standard protocol stacks such as IP became common practice in NewSpace. With this approach, meaningful data is carried as payload on top of several protocol layers, meaning that not every bit modulated in the channels is meaningful (from an application perspective) anymore; in other words, some overhead in the space link is accepted as a reasonable "price" to pay for the rich ecosystem of well-proven service the IP stack provides. For example, in the past, in order to transfer a file from ground to spacecraft, a file transfer ad hoc protocol had to be devised by the engineering team. In an IP-based link, the ground and space segment can send files using secure file transfer protocol (SFTP). Other services such as SSH (secure-shell), rsync (file system sync), and netcat facilitate the administration tasks of the remote spacecraft. With an IP-based link, a great deal of the concept of operations boils down to something very similar to operating a server across a network; i.e. operating a spacecraft becomes a sysadmin task. This also enables the use of sysadmin automation tools and methods which eases the operation of multisatellite constellations in an automated manner.

But IP datagrams cannot directly be modulated on a space radio link, since it lacks the underlying layers to deal for example with the physical layer (layer 1) and data link (layer 2). Typically, IP datagrams are encapsulated in CCSDS space data link protocols (SDLPs): telecommand (TC), telemetry (TM), advanced orbiting systems (AOS), and proximity-1 (Prox-1). IP datagrams are transferred by encapsulating them, one-for-one, within CCSDS encapsulation packets. The encapsulation packets are transferred directly within one or more CCSDS SDLP transfer frames (CCSDS 2012). CCSDS SDLPs are supported by most ground station providers, which makes it a "safe bet" for designers, but it is not the only option. Companies which own end-to-end TTC capabilities (meaning owning also the software-defined-radios at the ground segment) could define their own layer 1 and 2 protocols. This is highly discouraged for NewSpace orgs since it can be very time-consuming and error-prone.

5.2.6.1 Outcome

Typically, a conceptual design of data links should include a link budget. This is typically a spreadsheet where all the characteristic values of a digital radio link must be assessed (SNR basically, with BER assessment and Eb/No), to the best knowledge at that stage. Candidate equipment (radios), antennae, and ground stations are evaluated, and a preliminary selection is presented. For a CCSDS-based link, an assessment on encoding/decoding strategy (either hardware, software, or both) needs to be assessed, since it impacts the avionics and the software architecture.

5.2.7 Fault Detection, Isolation, and Recovery (FDIR)

As we discussed before (see Sect. 3.10), failure tends to happen. Since failure is rarely an atomic effect (unless a catastrophic event), but a combination of constitutive faults which combine toward a damaging event, a functionality must be included on-board to monitor occurrences of these faults and prevent them to find a path through the "Swiss cheese" toward disaster. For any fault-handling functionality to be effective, first it is required to have a thorough understanding of how the system can fail; i.e. a failure analysis must be performed. To understand how a system can fail, the failure analysis must first understand how the system operates. There are several ways to do this, but first, one needs to recognize that knowing how the system is supposed to operate does not mean one will know how it can fail. In fact, systems designers and development engineers (while helpful in defining how the system is supposed to operate) are sometimes not very helpful in defining how it can fail. Designers and development engineers are trained to think in terms of how the system is supposed to work, not how it is supposed *not* to work (Berk 2009). Failure analysis requires multidisciplinary brainstorming, and a well-documented output for easing the implementation: this is usually in the form of fault-trees, Ishikawa diagrams, or concept maps. Faults can happen at every layer of the system hierarchy, but frequently it is the software discipline to be informed about them (detect them), and the one to apply the logic once the fault has been detected.

In short, on-board fault handling is usually a software routine which runs in the on-board computer. This software routine continuously observes a set of variables and performs some action if those variables meet some criteria. Defining what variables to observe and what actions to take is a designer's decision, fed by the failure analysis. FDIR configuration can be very complex for big spacecraft, with thousands of variables to monitor, depending on particular operation modes. NewSpace must adopt a very lightweight approach to it; only the most critical variables must be monitored for faults. Moreover, since typically NewSpace cannot afford to spend months running failure analyses, this means knowledge about how the system can fail is partial. Therefore, FDIR capabilities must be progressively "grown" on orbit as more understanding about how the system performs is gained. When the failure modes are not well understood, it is recommended not to add many automatic actions in order to prevent unintended consequences.

5.2.7.1 Outcome

At conceptual stage, the FDIR strategy will be highly preliminary since the architecture remains fluid and failure modes will not be entirely understood. But as preliminary as this analysis can be, it can provide a good information on the complexity of the FDIR strategy and its different configurations and modes, and help the software team kickstart the development of the FDIR module.

5.2.8 *Further Design*

The boundaries between conceptual, preliminary, and critical design are blurry and arbitrary. The previous section on conceptual design stayed at a very high level of abstraction: spreadsheets, some documents, and perhaps some code. At the conceptual stage, things are fluid and there are still big unknowns to be sorted, such as a general avionics decomposition, suppliers, subsystems decompositions, make vs buy, etc. Many learning curves are at the beginning. The conceptual stage is very necessary to define directions for all these unknowns. Then, as design matures, some prototypes and early iterations of the work start to emerge, and this is what is usually called preliminary design.

Preliminary design evolves until a point when a "snapshot" of such design needs to be put under scrutiny before moving further ahead. The purpose of reviewing this snapshot is to make sure no further steps will be taken if the design does not seem (at least at this stage) to fulfill the requirements.

At the preliminary design review, a set of information "pieces" are expected to be defended against a board of people who are supposed to know

- Master WBS and a schedule
- Current physical configuration:
  - The size and shape of the spacecraft body (bus)
  - Solar panel locations
  - Payload location
  - AOCS sensors and actuators locations and orientations
- A system block diagram
- Detailed subsystems block diagrams
- Thermal design
- Power subsystem dimensioning: battery, solar panels.
- Electromagnetic compatibility considerations
- Fabrication and manufacturing considerations
- Preliminary supply chain breakdown structure (Bill of materials)
- All technical budgets (mass, power, pointing, propulsion, link, thermal)
- Software development plan
- Risk management strategy

- Reliability studies/risk analyses: what are the biggest concerns in terms of reliability, and what are the single points of failure.
- Safety and environmental strategy
- Cost breakdown.

For the design to move ahead, an approval of the preliminary proposal needs to be granted. For this, the main stakeholders of the mission (including the payload) are brought around a table where all the IPTs present their rationales behind the design decisions taken. If the preliminary design is accepted, the design moves forward to synthesize the different elements of the architecture. By the time the design reaches a critical stage, no major unknowns or learning curves must remain unsorted.

5.3 Modular Avionics Design

Engineers are extremely good at reinventing the wheel. They (alas, we) tend to believe that everything is better if designed and developed from scratch. We typically run very biased and partial *make vs buy* analyses and we consistently rig them to look like going in-house is the best option. For space startups this is particularly damaging, since developing things from ground up is (of course) very time-consuming but at the same time it generates non-recurring costs that can be directly suicidal. A NewSpace project should only insist on developing from scratch the technology that represents the strategic differentiating "core" of whatever the company is trying to do, and all the rest should come from off-the-shelf. NewSpace orgs should (must) put themselves in the role of system integrators and minimize non-recurring costs; i.e. avoid burning money. Of course, off-the-shelf is not such an easy option for every discipline or domain. Mechanics, for example, is often designed in a very ad hoc way, for specific mission requirements. This is understandable since the payload defines many factors (as we say in the Conceptual Design section) such as sensor and actuator placing, solar panels, battery size, etc. But avionics is an area where options in the market are many and increasing. Computing is probably the one area that benefits the most from a rich ecosystem of vendors and options. Today, there are fewer on-board computing requirements that cannot be met with off-the-shelf commodity embedded computers that are very capable.

A very generic avionics architecture of a spacecraft looks as depicted in Fig. 5.20.

The green boxes are different functional *chains* or blocks that provide some essential capability for the spacecraft. Regardless of what type of application or mission the spacecraft is supposed to perform, those functional blocks are mostly always present; in other words, you cannot do space without them. The "payload" yellow box encloses the functionality of the actual application which gives a purpose to the mission, which can be:

- Connectivity:
  - IoT

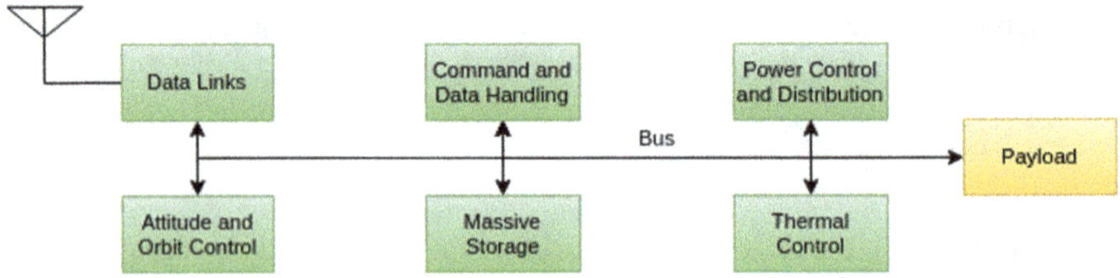

Fig. 5.20 A generic avionics block diagram

  - Satcom
  - LaserComm
- Earth Observation:
  - Optical
  - Radar
  - Infrared
- Other:
  - In-orbit robotics
  - Debris removal, in-orbit assembly, inspection, etc.
  - Pure science: Atmosphere, astronomy, etc.

Some of these functional chains or blocks do not need to have a computer inside every time; they can be passive as well. For example, thermal control can be passive (using insulation, paints, and radiators), hence computing will not be present there.

What stands out from the figure is that spacecraft avionics needs a lot of interconnection. This means the functional chains must exchange data with each other to couple/combine their functionalities for the global function of the spacecraft to emerge. That data is in the form of commands, telemetry, or generic data streams such as files, firmware binaries, payload data, etc. The architecture is recursive which means the functional chains have internal composition as well which will (in most cases) also require interconnection. For example, for the attitude control subsystem, the internal composition is depicted in Fig. 5.21.

Spacecraft function coupling highly depends on an aggregation of communication buses. It is clear that interconnection is probably the most important functional coupling requirement of any spacecraft avionics architecture. A spacecraft with poor interconnection between functional chains will see its performance and concept of operations greatly affected. This is a factor that is usually overlooked by space startups: low-speed, low-bandwidth buses are chosen at the early stages of the project, just to find out later that the throughputs are insufficient for the overall performance. Changing interconnection buses at later stages can be costly, both in money and time. With high-speed serial buses and high-performance processors dropping prices and being more and more accessible, there is no reason not to design the avionics to be highly interconnected and using high-speed connections.

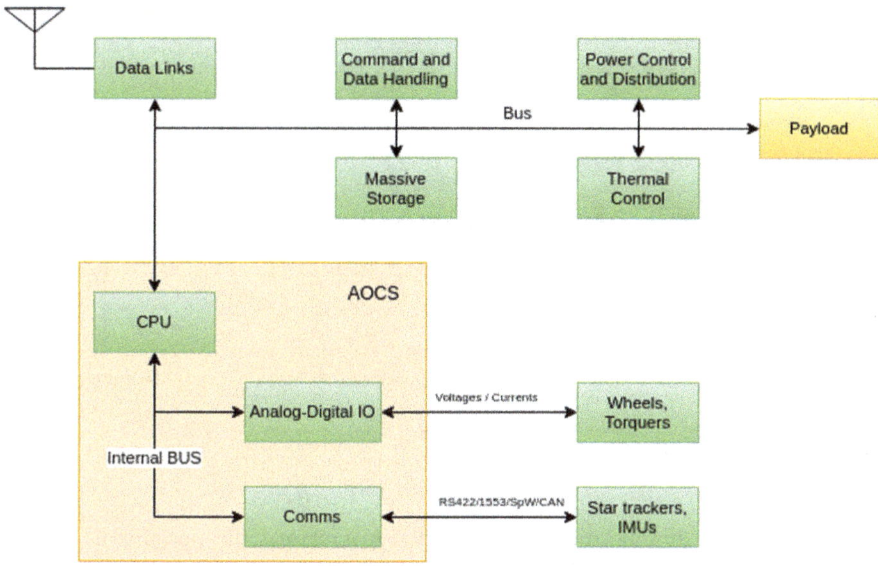

Fig. 5.21 AOCS functional chain as a member of the avionics architecture

Historically, spacecraft avionics has used hybrid interconnect approaches; most typically, ad hoc, daisy-chain-based topologies, where cables come out from a box and go inside the next one. Legacy spacecraft avionics feature a fair deal of "private" buses; i.e. buses that are only accessible by some subsystems and not from the rest of the architecture. When discussing interconnection, there are two different levels we will discuss:

- **Subsystem level**: how a subsystem chooses to connect and functionally couple with its internal components.
- **System level**: how different subsystems can connect to each other to provide the spacecraft "global" function. For example, how the command and data-handling subsystem connect to the power subsystem, and vice versa.

At the subsystem level, the approach has been hybrid as well. Typically:

- Box-centric "star" approach: the subsystem main unit (which usually hosts its CPU) resides in a box of customized form factor and this box is the central "concentrator" of the subsystem. Everything flows toward it. The box exposes a set of connectors. Then, different harnesses and cables come in and out from those connectors, toward external peripherals. These peripherals can be either point-to-point or connected through a bus, or both (Fig. 5.22).

In this type of design, the mechanical functional coupling between different parts of the subsystem is likely different for the different peripherals; i.e. different types of connectors, pinouts, harness, etc.

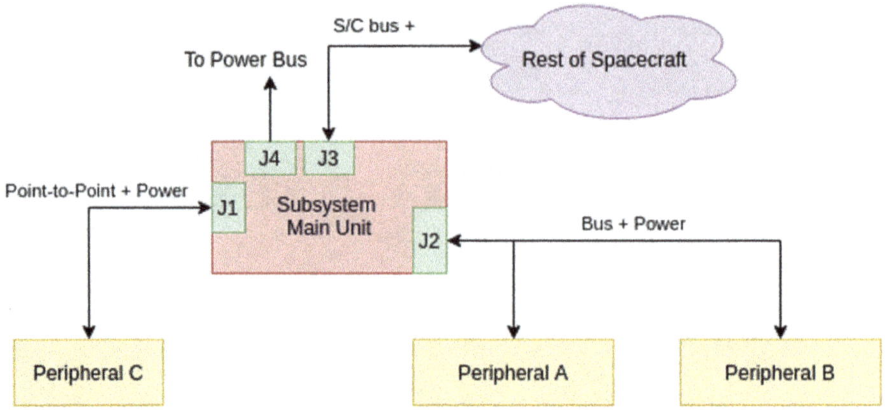

Fig. 5.22 Subsystem federated architecture

- Backplane: In this approach, the computing unit and the peripherals share a mechanical interface which allows them to connect to a board (called backplane) acting as the mechanical foundation. The peripherals connect by sliding in through slot connectors and mating orthogonally to the backplane. The backplane not only provides the mechanical coupling but also routes signal and power lines between all the different modules connected to it (Fig. 5.23).

Fig. 5.23 A backplane connecting 1 CPU unit and 2 peripheral boards

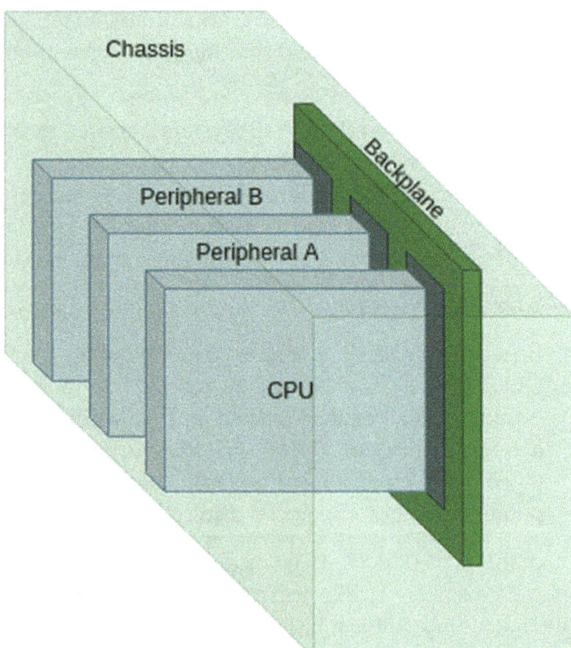

How to route the signals in the backplane is a design decision, since the backplane is basically a printed circuit board like any other. The backplane approach quickly gained popularity among designers: the advantage of routing signals in a standardized way quickly became an advantage. This made it possible for multiple vendors to be able to interconnect their products in backplanes and achieve interoperability. Several backplane standards proliferated, but one stood out as probably the most popular standard from those years: Versa module Europe (VME), and it is still in use today in some legacy applications. VME is one of the early open-standard backplane architectures. It was created to enable different companies to create interoperable computing systems, following standard form factors and signal routing. Among the typical components in the VME ecosystem, you can find processors, analog/digital boards, etc., as well as chassis (housings), backplanes, power supplies, and other subcomponents. System integrators benefited from VME in the next aspects:

- It provided multiple vendors to choose from (supply chain de-risk)
- A standard architecture vs costly proprietary solutions
- A tech platform with a known evolution plan
- Shorter development times (not having to start from scratch)
- Lower non-recurring costs (by not starting from scratch)
- An open specification to be able to choose to do subsystems in-house if needed.

The VME specification was designed with upgrade paths so that the technology would be usable for a long time. VME was based on the Eurocard form factor, where boards are typically 3U or 6U high. Design was quite rugged; with shrouded pins and rugged connectors, the form factor became a favorite for many military, aerospace, and industrial applications. VME was upgraded to VME64x (VITA 1.1), while retaining backwards compatibility. Over the years though, even these upgrades weren't enough bandwidth for many applications. Then, the switch fabrics entered the game.

5.4 VPX

VME was (is) based on a parallel bus, with signals daisy-chained at each slot. In contrast, VPX is based on switched fabrics. Parallel buses convey many signals simultaneously, i.e. concurrently. Signals physically run close to each other (hence the name); strictly speaking, they do not need to run parallel all the way from transmitter to receiver, but each signal in a parallel bus has its own private physical track all the way. In crowded parallel buses, arbitration needs to take place to avoid multiple users of the bus claiming it and disrupting the flow of other nodes, causing data corruption and delays. Also, if a node breaks down in a parallel bus, it can take down the whole bus altogether, affecting the overall function of the system. In switched fabrics, there are multiple paths for a node to reach a destination. Thus, fabrics allowed the data rates to go much higher and are inherently much more reliable. Switches are a central part of the network topology, and the nodes can be connected to switches in different ways. For example, in a centralized topology, there's one switch in a star configuration or two switches in a dual-star configuration. In a distributed topology, there can be multiple connection options with a mesh configuration (where each node is connected to every other node) being quite common. VME also started to face some concerns in terms of reliability, due to its parallel nature, but also as speeds were consistently increased, VME faced limitations on the performance of the connector. With VME quickly reaching its retirement, a retirement that has never fully materialized but its use for high-performance applications has declined. VPX was next in the evolution. First, VPX adopted high-speed connectors—the MultiGig RT2. By changing connectors, pinouts had to change as well. So, the VITA 46 standard was created to specify how to route the pins across the new connector, but the pinout choice remained on the designers' side. VPX quickly found adoption in computing intensive applications such as defense, aerospace, acquisition, test and measurement, space, radar, software-defined radio, and other markets. Today's highly interconnected military systems and their related applications require a great amount of very-high-speed computing performance across a reliable system. These applications became VPX's main driver and sponsor. As VPX grew, with all the multiple options about signals and speeds it offered (from single ended to multiple serial pipes), it became extremely difficult to ensure interoperability, so the OpenVPX (Vita 65) specification was created as the primary architecture *toolbox* to make different bits from different suppliers seamlessly work with each other. VPX supports a whole string of fast serial fabrics. However, these are described in extra specifications complementing the base specification which only defines the mechanics and the electrics of the standard.

VPX/OpenVPX offers many benefits, including:

- Increased reliability due to the choice of fabric approach
- High-speed MultiGig RT2 connector rated to about 10 Gb/s
- Support of standard SERDES data rate options of 3.125, 5.000, and 6.250 Gb/s (with 10 Gb/s and beyond becoming more popular)
- Defined areas for control plane, data plane, utility/management plane, etc.
- Options for system management per VITA 46.11
- Fully tested and rugged differential connectors
- Guide pins on the backplane for easy blind-mating (Fig. 5.24).

5.4.1 OpenVPX

Note: This chapter contains material that is reproduced and adapted with permission from VITA.

For NewSpace organizations planning to launch their own spacecraft designs, they must ruthlessly focus on system integration and discard complex, endless hardware designs from scratch, unless those things are considered part of their "core" technology, or essential for their business. In most cases this is a payload or an element of the payload, but the bus architecture components can (and should) be

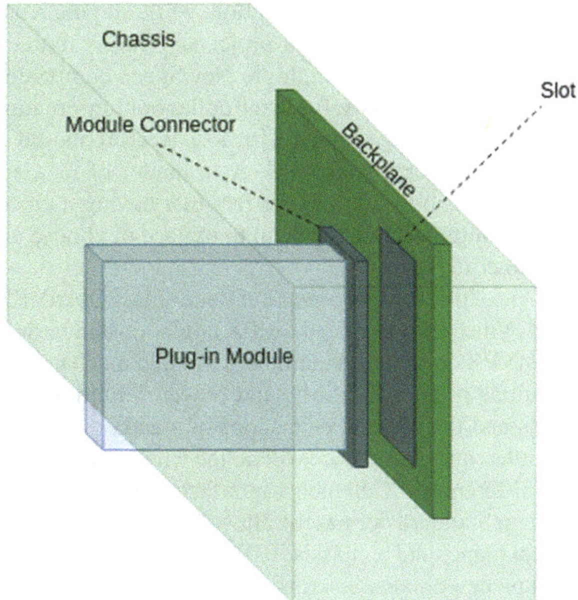

Fig. 5.24 VPX system main building blocks: plug-in module, slot, backplane, chassis

largely commoditized. NewSpace projects need to get to space as soon as possible. This means the architects need to close the EDA tool, take the shopping cart, and go scan the marketplace for bits and parts off-the-shelf. This approach contributes to: (a) de-risk the supply chain (buying the whole bus from one single supplier can be too high risk); (b) minimize pointless non-recurrent costs associated with complex R&D. In short, own the architecture, do now own all the parts of it. There is very little point in developing whole subsystems in-house, at least in the proof-of-concept stage. Engineers are usually lured by the celestial siren of R&D singing them to shipwreck. Too often, space startups embark on long R&D endeavors which have zero strategic value, which end up being expensive distractions misleadingly disguised as victories. This also contributes to integral and "closed" architectures which are very difficult to interface to other architectures, modify and extend further on. Developing subsystems from zero is an attractive challenge for any engineering team, but never an easy one. Everything is more complex than expected, everything is more expensive than expected, and (as Hofstadter's law states), everything takes longer than expected.

> Disclaimer: Research and Development is great, and very needed in order to stay innovative and differentiated in crowded markets; but, that collective brain power should be put to develop technologies that are truly strategic; what special single-board computer could you design that hasn't already been designed?

But, to be able to *commoditize* subsystems and components while defining the architectural "puzzle", standardized form factors and buses are a must. Working with a standard form factor puts the emphasis on system design and interconnection, with a stronger focus on the application, while staying reasonably away from high-speed, high-density digital design projects. What is more, standardized form factors for avionics could enable multiple NewSpace companies to interoperate their components more seamlessly, fostering collaboration, creating consortia, joint missions, etc. In short, choosing elements for a spacecraft should not be different than choosing components for a gamer PC. As a matter of fact, space industry is not very fond of interoperability due to the fact that having a customer captive of a product and making it impossible for them to replace or choose another one is a very interesting customer to have.

One of these interconnection standards is OpenVPX: The OpenVPX System Standard (Vita 65). Based on the VPX family of standards we briefly discussed before, the OpenVPX System Standard uses module mechanical, connectors, thermal, communications protocols, utility, and power definitions provided by specific VPX standards and then describes a series of standard profiles that define slots, backplanes, modules, and chassis (we will define what each of these are). Figure 10 illustrates how the different VPX elements interface with each other. In a nutshell, the OpenVPX System Standard defines the allowable combinations of interfaces between a module, a backplane, and a chassis. In the VPX terminology, a chassis hosts a backplane. A backplane contains a series of slots, and modules connect to those slots by means of their on-board connectors. The main task (and challenge) for systems integrators is to ensure the right signals are routed across the design, and for that OpenVPX specifies a naming convention as an abstraction (just as domain names abstract from lengthy IP network addresses), which can be used while selecting elements from the market (Fig. 5.25).

5.4.1.1 OpenVPX Key Element Descriptions

It is important first to discuss some key elements and terminology that are central to the VPX/OpenVPX architecture: port, plane, pipe, module, and profile.

Port: A physical aggregation of pins for a common I/O function on either a plug-in module's backplane connectors or a backplane slot's connectors.

Plane: A physical and logical interconnection path among elements of a system used for the transfer of information between elements. How the planes are used is a guideline. The following planes are predefined by OpenVPX:

- Control plane: A plane that is dedicated to application software control traffic.
- Data plane: A plane that is used for application and external data traffic.
- Expansion plane: A plane that is dedicated to communication between a logical controlling system element and a separate, but logically adjunct, system resource.
- Management plane: A plane that is dedicated to the supervision and management of hardware resources.
- Utility plane: A plane that is dedicated to common system services and/or utilities. The utility plane (UP) in an OpenVPX backplane includes common power

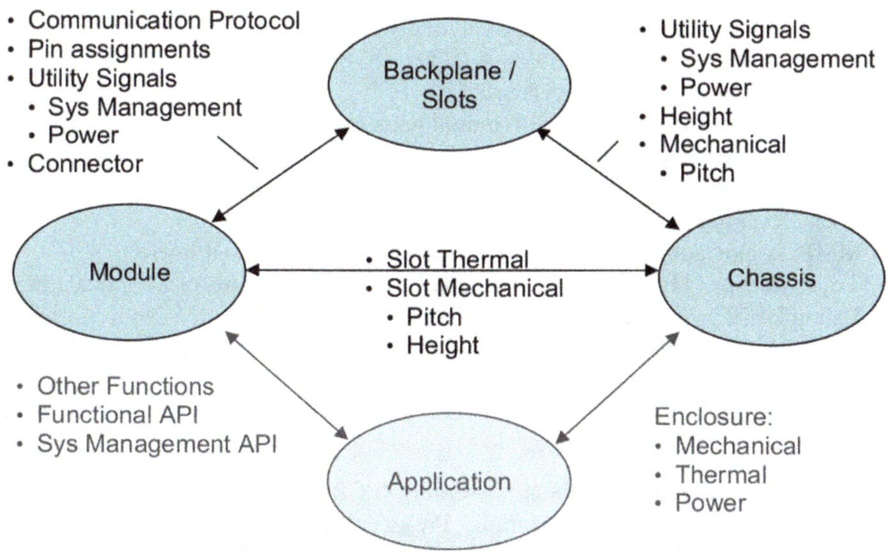

Fig. 5.25 System interoperability diagram (*Courtesy* VITA, used with permission). The OpenVPX System Standard acknowledges but does not define the interfaces between the application and the module or chassis (grayed out text and lines)

distribution rails, common control/status signals, common reference clocks, and the system management signals.

Typically, the plane speeds increment from utility (low-speed), control plane (medium), and data plane (high). Utility planes are used for low-speed communication between modules (control signals, configuration of low-level devices, etc.). Control planes are typically used for commands and telemetry for the application software. The data plane is used for high-bandwidth data transfers between application software, typically payload data.

Pipe: A physical aggregation of differential pairs or optical fibers used for a common function that is characterized in terms of the total number of differential pairs or optical fibers. A pipe is not characterized by the protocol used on it. The following pipes are predefined by OpenVPX:

- Single pipe (SP): A pipe consisting of a single differential pair or optical fiber.
- Ultra-thin pipe (UTP): A pipe comprising two differential pairs or two optical fibers. Example: 1000BASE-KX Ethernet, 1 × Serial RapidIO, × 1 PCIe, and 10GBASE-SR interfaces.
- Thin pipe (TP): A pipe composed of four differential pairs or four optical fibers. Example: 1000BASE-T interfaces.
- Triple ultra-thin pipe (TUTP): A pipe comprising six differential pairs or six optical fibers.

- Fat pipe (FP): A pipe composed of eight differential pairs or eight optical fibers. Example: 4 × Serial RapidIO, × 4 PCIe, 10GBASE-KX4, 40GBASE-SR4 (40 Gbit Ethernet over fiber) interfaces.
- MP: A pipe composed of 10 differential pairs or 10 optical fibers.
- WP: A pipe composed of 12 differential pairs or 12 optical fibers.
- Double-fat pipe (DFP): A pipe composed of 16 differential pairs or 16 optical fibers. Example: × 8 PCIe interface.
- MMP: A pipe composed of 20 differential pairs or 20 optical fibers.
- Triple-fat pipe (TFP): A pipe composed of 24 differential pairs or 24 optical fibers. Example: 12 × InfiniBand interface.
- Quad-fat pipe (QFP): A pipe composed of 32 differential pairs or 32 optical fibers. Example: × 16 PCIe interface.
- Octal fat pipe (OFP): A pipe composed of 64 differential pairs or 64 optical fibers. Example: × 32 PCIe interface.

Module: A printed circuit board assembly (PCBA), which conforms to defined mechanical and electrical specifications. Pre-existing examples of modules that are applicable to OpenVPX include 3U plug-in modules; 6U plug-in modules; backplanes, mezzanine modules such as XMC, PMC, or FMC modules; and rear transition modules. Additionally, the following module types are defined by OpenVPX:

- Bridge module: A plug-in module in an OpenVPX system that might be required to provide communication paths between multiple plug-in modules that support different plane protocols and/or implementations. When the transfer of information is necessary between plug-in modules utilizing dissimilar interfaces for communication, the bridge module terminates the channel and/or bus from the plug-in module(s) communicating via the initial protocol and transmits the information along the plug-in module(s) communicating via the second protocol on a separate channel or bus.
- Payload module: A plug-in module that provides hardware processing and/or I/O resources required to satisfy the needs of the top-level application. Example: A payload module might be an embedded processor or an I/O controller module.
- Peripheral module: A plug-in module such as an I/O device interface that is usually subservient to a payload module.
- Plug-in module: A circuit card or module assembly that is capable of being plugged into the front side of a backplane.
- SpaceUM: Space utility management module contains the utility management selection circuitry for the SpaceVPX module. The SpaceUM module receives redundant utility plane signals through the backplane and selects one set to be forwarded to the standard slot utility plane signals for each slot it controls.
- Storage module: This module provides the functionality of a disk drive. An example is a SATA HDD/SSD (hard disk drive/solid-state drive) carrier.
- Switch module: A plug-in module in an OpenVPX system that minimally serves the function of aggregating channels from other plug-in modules. These channels might be physical partitions of logical planes as defined by a backplane profile. This module terminates the aggregated channels and provides the necessary switch

fabric(s) to transfer data frames from a source plug-in module to a terminating plug-in module as defined by the assigned channel protocol. This module is typically used in systems that implement centralized switch architectures to achieve interconnection of their logical planes. Distributed switch architectures typically do not include a switch module.

Profile: A profile is a specific variant of a possible set of many combinations. In the VPX context, it applies to backplanes, chassis, modules, and slots.

- Backplane profile: A physical definition of a backplane implementation that includes details such as the number and type of slots that are implemented and the topologies used to interconnect them. Ultimately, a backplane profile is a description of channels and buses that interconnect slots and other physical entities in a backplane.
- Chassis profile: A physical definition of a chassis implementation that includes details such as the chassis type, slot count, primary power input, module cooling type, backplane profile, and supplied backplane power that are implemented in the standard development chassis profile.
- Module profile: A physical mapping of ports onto a given module's backplane connectors and protocol mapping(s), as appropriate, to the assigned port(s). This definition provides a first-order check of operating compatibility between modules and slots as well as between multiple modules in a chassis. Module profiles achieve the physical mapping of ports to backplane connectors by specifying a slot profile. Multiple module profiles can specify the same slot profile.
- Slot profile: A physical mapping of ports onto a given slot's backplane connectors. These definitions are often made in terms of pipes. Slot profiles also give the mapping of ports onto plug-in module's backplane connectors. Unlike module profiles, a slot profile never specifies protocols for any of the defined ports.

Slot:

A physical space on a SpaceVPX backplane with a defined mechanical and electrical specification intended to accept a plug-in module. Pre-existing examples of slots that are applicable to OpenVPX include 6U and 3U. Additionally, the following slot types are defined by a SpaceVPX:

- Controller slot: A slot in a SpaceVPX system that will accept a controller plug-in module. A controller slot always has the control plane switch and system controller function. It can be combined with the switch function for the data plane.
- Payload slot: A slot in a SpaceVPX system that will accept a payload plug-in module such as, but not limited to, a hardware processing and/or I/O plug-in module.
- Peripheral slot: A slot in a SpaceVPX system that will accept a peripheral plug-in module that is usually subservient to a payload or controller module. It can also serve to bridge an interface such as PCI from the payload or controller slot.
- Switch slot: A slot in a SpaceVPX system that will accept a switch plug-in module.

5.4.1.2 Profiles

Backplane profiles:

At the center of each OpenVPX architectural definition is the backplane profile. This profile contains two important elements: a backplane topology for each communication plane and a slot interconnection definition for each slot type used. Each backplane profile references a slot profile for each slot position on the backplane and then defines how each pipe in each slot is interconnected and each pipe's electrical performance. The backplane profile defines which pins or set of pins are routed in the backplane. The backplane profile also defines allowed slot-to-slot pitch.

Slot profiles:

Slot profiles define the connector type and how each pin, or pair of pins, is allocated. Single pins are generally allocated to the utility plane for power, grounds, system discrete signals, and system management. Differential pin/pairs are generally allocated for the three communication planes (control, data, and expansion). Differential paired pins are grouped together to form "pipes" (the definition of planes, pipes, and profiles has been specified before). Slot profiles also specify which pins are user-defined. Slot profiles are divided into categories, but not limited to: payload, switch, peripheral, storage, bridge, and timing.

Chassis profiles:

Within the context of OpenVPX, a chassis is targeted for plug-in module system integration and test. OpenVPX defines three variants of standard development chassis: small, medium, and large.

Module profiles:

The module profile defines what communication protocol can be used on each pipe as defined in a corresponding slot profile. Each module profile specifies a particular slot profile, which in turn specifies the connector types. The module profile also specifies the module height (6U/3U), see Fig. 5.26.

Module profiles and backplane profiles guide the system integrator in selecting plug-in modules and backplanes that can work together. However, when user-defined

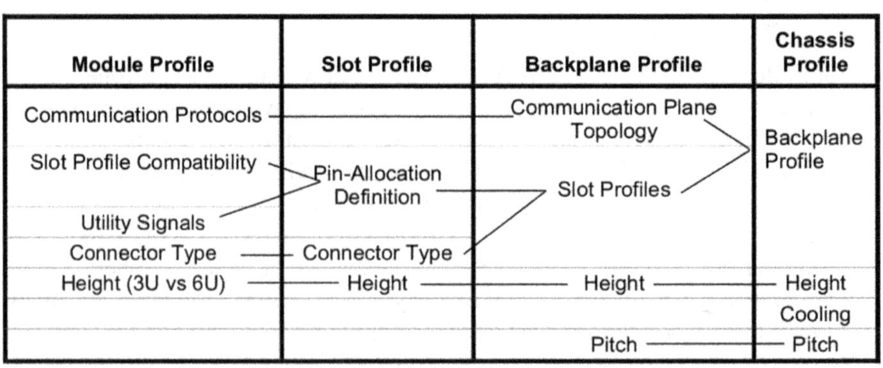

Fig. 5.26 OpenVPX profile relationships (*Courtesy* VITA, used with permission)

pins are used, the system integrator needs to ensure that the user-defined pins on the backplane are routed accordingly.

5.4.1.3 Slot Vs Module Profiles

Slot profiles are used to specify how ports are mapped to pins of a backplane slot. A guiding principle in the way the OpenVPX standards are organized is that things that affect the implementation of the backplane are in slot and backplane profiles. Module profiles are specifying the protocols running over the physical connections specified by slot and backplane profiles.

5.4.1.4 Profile Names—Use and Construction

OpenVPX implements a detailed naming convention for all its profiles. This provides a great abstraction for the system architect when it comes to choose VPX components off-the-shelf. Readers are invited to check the standard (VITA 2019) for further details on OpenVPX naming convention.

5.4.1.5 SpaceVPX

The OpenVPX (VITA 65) backplane standard has been chosen as the base for the SpaceVPX backplane standardization effort. The SpaceVPX Systems Specification was created to bridge the VPX standards to the space market. SpaceVPX enhances the OpenVPX industry standard by adding spacecraft interfaces and balanced fault tolerance. SpaceVPX was developed as a standardized element of the Next-Generation Space Interconnect Standard (NGSIS) working group breadth of products. The primary focus is to enhance the OpenVPX standard with features required for space applications, such as single-point failure tolerance, spacecraft interfaces, spare module support, redundancy management, and status and diagnostic support.

Figure 5.27 captures the essence of a SpaceVPX system. Each box represents one module function. The main data flow in most space systems is shown by the blue boxes, typically implemented as payload modules. Any of the four functions may be combined into a specific payload implementation. The properties of this use case follow the circled letters in the figure.

- A: Data enters the system through a data in module. The next four letters represent potential data transfers between the main payload functions within a SpaceVPX system.
- B: Data from the data in module can be transmitted to the processing module. Processed data from the processing module can be stored in the storage module. Data from the storage module can be routed to the data out module.
- C: The processor module can access data from the storage module.

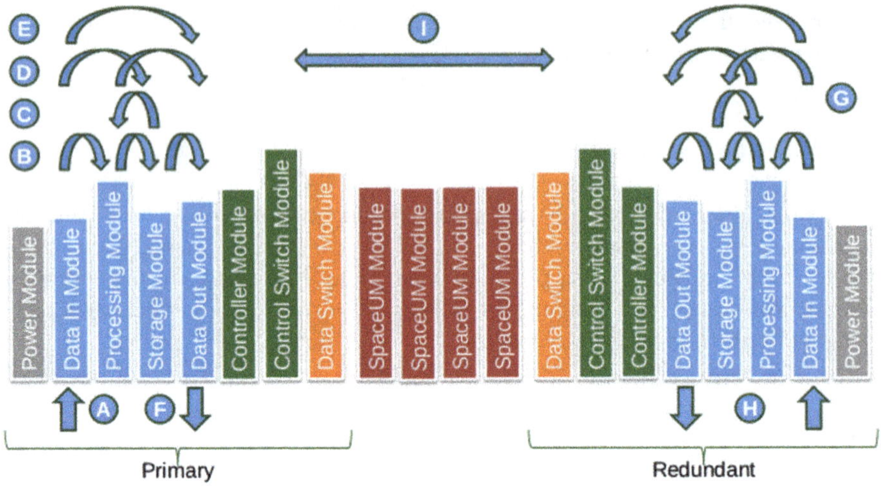

Fig. 5.27 SpaceVPX system use case (*Courtesy* VITA, used with permission)

- D: Data from the data in module can be routed directly to the storage module. Data from the processing module can be routed directly to the data out module.
- E: Data from the data in module can be routed directly to the data out module.
- F: Data routed to the data out module can then be transmitted out of the system.
- G: For fault tolerance, there will be at least one spare of each function. These redundant payloads will have the same data routing as the primary payloads so that no error will reduce capabilities. Though not shown, more than one of each function type may be present and M of N sparing (minimum 1 of 2) may be implemented for any payload type.
- H: There will also be redundant external inputs and outputs to and from the SpaceVPX system. These are typically cross-strapped between redundant elements.
- I: Finally, there will be typically cross-strapping between the primary and redundant payload groups so that any one payload can fail and the remaining payloads can operate with sufficient interconnects for any combination of primary and redundant elements.

This base use case can apply to both OpenVPX and SpaceVPX systems.

5.4.1.6 Mesh Data Backplane (Switchless Approach)

A mesh data backplane specifically provides a maximum simultaneous data connectivity between payload modules within the system—it is worth noting that in these cases the modules often are of similar configurations, i.e. processors, I/O, storage, or other peer-to-peer modules. Given the topology of the backplane, no data switches are required although a switch can be embedded in any of the payload modules; in

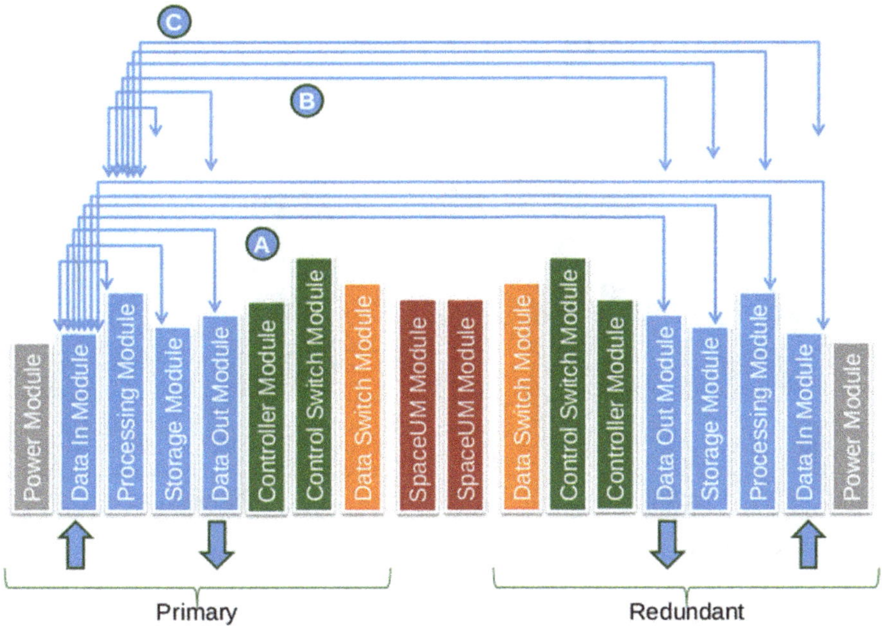

Fig. 5.28 SpaceVPX system use case—mesh data backplane (*Courtesy* VITA, reproduced with permission)

addition, each of the payload slots is connected to all the other payload slots with one or more data plane connections. Redundancy is managed at the module level, i.e. N + M so there is at least one spare of each module type. As such, a loss of a module disables only all the connections to that module.

As shown in Fig. 5.28, the payloads of the SpaceVPX system use case are interconnected by a mesh data backplane:

- A: The data in module has a direct data plane connection to every other payload module.
- B: The processing module has a direct data plane connection to every other payload module.
- C: Similar connections are provided between all other payload modules. Data can be transferred simultaneously on all links.

5.4.1.7 Switched (Star/Redundant Star) Data Backplane

A switched (star/redundant star) data backplane provides a maximum amount of connectivity between associated payload modules that reduces cost for the overall system. Given that it is a switched backplane, the data that is transmitted to and from each of the backplanes modules is limited to its switched connections only—typically the number of switch ports in a given system is higher than the number of available

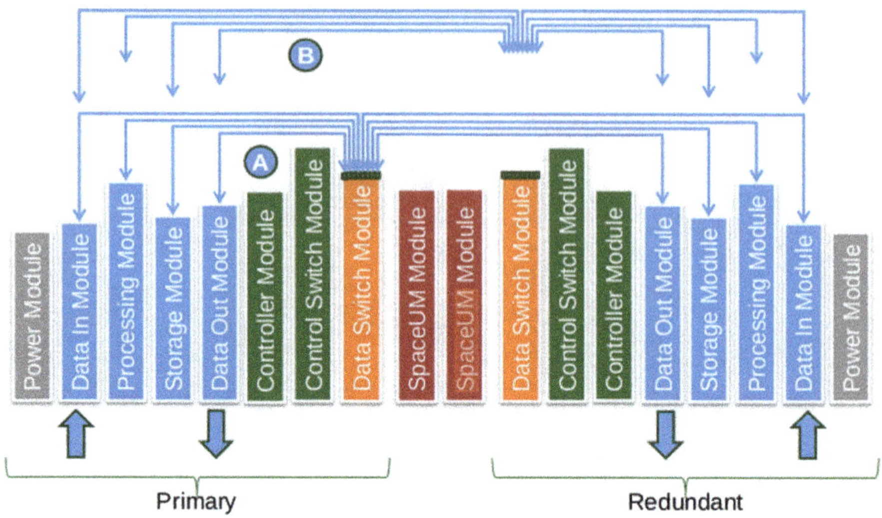

Fig. 5.29 SpaceVPX system use case—switched data backplane (*Courtesy* VITA, used with permission)

payload module ports. As shown in Fig. 5.29, the primary switch is connected to each payload slot on one or more data plane connections per slot. For SpaceVPX, the inclusion of a redundant switch increases fault tolerance of the overall system. All connections for the redundant switch mimic those of the primary switch.

As shown in Fig. 5.29, the payloads of the SpaceVPX system use case are connected to data switch modules.

- A: The primary data switch module has a connection to every primary and redundant payload module. The data switch allows multiple simultaneous connections.
- B: The redundant data switch module has a connection to every primary and redundant payload module.

5.4.1.8 System Controller and Control Plane Connections

System controller is an essential part of the SpaceVPX standard. As shown in Fig. 5.30, the system controller can use the control plane to communicate with all other payload and switch modules at rates 10 to 1000 times faster than what is possible on the utility plane, yet 10 to 1000 times slower than what is possible on the data plane. The system controller can use the control plane to perform command processing, detailed telemetry collection, module, control and data network setup and monitoring, or medium-speed data transport. The control switch module routes the redundant control plane to all other modules in the SpaceVPX system. The control switch function may be combined with the system controller, data switch, or any

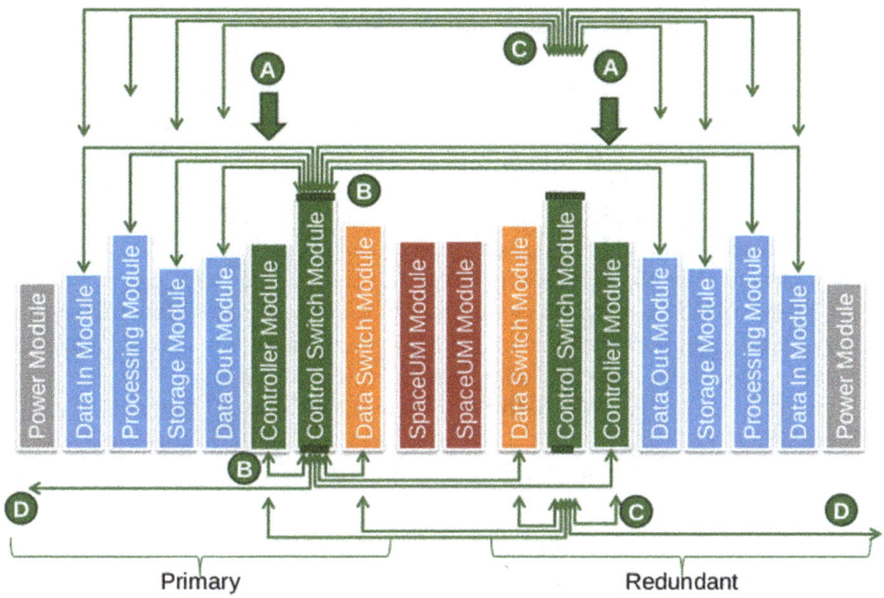

Fig. 5.30 SpaceVPX system use case showing system controller and control plane connections (*Courtesy* VITA, used with permission)

payload module function. Fault tolerance is provided with the addition of a redundant system controller cross-strapped to the primary system controller through the control switch. Additional control plane links can be provided for external control or to extend the control plane to other boxes with or without cross-strapping. Note that SpaceVPX utilizes SpaceWire for all control plane links and switches. OpenVPX supports similar structures for the control plane using either PCI Express or Ethernet.

From Fig. 5.30, the system controller uses control plane connections to all active logic modules.

- A: Controller modules provide a fault-tolerant selection for plug-in modules that could be system controllers.
- B: The system controller module utilizes a control switch function (either integrated with the system controller or on separate modules) to connect to every other payload or data switch module. This switched network provides a full network fabric for the control plane.
- C: A redundant system controller function provides a redundant switch fabric for the control plane with identical connections to all payload and switch modules.
- D: External control plane connections to the control plane of other systems are provided through the control switch. The external control plane connections are typically cross-strapped.

5.4.1.9 System Controller and Utility Plane Connections

The system controller also controls utility plane signal distribution to all the logic modules. As shown in Fig. 5.31, each system controller routes identical copies of utility plane signals (e.g. resets, reference clocks, etc.) to each active space utility management (SpaceUM) module. Since each SpaceUM module supports the utility plane selection of up to 2 (3U) or 8 (6U) logic modules, a SpaceVPX system supports up to four SpaceUM modules. The SpaceUM modules then select which set of utility plane signals to route to each logic module. The default OpenVPX routing of these utility plane signals is as buses which introduce single point of failures between the modules. Fault tolerance is enhanced with the separate radial distribution of the selected utility plane signals to each module. Some of these SpaceVPX extensions are optional in OpenVPX [VITA 46.11]. However, the SpaceUM module is unique to SpaceVPX.

From Fig. 5.31, the system controller controls the utility plane signals in a SpaceVPX system.

- A: Controller modules provide a fault-tolerant selection for plug-in modules that could be system controllers.
- B: Controller modules distribute copies of the system control and reference clocks to each active SpaceUM module.
- C: SpaceUM modules use the system controller select signals (either external or internally generated) to select which set of system control and reference clocks to distribute to logic modules.

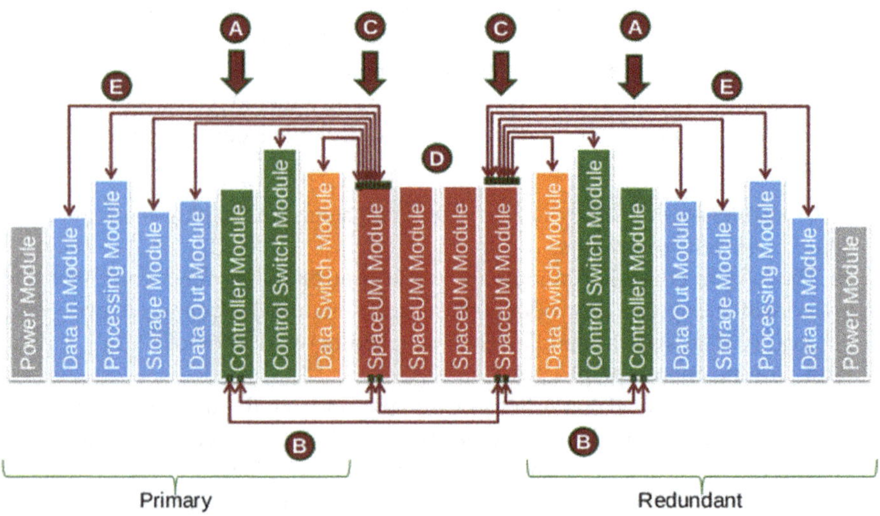

Fig. 5.31 SpaceVPX system use case showing system controller and utility plane connections (*Courtesy* VITA, used with permission)

- D: SpaceVPX supports switching of up to 32 6U logic modules using up to four SpaceUM modules, each handling eight logic modules. In this figure, only two SpaceUM modules are needed to support the 14 logic modules. If more than two logic modules were added to this system, the third SpaceUM would be required for the 17th through 24th module and the fourth SpaceUM would be required for the 25th through 32nd modules. Only two logic modules are supported by a 3U SpaceUM module. Since the SpaceUM modules are simply extensions of the controller, payload, and power modules, no redundancy is required.
- E: The SpaceUM module distributes the utility signals to all other logic modules including payloads, data switches, and control switches.

5.4.1.10 Utility Plane Power Connections

Utility plane power connections are also routed through the SpaceUM modules in a SpaceVPX system. The utility plane power connections consist of at least dual power modules to provide power to the SpaceVPX system. Resulting module power selection are individually selected by the SpaceUM modules and separately routed for each logic module. The A/B selection can be internal (autonomous) or external. Fault tolerance is provided by fault containment between redundant sections of each SpaceUM module. OpenVPX supports a single power mode for each plug-in logic module. SpaceVPX maintains this by moving the power selection into the SpaceUM module.

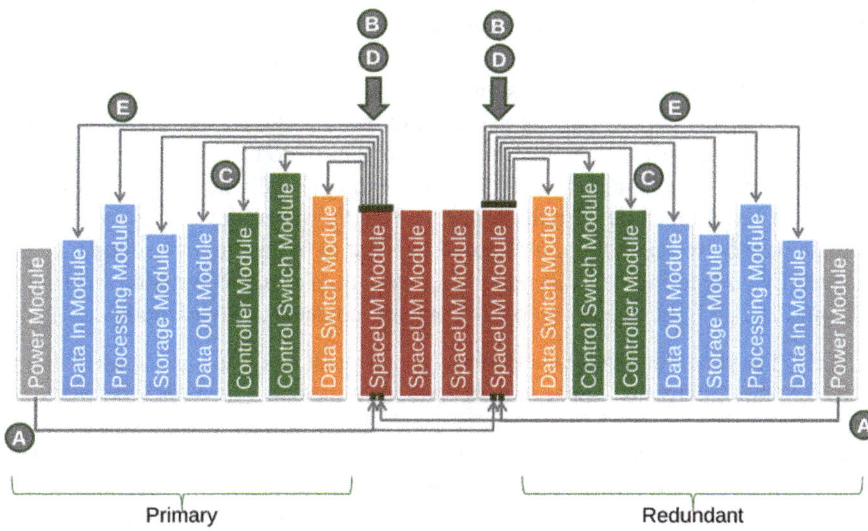

Fig. 5.32 SpaceVPX use case—utility plane power routing and control (*Courtesy* VITA, used with permission)

From Fig. 5.32, SpaceVPX modules receive switched fault-tolerant power through an attached SpaceUM module.

- A: Two power modules provide A and B power for the system.
- B: System controller selection identifies which controller modules to power initially.
- C: Switched power is routed to the system controller(s) that are marked to receive power.
- D: Power supply select is used to select power for each logic module. This may also be done by the system controller using the controller utility signals.
- E: Power is routed separately to each logic module in the SpaceVPX system. Each 6U SpaceUM supports routing power up to eight logic modules. As shown, the 14 logic modules require two SpaceUM modules. If more than two logic modules were added to this system, a 3rd SpaceUM module would support the 17th through 24th logic modules and a 4th SpaceUM module would support the 25th through 32nd logic modules.

5.4.1.11 Expansion Plane Extensions

The expansion plane provides additional module connections using a separate connector segment. These can be used for connecting on the backplane to heritage or special I/O not defined elsewhere. Another use of the expansion plane is for additional control or data plane connections or lower latency data connections in addition to the dominant switch (star/redundant star) or mesh topology previously discussed. Any size data connections can be used across the 32 differential pairs of signals on the expansion plane connector. Connections are made between adjacent physical modules with matching expansion plane connectors with a typical distribution of half of the pairs on the module assigned to the module on the right and half to the module on the left. Logically, these expansion plane extensions form a string of connections. The end modules of the string can optionally be connected together through the backplane or externally to form a ring, however, the latency to travel the alternate path between end modules could turn out to be too large. Redundancy can be added by connecting twice as many data pairs than required for bandwidth.

From Fig. 5.33, the expansion plane connects adjacent payload modules.

- A: Each payload module connects to its two nearest neighbors using up to half of the 32 pairs of signals. These connections have the shortest latency between slot positions.
- B: Peripheral modules typically contain payload functions without a data plane connection. The expansion plane connects them to the nearest payload module(s).
- C: Any peripheral module will also include at least one second copy for redundancy.
- D: If the peripheral module has utility plane connections, it is treated as an independent module. If it must derive its utility plane connections from the adjoining

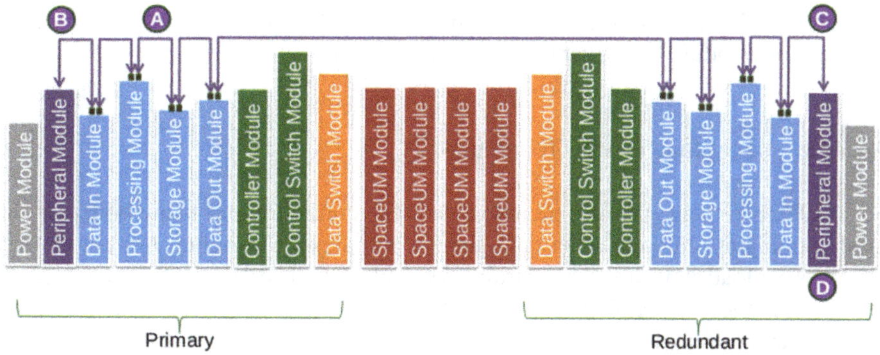

Fig. 5.33 SpaceVPX system use case—expansion plane extensions (*Courtesy* VITA, used with permission)

module through user-defined I/O, it is considered an attached peripheral module and is grouped with its payload module bridge for redundancy.

5.4.1.12 SpaceVPX Reliability Model

Figure 5.34 shows the reliability model diagram for the SpaceVPX use case system. The diagram includes the contents of two SpaceUM modules, dual power supply modules, dual system controller modules, dual control switch modules, dual data switch modules, four sets of dual payload modules, and one set with an optionally attached peripheral module. The diagram also includes a peripheral module that is independent. The utility, data, and control planes are all dual redundant. The utility plane is indicated by P (Power) and S (System) connections. Data plane redundancy is shown as dual redundant but is application-dependent. Additional payloads, peripherals, and switches can be added according to the application and the redundancy.

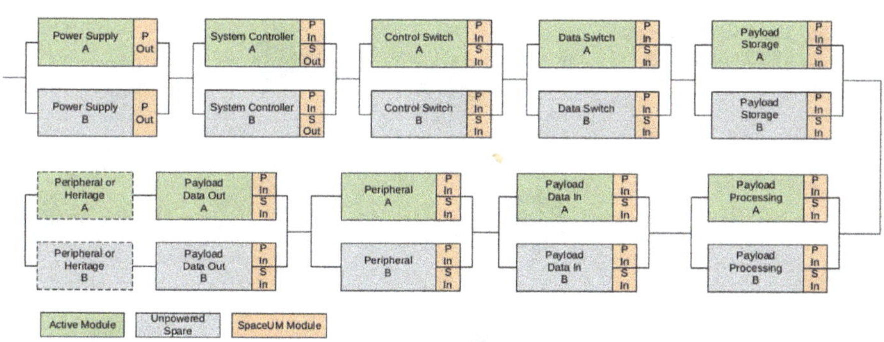

Fig. 5.34 Typical SpaceVPX reliability model diagram (*Courtesy* VITA, used with permission)

Note that the SpaceUM module is distributed across the various modules and not as a standalone module that must be spared (Fig. 5.34).

5.4.1.13 Backplane Profile Topologies

SpaceVPX backplane profiles include data plane configurations using both star and mesh topologies although other data plane topologies are possible. A star topology connects payload slots through centralized switch slots. The SpaceVPX control plane always uses the star topology while the SpaceVPX data plane can be either a star topology or a mesh topology. Note that when the control and data planes are both star topology, the control and data plane switches can be combined on a single module. In star topologies with dual redundant fault tolerance, each switch topology has two switch slots and all payload slots are connected to each switch slot. A mesh topology connects each of the payload slots directly to other payload slots, without the use of a centralized switch slot. The SpaceVPX data plane is implemented as a full mesh where each payload slot data plane interface is connected to each of the other payload slots. In a full mesh, failure of one module does not adversely affect the ability of any working module to communicate with any other working module in the mesh. Also, modules in the mesh can be in a standby redundant off-state without adversely affecting communication between modules turned on. Any mesh configuration requiring one or more modules to be alive and working for other modules in the mesh to communicate with others are not conducive to space applications. Figure 5.35 gives examples of star topology. A star topology connects payload slots through centralized switch slots. Note: For the purpose of the discussion in this section, a peripheral or controller slot is considered as a type of payload slot.

- A dual star (Fig. 5.35), using two switch slots (or redundant star), is where there are two switch slots and every payload slot is connected to both switch slots. See Fig. 5.35.

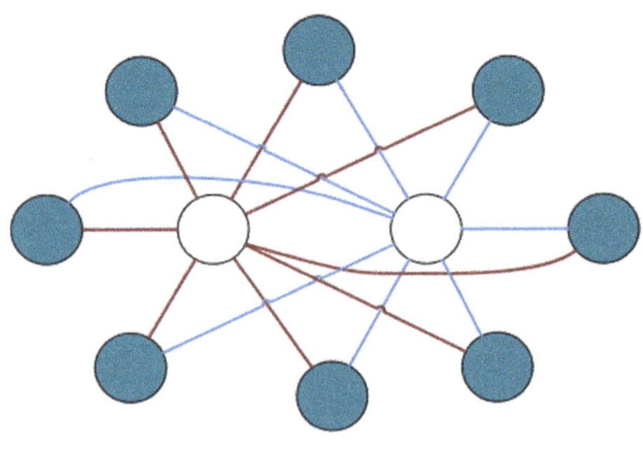

Fig. 5.35 Dual star topology (*Courtesy* VITA, used with permission)

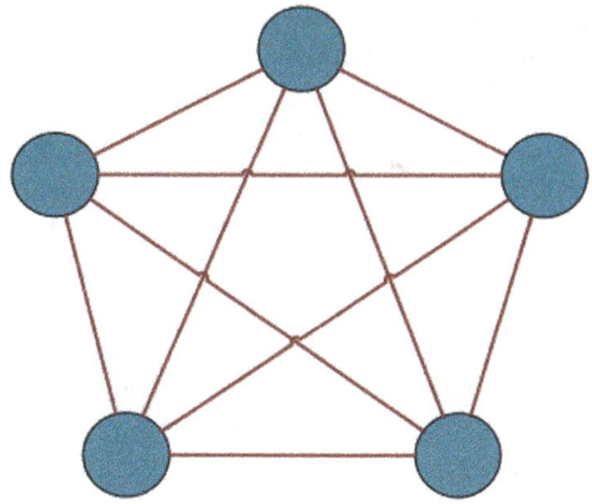

Fig. 5.36 Full mesh topology (*Courtesy* VITA, used with permission)

- A quad star, using two switch slots, is where there are two switch slots and every payload slot is connected to each switch slot with two links (pipes).

A mesh topology is where payload slots are directly connected to other payload slots, without the use of a centralized switch slot.

- A full mesh is a mesh topology in which every payload slot is directly connected to every other payload slot. An example is shown in Fig. 5.36.

Centralized

Pictorially, this topology is a star, with two or more centralized switches. One or more switch slots is the center of the star with payload slots as the points. In these topologies, at least one switch provides a redundant connection to each payload module. This redundant connection must not be included as providing additional bandwidth since it may not be available due to failure or it can be off in the case of standby redundancy.

Distributed Topology

These are topologies that are intended to be used by plug-in modules that each contribute to the switching as opposed to having centralized resources for switching. These can be mesh, daisy-chain, ring topologies, or some combination. For generic topologies of this type, see Fig. 5.37 above. SpaceVPX specification restricts distributed topologies to the data plane.

Hybrid Topology

A hybrid topology has slots defined by a different standard, such as PCI or VME, on the same backplane with VPX slots. A bridged slot profile then provides a pin allocation allowing connection between them.

5.4.1.14 SpaceVPX Fault Tolerance

The goal of SpaceVPX is to achieve an acceptable level of fault tolerance while maintaining reasonable compatibility with OpenVPX components, including connector pin assignments. For the purposes of fault tolerance, a module is considered the minimum redundancy element. The utility plane and control plane are all distributed redundantly and in star topologies to provide fault tolerance. For space applications, the major fault tolerance requirements are listed below:

1. Dual-redundant power distribution (bussed) where each distribution is supplied from an independent power source.
2. Dual-redundant utility plane signal distribution (point-to-point cross-strapped) where each distribution is supplied from an independent system controller to a SpaceUM module that selects between the A and B system controllers for distribution to each of the slots controlled by the SpaceUM module.
3. Card-level serial management
4. Card-level reset control
5. Card-level power control
6. Matched length, low-skew differential timing/synchronization/clocks
7. Fault-tolerant power supply select (bussed)
8. Fault-tolerant system controller signal selection (bussed)
9. Dual-redundant data planes (point-to-point cross-strapped)
10. Dual-redundant control planes (point-to-point cross-strapped).

5.5 CompactPCI Serial Space

Another modular interconnection standard which is based on backplanes and plug-in modules is CompactPCI Serial Space, which is based on CompactPCI Serial. CompactPCI Serial will be introduced in the next section. Since CompactPCI Serial is considerably less complex than VPX, this section will be shorter. This does not imply per se an inclination for VPX (this book stays neutral with respect to that choice), it is just that adding more information would not add to understanding better the standard, which is pretty straightforward.

5.5.1 CompactPCI Serial

CompactPCI Serial is a standard managed by PICMG.[4] PICMG is a consortium of roughly 50 companies, notably Intel, Airbus, BAE, Advantech, National Instruments, and others. PICMG works on defining and releasing open standards. An open standard is a definition of everything a vendor needs to know to build equipment (and write software) that will work with compatible products offered by other vendors. CompactPCI Serial combines the simplicity of VME with the speed and bandwidth of serial pipes. Technically speaking, CompactPCI Serial cannot be strictly considered

[4] https://www.picmg.org/openstandards/compactpci-serial/.

a "switched fabric" since basically no switching infrastructure is required, provided the slot count remains within some limits.

CompactPCI Serial is a design evolution for the proven parallel CompactPCI. The new base standard (CPCI-S.0) replaces the parallel signals by fast serial point-to-point data links and introduces a new connector type (Airmax connector). CompactPCI Serial supports all modern high-speed serial data connections while keeping mechanical compatibility with the IEEE 1101 and Eurocard format (Figs. 5.37 and 5.38). CompactPCI Serial is based on a star topology for PCI Express (also Serial RapidIO), SATA, and USB. One slot in the backplane is reserved as the system slot. The system slot supports up to eight peripheral lots. The system slot board (CPU) is the root complex (source) for up to 8 × PCIe Links, 8 × SATA, 8 × USB, and 8 × GbE distributed across the backplane to the peripheral slots.

In principle all peripheral slots are the same. The pin assignment of all peripheral slots is 100% identical, so there are no slot nor backplane profiles as in VPX. The backplane remains also the same all along. Two slots are especially connected to the system slot using an extra wide PCI Express link called a fat pipe (see Fig. 5.39). These slots can be used for high-end processing-intensive applications which need high throughput back and forth from the computing unit. CompactPCI Serial architecture is pretty simple and literally plug&playable. CompactPCI Serial does not require any switches or bridges in systems with up to nine slots.

Ethernet is wired as a full mesh network. In full mesh architectures each of the nine slots is connected to each of the other eight slots via a dedicated point-to-point connection. In Fig. 5.39, a detailed description of the different connectors on the backplane for the different slots and board roles can be seen. System slots use all

Fig. 5.37 CompactPCI Serial system card and backplane (*Courtesy* EKF, used with permission)

Fig. 5.38 CompactPCI Serial backplanes (*Courtesy* EKF, used with permission)

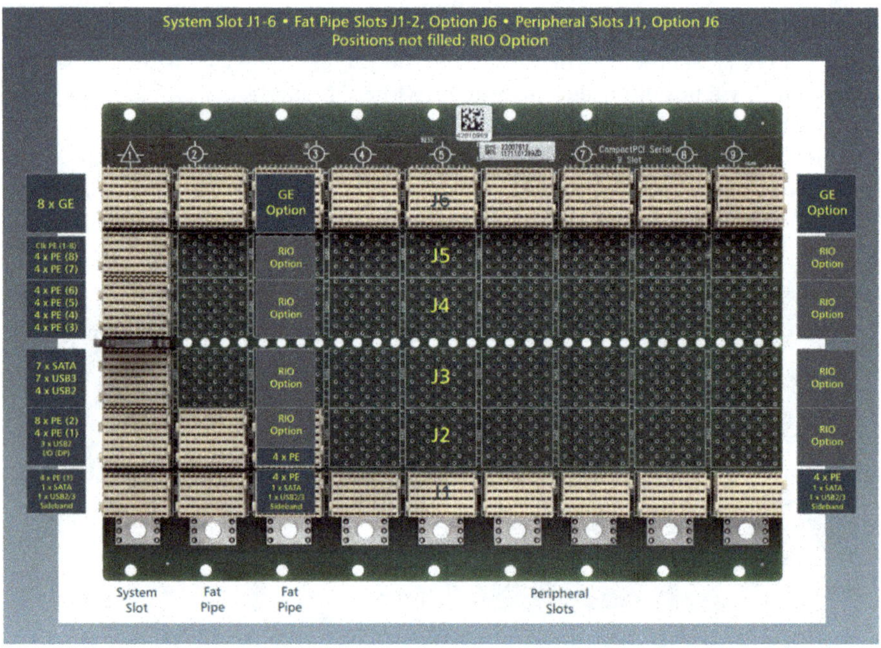

Fig. 5.39 Slots in a CompactPCI Serial system (*Courtesy* EKF, used with permission)

six-connectors (J1-J6), whereas fat pipe slots use J1-2 with J6 being optional for GigE. For regular peripheral slots, J1 and optionally J6 (for GigE connection) are used. Note that all connectors not used can be used for SRIO.

One pin signals to the plugged-in module/board whether it is located in a system slot or peripheral slot. This allows to plug a system-slot board (normally a CPU card) also into any peripheral slot. The next batch of figures depict the different board "roles" and connectors and the interfaces used from a module perspective: system slot card, fat-pipe peripheral slot card, and regular peripheral slot card (Figs. 5.40, 5.41, and 5.42, respectively). This can be matched with Fig. 5.40 to observe how the interconnect takes place.

In addition, there are a number of "utility" signals to support the slots and for general system management, such as reset, hot plug, geographical addressing, etc. 12 V are available for power supply, allowing a maximum power consumption of 60 W for one 3U slot. This includes the peripheral slots. In case the system architecture requires more than the eight peripheral slots CompactPCI Serial allows, it is possible to use backplane couplers (Fig. 5.43).

As shown before, CompactPCI Serial supports a full meshed Ethernet network. The backplane wiring creates a dedicated connection of every slot to every other slot. Each of the nine slots in a CompactPCI Serial system is connected with each of the other eight slots via the backplane, through the J6 connector. CompactPCI Serial allows using a system-slot CPU also as a peripheral card without any problems, which makes it very straightforward to create modular multi-CPU architectures.

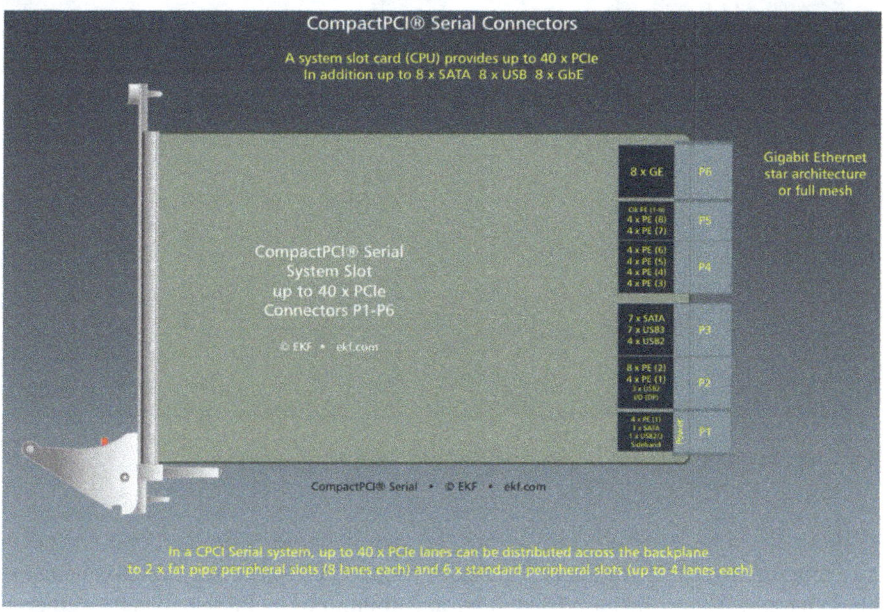

Fig. 5.40 System-slot CompactPCI Serial card (*Courtesy* EKF, used with permission)

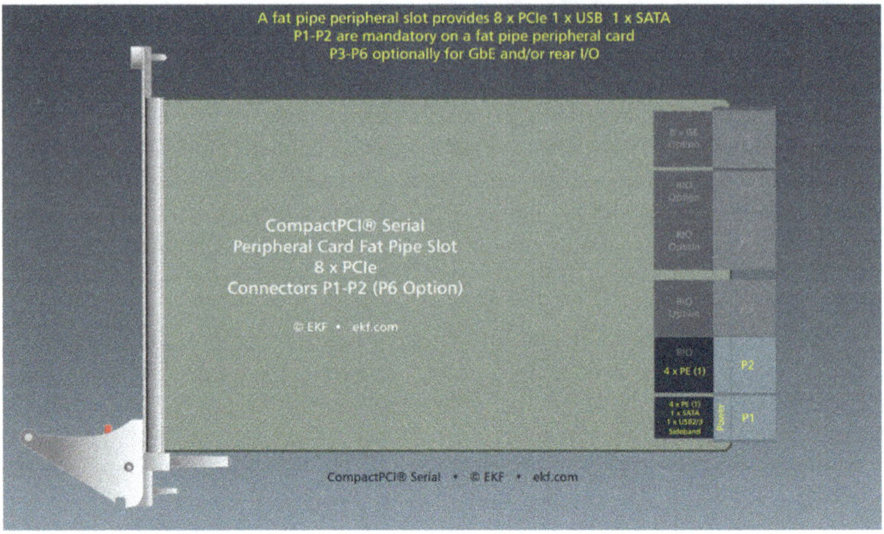

Fig. 5.41 Fat pipe peripheral board/slot (*Courtesy* EKF, used with permission)

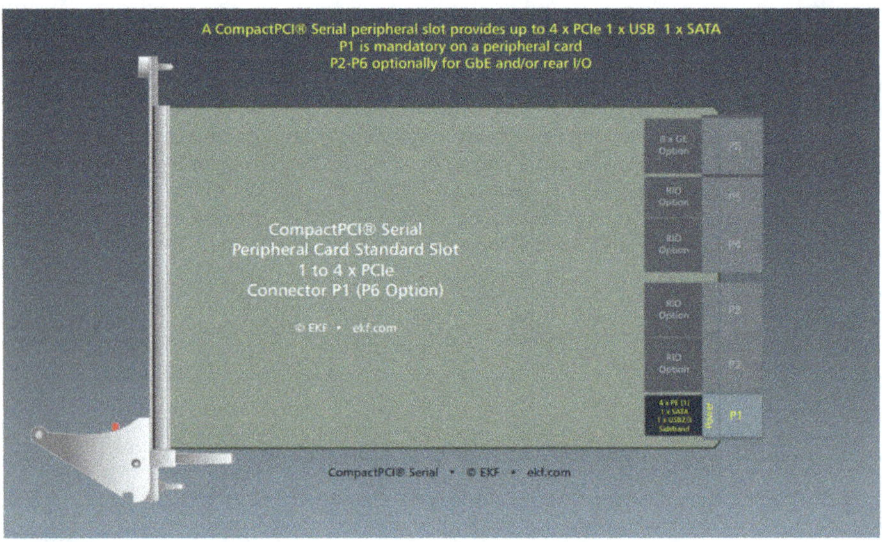

Fig. 5.42 Peripheral board/slot (*Courtesy* EKF, used with permission)

As these are point-to-point connections, no switch board is needed, and no special infrastructure or configuration is required. CompactPCI Serial enables easy and cost-efficient multiprocessing based on known and ubiquitous technology such as Ethernet communication.

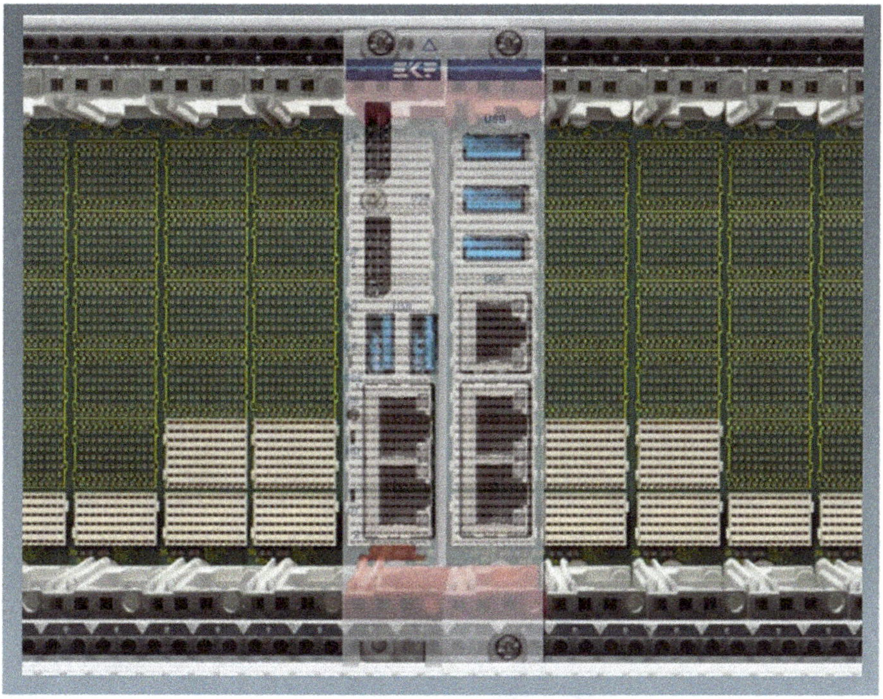

Fig. 5.43 Backplane coupler (*Courtesy* EKF, used with permission)

5.5.2 CompactPCI Serial for Space Applications

CompactPCI Serial Space initiative started in 2015, triggered by DLR project OBC-SA and conformed by companies such as Thales Alenia Space (Germany), SpaceTech, Airbus Defence and Space, MEN, and supported by EBV, EKF, FCI, Fraunhofer, Heitec, TTTech (Fig. 5.44).

Regular CompactPCI Serial products can be combined with Space CompactPCI Serial Space products to realize on-board functional models, test and simulation systems. One of the main differences compared to CompactPCI Serial is that it contains two system slots compared to one of the former. This means CompactPCI Serial Space has slots for seven peripheral slots (Fig. 5.45). Another key difference is that CompactPCI Serial Space does not include SATA and USB as standard interfaces, but it adds CAN and SpaceWire. As for networking, CompactPCI Serial Space offers dual star topology for PCIe, Ethernet, and SpaceWire. CompactPCI Serial Space defines four slot profiles: power, system, peripheral, and utility.

In redundant CompactPCI Serial Space configurations, the second system slots can host a network module. This way, all boards in peripheral slots are connected to the primary and to the redundant command and data handling link. In CompactPCI Serial Space the maximum number of slots is limited as well. In case an architecture

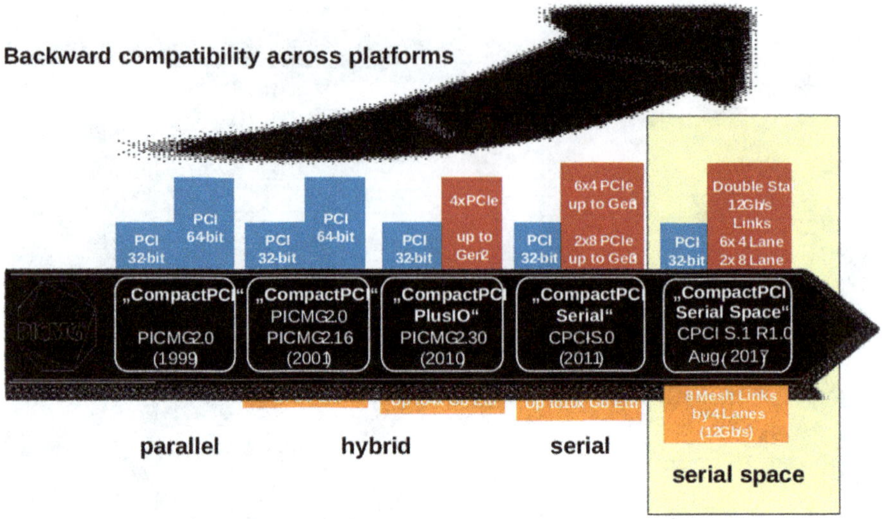

Fig. 5.44 Evolution of the PCI standard throughout the years (*credit* PICMG, used with permission)

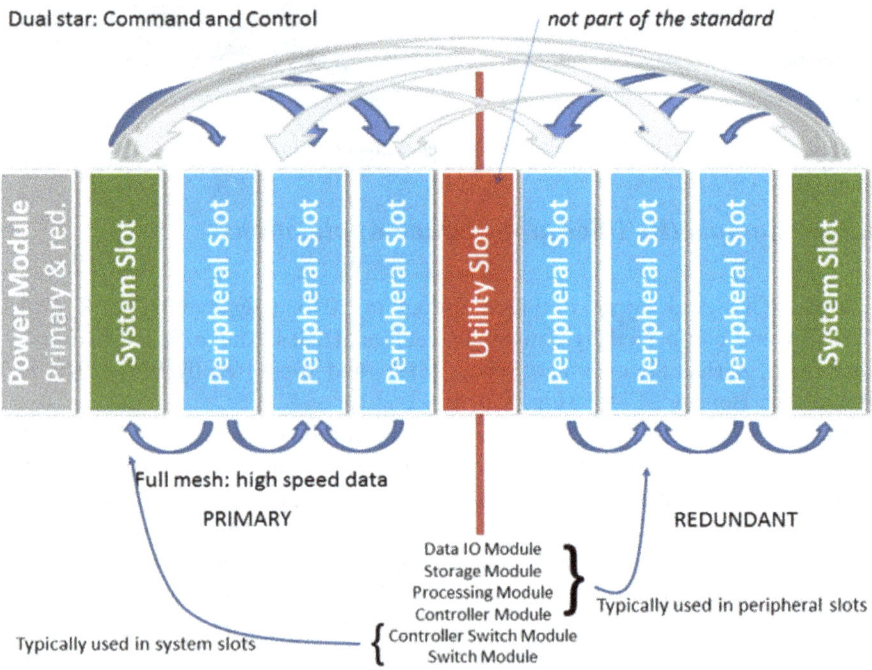

Fig. 5.45 CompactPCI Serial Space backplane profiles (*credit* PICMG, used with permission)

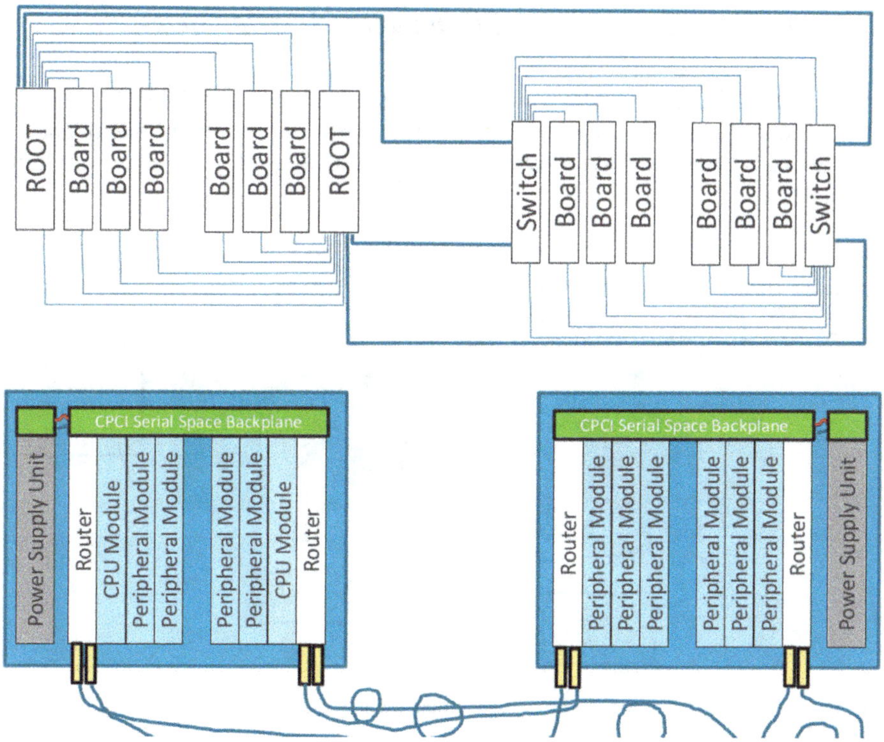

Fig. 5.46 Interconnecting two CompactPCI Serial Space chassis together (*credit* MEN, used with permission)

requires a higher amount of boards than the maximum allowed, more chassis need to be allocated and interconnected (Fig. 5.46).

5.5.3 Power Distribution

Neither the power supply itself nor the connecting of the power supply to the backplane is a part of the specification. Redundancy concepts for power supplies are also not specified but of course common practice. The power distribution over the backplane can be implemented in different ways depending on the application.

5.5.4 Typical Applications of CompactPCI Serial Space

A typical avionics architecture using cPCISS can be seen in Fig. 5.47.

This could be interfaced with other cPCISS boxes implementing functionalities such as mass memory (Fig. 5.49), input/output modules (Fig. 5.48), or a GNSS payload (Fig. 5.50).

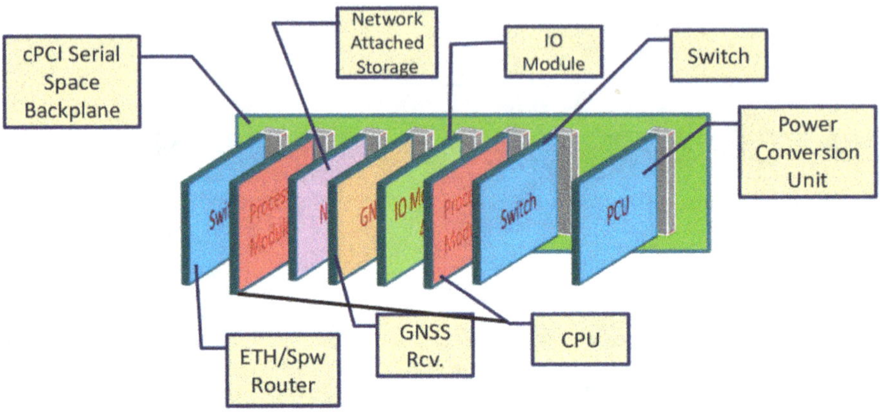

Fig. 5.47 An avionics architecture using CompactPCI Serial Space (*credit* MEN, used with permission)

Fig. 5.48 A cPCISS input/output 6U backplane (*credit* MEN, used with permission)

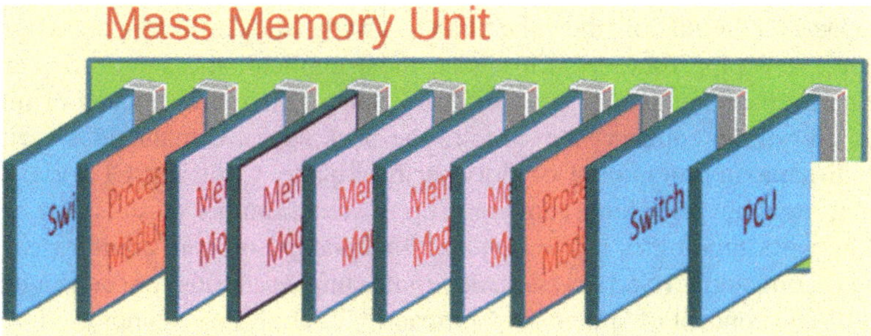

Fig. 5.49 A MMU backplane (*credit* MEN, used with permission)

Fig. 5.50 A GNSS payload backplane example (*credit* MEN, used with permission)

5.6 CompactPCI Serial or VPX?

One of the most remarkable aspects of comparing VPX with CompactPCI Serial is that it illustrates very precisely the nuances of restricted or unrestricted degrees of freedom for designers and architects.

Although these two standards can be perceived as quite similar from a distance (after all, it's about boards connected to a backplane) they are quite different. VPX adopted an approach of giving the system architect almost full freedom to choose topologies and pipe routings. This did not come without penalties: the possibilities are so many that the specification had to prescribe some "variants" for slots and modules in order to facilitate interoperability (OpenVPX). This also means VPX backplanes can be quite ad hoc (in other words, backplanes are application-specific), making the interoperability somewhat more laborious; it can be challenging to exchange plug-in boards of different manufacturers one-to-one. VPX and all its derivative specifications (OpenVPX, SpaceVPX) are not strictly *plug and play*. Also, VPX systems often require switching equipment in the backplane (except for the switch-less approach described before, see Sect. 5.4.2.1). Designing a VPX-based system requires a considerable amount of non-recurring engineering. The impossible-to-deny gain is full freedom for the system architect to choose among the topologies, protocols, and layouts that suit the requirements the best.

CompactPCI Serial, on the other hand, strictly standardized pin assignment, topologies, and pipe routings. This way, most applications, simple or complex, can be built up of standard boards and backplanes (there are no slots or modules profiles, everything connects the same to the backplane). CompactPCI Serial also restricts the maximum amount of slots per backplane. CompactPCI Serial can do completely without switching hardware (up to some number of slots), by creating default network topologies in the standard backplane. Software complexity can be lower in a CompactPCI Serial system due to its simpler architecture and switch-less approach. The concept of "plane" in CompactPCI Serial Space is not used.

By pre-defining topologies and pipe routing, CompactPCI Serial (and Serial Space) reduces complexity but the system architect sees the design degrees of freedom constrained. CompactPCI Serial Space has been selected by several European players in the space industry, making it the standard of choice for European space systems in the incoming years. VPX, on the other hand, is the standard of choice for interconnected US military systems, electronic warfare, and industrial high-performance computing industry.

5.7 Integrated Modular Avionics (IMA)

A modular subsystem interconnect would be only partially exploited without a software architecture on top which matches and exploits this modular approach. A modular software architecture will be discussed in this section.

We have assessed so far interconnection approaches which generally relate to the way signals are routed from modules to slots and backplanes, but not much has been said about the software running on the CPUs connected to those backplanes.

Space is a conservative industry, and for valid reasons. Spacecraft designers don't get the chance to push the reset button in space if something hangs, so things must be done in special ways to minimize the chance of a failure which could turn the mission useless, with all the losses and damages associated with that. Surprisingly, the aerospace industry is less conservative when it comes to exploring new architectural approaches, despite its extremely strict safety requirements. Main reason for this is a never-ending search for fuel efficiency which means weight reduction. Due to this, in the last 15 or 20 years the architecture behind aerospace avionics development has shifted its paradigm considerably. The federated architecture (as in one computer assigned to one functionality) that has been popular up to the end of the century is being replaced by a different approach, which is called integrated modular avionics (IMA). Some sources point that the origins of the IMA concept originated in the United States with the new F-22 and F-35 fighters and then migrated to the commercial airliner sector. Others say the modular avionics concept has been used in business jets and regional airliners since the late 1980s or early 1990s. The modular approach also is seen on the military side in tankers and transport aircraft, such as KC135, C-130 as well as in the Airbus A400M (Ramsey 2007). In a federated architecture, a system main function is decomposed into smaller blocks that provide certain specific functions. Each black box—often called line replaceable unit (LRU)—contains the hardware and software required for it to provide its function. In the federated concept, each new function added to an avionics system requires the addition of new LRUs. This means that there is a linear correlation between functionality and mass, volume, and power; i.e. every new functionality proportionally increases all these factors. Moreover, for every new function added to the system there is a consequent increase in multidisciplinary configuration control efforts, updates, iterations, etc. This approach quickly met a limit. The aerospace industry understood that the classical concept of "one function maps to one computer" could no longer be maintained. To tackle this issue, the integrated modular avionics (IMA) concept emerged. Exploiting the fact that software does not weigh anything in and of itself, IMA allowed retaining some advantages of the federated architecture, like fault containment, while decreasing the overhead of separating each function physically from others. The main architectural principle behind IMA is the introduction of a shared computing resource which contains functions from several LRUs. This means function does not directly map 1:1 to the physical architecture, but one physical computing unit (CPU) can share its computing resources to execute more than one function.

A sort of contradiction surrounds the IMA concept. It could be argued that IMA proposes an architecture that technology has already rendered obsolete: centralized architectures. With embedded processors, memories and other devices becoming more reliable and less expensive, surely this trend should favor *less* rather than more centralization. Thus, following this argument, a modern avionics architecture should be more, not less, federated, with existing functions "deconstructed" into smaller components, and each having its own processor (Rushby 1999). There is some plausibility to this argument, but the distinction between the "more federated" architecture and centralized IMA proves to be debatable on closer inspection. A federated architecture is one whose components are very loosely coupled—meaning that they can operate largely independently. But the different elements of a function—for example, orbit and attitude control—usually are rather tightly coupled so that the deconstructed function would not be a federated system so much as a *distributed* one—meaning a system whose components may be physically separated, but which must closely coordinate to achieve some collective purpose. Consequently, a conceptually centralized architecture will be, internally, a distributed system, and the basic services that it provides will not differ in a significant way from those required for the more federated architecture (Rushby 1999).

Then, IMA quickly gained traction in aerospace, and its success caught the attention of the space industry, mainly of the European Space Agency (ESA). IMA became of interest for space applications due to the fact that it allowed for mass, volume, and power savings by pushing more functionalities to the software (Fig. 5.51).

Combining multiple functions on one processing/computing module requires specific considerations and requirements which are not relevant for regular federated LRUs. The physical separation that existed between LRUs has to be provided now virtually for applications running in the same core processing module (CPM). Furthermore, sharing of the same computing resource influences the development

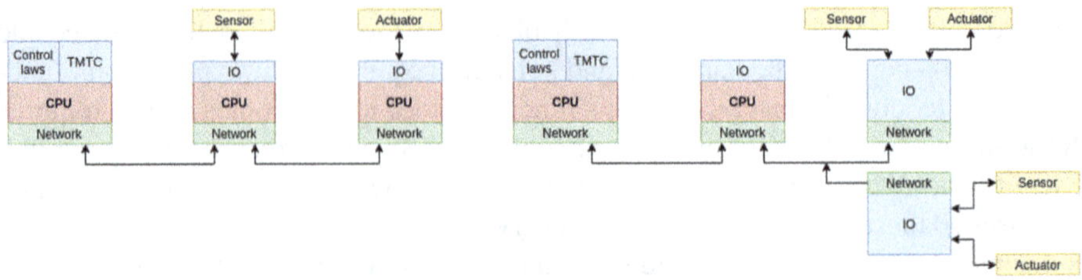

Fig. 5.51 Federated (left) vs integrated (right) architectures

process, because new dependencies appear among different elements. Input/output common resources are provided by I/O modules. These I/O modules (IOMs) interface with sensors/actuators, acting as a bridge between them and core processing modules (CPMs). A core processing module that also contains I/O capabilities is named a core processing input/output module (CPIOM). If the architecture does not make use of CPIOMs, the IO layer remains as *thin* as can be. Removing all software from the IOMs removes complexity and therefore verification, and configuration control costs considerably. The physical separation which was inherent in the LRUs for the federated architecture from a software point of view must be virtually enforced in an IMA platform. The performance of each application shall be unaffected by the presence of others. This separation is provided by partitioning the common resources and assigning those partitioned resources to an application. The partitioning of processing power is enforced by strictly limiting the time each application can use the processor and by restraining its memory access. Memory access is controlled by hardware thus preventing partitions from interfering with each other. Each software application is therefore partitioned in space and time.

Avionic software applications have different levels of criticality based on the effects that a failure on a given application would cause to the system. Those criticality levels are specified in standards such as the RTCA/DO178C[5] which defines five different criticality levels (from A: catastrophic, to E: no safety effect). The software development efforts and costs grow exponentially with the criticality level required for certification, since the process of testing and validation becomes more complex. In an integral, non-partitioned architecture all the software in a functional block has to be validated under the same criticality level. The IMA approach enables different software partitions with different criticality levels to be integrated under the same platform and certified separately, which eases the certification process. Since each application is isolated from others, it is guaranteed that faults will not propagate, provided the separation "agent" is able to do so. As a result, it is possible to create an integrated system that has the same inherent fault containment as a federated system. For achieving such containment, Rushby specifies a set of guidelines (Rushby 1999):

[5]DO-178C, Software Considerations in Airborne Systems and Equipment Certification is the primary document by which the certification authorities such as FAA, EASA and Transport Canada approve all commercial software-based aerospace systems. The document is published by RTCA, Incorporated.

- Gold standard for partitioning: A partitioned system should provide fault containment equivalent to an idealized system in which each partition is allocated an independent processor and associated peripherals and all inter-partition communications are carried on dedicated lines.
- Alternative gold standard for partitioning: The behavior and performance of software in one partition must be unaffected by the software in other partitions.
- Spatial partitioning: Spatial partitioning must ensure that software in one partition cannot change the software or private data of another partition (either in memory or in transit) nor command the private devices or actuators of other partitions.
- Temporal partitioning: Temporal partitioning must ensure that the service received from shared resources by the software in one partition cannot be affected by the software in another partition. This includes the performance of the resource concerned, as well as the rate, latency, jitter, and duration of scheduled access to it.

The mechanisms of partitioning must block the spatial and temporal pathways for fault propagation by interposing themselves between avionics software functions and the shared resources that they use.

5.7.1 Application Interfaces and ARINC-653

From the previous section, we still have not discussed the entity in charge of guaranteeing that different applications will run on different partitions on top of a shared processing platform and with enough isolation. ARINC 653 defines a standard interface between the software applications and the underlying operating system. This middle layer is known as application executive (APEX) interface. The philosophy of ARINC 653 is centered upon a robust time and space partitioned operating system, which allows independent execution of different partitions. In ARINC 653, a partition is defined as portions of software specific to avionics applications that are subject to robust space and time partitioning. They occupy a similar role to processes in regular operating systems, having their own data, context, attributes, etc. The underlying architecture of a partition is similar to that of a multitasking application within a general-purpose computer. Each partition consists of one or more concurrently executing processes (threads), sharing access to processor resources based upon the requirements of the application. An application partition is limited to use the services provided by the APEX defined in ARINC 653 while system partitions can use interfaces that are specific to the underlying hardware or platform (Fig. 5.52).

Each partition is scheduled to the processor in a fixed, predetermined, cyclic basis, guaranteeing temporal segregation. The static schedule is defined by specifying the period and the duration of each partition's execution. The period of a partition is defined as the interval at which computing resources are assigned to it while the duration is defined as the amount of execution time required by the partition within

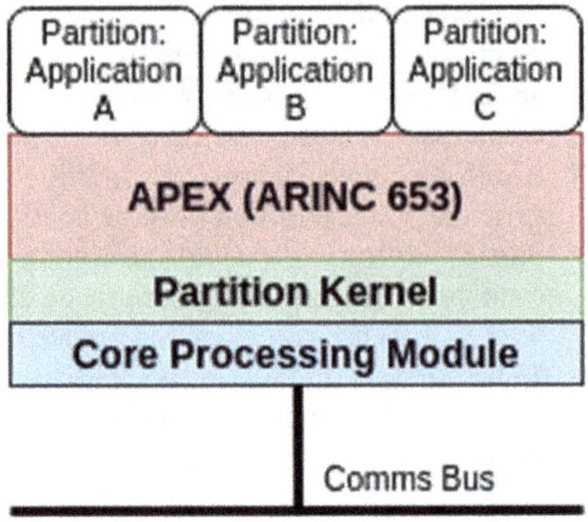

Fig. 5.52 Building blocks of a partitioned system

one period. The periods and durations of all partitions are used to compose a major timeframe. The major timeframe is the schedule basic unit that is cyclically repeated.

Each partition has predetermined areas of memory allocated to it. These unique memory areas are identified based upon the requirements of the individual partitions and vary in size and access rights.

5.7.2 *Benefits of Using Integrated Modular Avionics (IMA)*

The expected benefits from implementing IMA solutions are:

- Optimizations and saving of mass, volume, and power consumption;
- Simpler assembly, integration, and verification (AIV) activities due to smaller number of physical units and simpler harness;
- Focused development efforts: the developer can focus wholly on their software, instead of focusing on the complete development of an LRU;
- Retaining federated system properties like fault containment;
- Incremental validation and certification of applications;
- IMA has already been applied in the development of several aircrafts, most noticeably on the development of the A380 and Boeing 787. The Airbus and Boeing programs reported savings in terms of mass, power, and volume of 25, 50, and 60%, respectively (Itier 2007). IMA has eliminated 100 LRUs from the Boeing 787 (Ramsey 2007).

5.7.3 IMA for Space (IMA-SP)

There are some major differences between the space and aeronautical domains which constrain the development of space avionic systems. While most aeronautical systems operate with human intervention, most space systems are unmanned. The IMA approach used in the aeronautics domain cannot be used directly in the space domain. The reasons are: (1) The processing platforms currently used in space are less powerful than the platforms in aeronautics, i.e. there is a technology gap. (2) there are strong requirements that hardware and software modules already developed for the current platform shall remain compatible with any new architecture in order to keep the costs of the architecture transition low (Herpel et al. 2016).

Radiation is a major threat to on-board electronic systems and software since it can cause electrical fluctuations and software errors. Additionally, space systems are very constrained in terms of available power, mass, and volume. It is expensive and impractical to develop and deploy complex platforms; therefore, most systems have very limited hardware. Satellites are usually limited to one or two main computers connected with the required transducers and payload equipment through robust data buses. When compared with commercial aviation the space market largely lacks standardization. Each major player in the industry designs and operates its systems using their own internal principles. The European Space Agency has invested a great amount of effort in the last decades to standardize space engineering across Europe, as well as the European Cooperation for Space Standardization[6] (ECSS, which includes ESA as a member). ESA has defined[7] several ground rules which guide the IMA-SP system platform specification. They intend to adapt IMA to space without totally breaking with the current space avionics approach.

To enable the usage of an operating system with the requirement to implement time and space partitioning, ESA defined a two-level software executive. The system executive level is composed of a software hypervisor that segregates computing resources between partitions. This hypervisor is responsible for the robust isolation of the applications and for implementing the static CPU allocation schedule. A second level, the application level, is composed of the user's applications running in an isolated environment (partition). Each application can implement a system function by running several tasks/processes. The multitasking environment is provided by a paravirtualized operating system which runs in each partition. These partition operating systems (POS) are modified to operate along with the underlying hypervisor. The hypervisor supplies a software interface layer to which these operating systems attach to. In the context of IMA for space, ESA selected the use of RTEMS as the main partition operating system.

[6]https://ecss.nl/.

[7]https://www.esa.int/Enabling_Support/Space_Engineering_Technology/Shaping_the_Future/IMA_Separation_Kernel_Qualification_preparation.

Three hypervisors are currently being evaluated by ESA as part of the IMA-SP platform: XtratuM,[8] AIR,[9] and PikeOS[10]. XtratuM is an open-source hypervisor available for the × 86 and LEON architectures that is developed by the Universidad Politécnica de Valencia. Despite providing similar services to the ARINC 653 standard, XtratuM does not aim at being ARINC 653 compatible. PikeOS is a commercial microkernel which supports many APIs and virtualized operating systems. Finally, AIR is an open-source operating system developed by GMV and based on RTEMS. The IMA-SP specification also leaves an option to have partitions without RTOS. These "bare metal" partitions can be used for very critical single threading code which doesn't require a full featured real-time operating system.

5.7.4 Networking and Input/Output Considerations for IMA-SP

In a partitioned system, the quantification of time spent in I/O tasks is even more critical, since it shall be known on whose behalf I/O tasks are performed. The costs for these tasks should be booked to the applications that actually benefit from it. Robust partitioning demands that applications use only those time resources that have been reserved for them during the system design phase. I/O activities shall, hence, be scheduled for periods when the applications that use these specific capabilities are being executed. Furthermore, safety requirements may forbid that some partitions are interrupted by hardware during their guaranteed execution time slices. In consequence, it must be ensured that I/O devices have enough buffering capabilities at their disposal to store data during the time non-interruptible applications are running. Segregating a data bus is harder than segregating memory and processor resources. I/O handling software must be able to route data from an incoming bus to the application to which that data belongs to. General-purpose operating systems use network stacks to route data to different applications. In a virtualized and partitioned architecture, the incoming data must be shared not only with applications in the same partition but among partitions themselves. Each partition operating system could manage their own devices, but this is only feasible if devices are not shared among partitions. If a device is used by more than one partition then there is the latent risk of one of the partitions leaving the shared device in an unknown state, therefore influencing the second partition. In aeronautical IMA this problem is approached by using partitioning-aware data buses like AFDX. AFDX devices are smart in the sense that they can determine to which partition of an end system the data belongs to. This point is critical since the I/O in the platform must be managed in such way that behavior by one partition cannot affect the I/O services received by another. From

[8] https://fentiss.com/products/hypervisor/.

[9] https://indico.esa.int/event/225/contributions/4307/attachments/3343/4403/OBDP2019-S02-05-GMV_Gomes_AIR_Hypervisor_using_RTEMS_SMP.pdf.

[10] https://www.sysgo.com/products/pikeos-hypervisor/.

the rationale exposed in the last paragraphs, we can sum up a set of characteristics the I/O system must have to respect the characteristics of a partitioned system:

- The I/O module shall be generic and therefore decoupled from the application.
- The I/O module shall be robust, in the sense that it can be used by more than one partition without interference.
- The I/O module shall be able to route data to its rightful owner (a given application in a given partition).
- The I/O module shall be quantifiable (i.e. its execution time must be bound and measurable).
- The I/O module shall not interrupt, disrupt, or have any kind of impact in the time and space partitioning of the applications.

Another option for networking is to use TTEthernet.[11] A TTEthernet network includes features as global time using clock synchronization and offers fault isolation mechanisms to manage channel and node failures. TTEthernet defines three types of data flow: Time-triggered (TT) data flow which is the higher priority traffic; rate-constrained (RC) traffic, which is equivalent to AFDX traffic; and best effort (BE) traffic. This makes TTEthernet suitable for mixed-criticality applications such as avionic and automotive applications where highly critical control functions such as a flight management system cohabit with less critical functions such as an entertainment system. By adding TTEthernet switches guaranteed hard real-time communication pathways can be created in an Ethernet network, without impacting any of the existing applications. It can be used for the design of deterministic control systems, fault-tolerant systems, and infotainment/media applications which require multiple large congestion-free data streams (Robati et al. 2014). The TTEthernet product family supports bandwidths of 10, 100 Mbit/s, 1 Gbit/s, and higher, as well as copper and fiber optics physical networks. It enables purely synchronous and asynchronous operation over the same network (Herpel et al. 2016). The combination of IMA and TTEthernet enables the error isolation provided not only at the level of the modules through the partitioning but also the level of the network using different data traffics and the concept of virtual links. Third, TTEthernet enables the safe integration of data traffics with different performance and reliability requirements (Robati et al. 2014).

5.8 Conclusion: Don't Make Your Sandwich from Scratch

Time to market and wheel reinvention do not combine very well. A while ago some guy on YouTube decided to make himself a sandwich totally from scratch; it took him six months and 1,500 dollars. NewSpace orgs cannot make their sandwiches from scratch, yet they often do. The only way to get to space quickly and reliably

[11] https://www.sae.org/standards/content/as6802/.

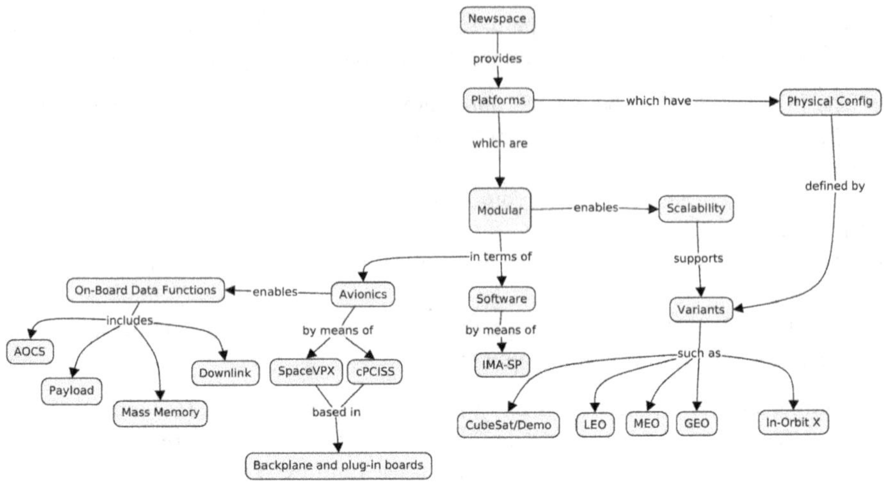

Fig. 5.53 Modularity, scalability, and interconnection concepts and how they relate

is to pick up things from the space shelf, use standardized form factors and high-speed interconnects, and set the mindset at the system level. Avionics and software modularity not only streamline interoperability between NewSpace suppliers and actors in general, but it also enables scalability. Thus, a baseline architecture can support small, medium, and large satellites. Modular avionics, by adopting high-speed wireless inter-satellite-links, can also leverage some advanced concepts such as in-orbit assembly, in-orbit servicing, and segmented architectures (Fig. 5.53).

References

Agrawal, B. N. (1986). *Design of geosynchronous spacecraft*. Upper Saddle River, NJ: Prentice-Hall.

Arianespace. (2016). *Ariane 5 User's Manual* (Issue 5, Rev 2 ed.).

Berk, J. (2009). *Systems failure analysis*. ASM International.

Brown, C. D. (1996). *Spacecraft propulsion* (J. S. Przemieniecki, Ed.). AIAA Education Series.

Brown, C. D. (2002). *Elements of spacecraft design*. AIAA Education Series. https://doi.org/10.2514/4.861796

CCSDS. (2012). IP over CCSDS Space Links—Recommended Standard CCSDS 702.1-B-1. CCSDS.

ECSS. (2002). ECSS-E-10-03A—Space engineering—Testing. European Cooperation for Space Standardization (ECSS).

Ghasemi, A., Abedi, A., & Ghasemi, F. (2013). *Propagation engineering in radio links design*. Springer. https://doi.org/10.1007/978-1-4614-5314-7.

Goebel, D. M., & Katz, I. (2008). *Fundamentals of electric propulsion: Ion and Hall thrusters*. JPL Space Science and Technology Series.

Herpel, H., Schuettauf, A., Willich, G., Tverdyshev, S., Pletner, S., Schoen, F., et al. (2016). Open modular computing platforms in space—Learning from other industrial domains. 1-11. 10.1109

Itier, J.-B. (2007). A380 Integrated Modular Avionics The history, objectives and challenges of the deployment of IMA on A380. Artist—European Network of Excellence in Embedded Systems Design. http://www.artist-embedded.org/docs/Events/2007/IMA/Slides/ARTIST2_IMA_Itier.pdf

Kuninaka, H., & Molina-Morales, P. (2004). Spacecraft charging due to lack of neutralization on Ion thrusters. *Acta Astronautica, 55*(1), 27–38.

Montenbruck, O., & Gill, E. (2001). *Satellite orbits models, methods, and applications*. Springer.

QY Research. (2020, Apr 17). *Global Lithium-ion power battery market insights, Forecast to 2026* (preview). https://www.marketstudyreport.com/reports/global-lithium-ion-power-battery-market-insights-forecast-to-2026

Ramsey, J. (2007, February 1). Integrated modular avionics: Less is more. *Aviation Today*. https://www.aviationtoday.com/2007/02/01/integrated-modular-avionics-less-is-more/

Robati, T., El Kouhen, A., Gherbi, A., Hamadou, S., & Mullins, J. (2014). An extension for AADL to model mixed-criticality avionic systems deployed on IMA architectures with TTEthernet. In *Conference: 1st Architecture Centric Virtual Integration Workshop (ACVI)*, ceur-ws 1233.

Rushby, J. (1999). Partitioning in avionics architectures: requirements, mechanisms, and assurance. NASA Technical Report DOT/FAA/AR99/58.

VITA. (2015). ANSI/VITA 78.00-2015 SpaceVPX System. ANSI.

VITA. (2019). ANSI/VITA 65.0-2019—OpenVPX System Standard. ANSI.

Wertz, J. (Ed.). (1990). *Spacecraft attitude determination and control*. Kluwer Academic Publishers.

Chapter 6
Simulating Cyber-Physical Systems

Computer science inverts the normal. In normal science, you're given a world, and your job is to find out the rules. In computer science, you give the computer the rules, and it creates the world.
—Alan Kay

The design of game playing machines may seem at first an entertaining pastime rather than serious scientific study and indeed, many scientists, both amateur and professional, have made a hobby of this fascinating subject. There is, however, a serious side and significant purpose to such work….
—Claude Shannon

Abstract Engineers create machines that interact with the surrounding physical reality for different purposes such as delivering a payload in orbit, controlling a chemical process, etc. These machines execute control laws in the cyber/computer domain and use sensors and actuators to gauge and affect the surrounding environment. In the process of developing such machines, we have to create synthetic realities which numerically re-create the physical laws in computers and in the laboratory, which we use to verify the control laws accordingly. Luckily, digital control computers are not able to tell if they are operating in their real environment or in synthetic ones, provided we supply them with consistent information which accurately mimics the behavior of the real processes, both qualitatively and quantitatively. Ultimately, *we* tell computers what type of laws they are ought to control; they can't tell by themselves. Developing cyber-physical systems such as space systems requires a layered simulation-centric approach which will also need two- and three-dimensional visualization capabilities.

Keywords Cyber-physical systems · CPS · Digital control · Digital twin · Digital engineering · Digital mock-up · Simulation

In conventional data-processing systems, a computer manipulates symbols and numbers while it operates exclusively in cyberspace. It interacts with an intelligent human being in the physical world across a human–machine interface (HMI). A human–machine interface accepts the input data from a human operator and delivers the output data to this human operator. The human operator checks these outputs for plausibility before he uses them to initiate an action in the physical world. The

I. Chechile, *NewSpace Systems Engineering*,
https://doi.org/10.1007/978-3-030-66898-3_6

human actor is the main decider on how to affect the physical surroundings in this scheme. On the other hand, a cyber-physical system (CPS) consists of two subsystems, the physical process in the environment and the controlling computer system (the cyber system) that observes selected state variables of this process and controls this process by providing set-points to actuators that change values of quantities in the physical world. Often there is no human in the loop who observes the environment with his diverse human sense organs, develops a mental model of the ongoing activities, and provides an intelligent buffer between cyberspace and physical space (Kopetz 2018). The first and most distinctive difference between a conventional data-processing system and a CPS relates to the operation of a CPS that has the capability to autonomously modify the state of the physical world. The available sensors of the cyber system determine and limit the view that the cyber system has of the physical world. The cyber model that controls the processes in the physical world is thus constrained by this limited view and cannot detect changing properties of the physical environment caused by environmental dynamics that are outside the scope of the sensors provided by the system designer. An unforeseen environmental event or a failure of a single sensor can cause a serious malfunction of a CPS. This is in contrast to a human operator with his diversity of redundant biological sensors and a large intuitive knowledge base in his conceptual landscape. Modern space systems necessarily have a tight coupling between on-board cyber (processing, communication) and physical (sensing, actuation) elements to survive the harsh extraterrestrial environment and successfully complete ambitious missions (Klesh et al. 2012). A NewSpace organization designing and building spacecraft is actually designing and building cyber-physical systems.

6.1 Real and Synthetic Environments

One big challenge (among many others) of working with space systems is that we never get the chance to realistically assess that everything works according to the plan before flinging them to space. Or, put in a different way, we only get to realistically evaluate them, *after* we insert them in the real environment. This peculiarity stems from the fact that doing space is about designing artifacts for an environment we cannot fully replicate at the ground level. There is no chance of having a "staging environment", as it is a common practice in software. A staging environment is a supposedly high-fidelity replica of the environment where the application will run, put in place for software testing purposes. Staging environments are made to test code, builds, patches, and updates to ensure a good level of quality under a production-like environment before the application is rolled-out to the field. Things breaking in a staging environment doesn't affect customers, it helps the engineer understand what went wrong, and the users of the application in the real environment are happier because they get better software, at least ideally. The truth is a staging environment can mean different things to different people; differences between staging and production environments are usually present. Another industry that has the advantage of evaluating their products in realistic environments is aerospace. Aerospace designs and builds very complex and safe aircrafts under strict requirements. During a new aircraft program development, extensive amounts of flight hours take place to evaluate that the prototypes comply with the design specifications. But also, during the production phase, every brand-new manufactured aircraft is evaluated in their real environment before roll-out.

For space projects, the "production" environment is in orbit, several hundred or thousands of kilometers away from the ground; and a very hostile environment, with particles of all kinds and energies impacting the spacecraft's subsystems from multiple directions.

While we do space engineering, we can only devise staging environments which are somewhat poor replicas of the real environments; in other words, *simulations* of the real environment. Bonini's paradox strikes again: the very true nature of simulations we are ought to create is to be less complex (i.e. less accurate) than reality, otherwise there is no reason for the simulation. And here is where budgets make a difference: classic space industry can afford better quality simulations whereas NewSpace not as much, for a variety of reasons we will discuss in this chapter. Simulation environment is yet another area where the fracture between classic space and NewSpace ways can be clearly seen. NewSpace, being more software-intensive and bold than classic space, can take approaches that can be reasonably cheap and bring a lot of value to the development process. Flatsats, as discussed in a previous section, are the closest to "staging environments" space companies can get to. But, since these environments cannot replicate many of the real-conditions spacecraft seen in orbit (thermal, power, etc.), deployments in these kinds of environments cannot be fully taken as "verified". Software tested on flatsats or testbeds can be verified as the realism of the testbed permits. More details about simulation environments will be discussed in this chapter.

The unspoken truth in most NewSpace companies is that development environments extend all the way into space. This is mainly because things are never ready before launch, and because of the poor verification environments during the development stage, things can be found in orbit which may need patching or rethinking. This is massively different from classic space. In most of the NewSpace missions I had the luck to take part of, the flight software was being patched after the second or third pass after launch. Having good simulations is great, but technically not a killer to launch something to orbit that works. The important bit is to think about the things that must NOT fail. Once you get those things sorted out, the rest of the development can continue in orbit.

In this chapter, we will discuss how the need of simulation grows as the system design grows, how the design can be aligned with the simulation infrastructure, and what types of technologies and approaches can make the design process more "digital". Here with "digital" we mean creating a representation of the system of interest in computer environments for evaluating, to the best of our knowledge and the precision of those environments, if the design is correct to fulfill the mission needs, finishing with a sort of hyperbole that designing space systems and video games are kind of two sides of the same coin.

6.2 Test, Verification, and Validation of Space Systems

Now mind that in the previous section, I repeatedly and purposely used the word "check" instead of "test", or "verify", or "validate". This is yet-another engineering semantic confusion among many other you can find out there. And the deeper you dig, the bigger the confusion gets. Let's see what bibliography says:

From INCOSE Handbook (INCOSE 2015):

- Verification: The purpose of the verification process is to provide objective evidence that a system or system element fulfils its specified requirements and characteristics.
- Validation: The purpose of the validation process is to provide objective evidence that the system, when in use, fulfills its business or mission objectives and stakeholder requirements, achieving its intended use in its intended operational environment.

PMBoK (Project Management International 2017):

- Verification: It is the evaluation of whether a product, service, or system complies with a regulation, requirement, specification, or imposed condition. It is often an internal process. This function may involve specific tests and inspections, as well as evaluating any test records and inspection logs completed during the course of the work. At the end of this process you have verified deliverables.
- Validation: It is the assurance that a product, service, or system meets the needs of the customer and other identified stakeholders. It often involves acceptance and suitability with external customers.

Few takeaways from above. INCOSE Handbook states that **validation** is about "*achieving its intended use in its intended operational environment*". Does this imply then that verification does not really need to be done in its *intended operational environment*? Example: We are called to design a deployable big antenna for a comms spacecraft. There will surely be a lot of requirements to fulfill, such as mass, stiffness, mechanical frequency response, thermal, etc. Since we cannot have access to the real environment, we can only partially verify requirements. For example, there is no way to re-create at ground level the zero-gravity conditions needed to properly assess a deployable appendage design works as intended, so engineers need to use jigs, helium balloons, etc., as "hacks" in order to re-create, to some extent, the conditions the design will meet in the real environment, and gain confidence (which is all they can hope for).

Now regarding how the word "test" enters the scene. PMBoK states (Project Management International 2017): "*[verification] may involve specific tests and inspections, as well as evaluating any test records and inspection logs completed during the course of the work*". There is a great deal of information in that otherwise anodyne sentence. This means that even if we had access to the real environment, that does not solve all our problems. We also need to come up with the strategy and the means to gauge our designs in order to see if they comply (or not) to what was initially required. By "coming up with scenarios" here I mean analyzing and synthesizing the evaluation cases, their scope, and the method of evaluating or gauging: inspection, analysis, or testing. This strategy requires design as well, which means everything we have said already about design in all previous chapters applies here as well.

This, of course, means we can make mistakes in designing the verification methods; then, the reader could in all fairness ask: who verifies the verification methods?

The amount of test, analysis, or inspection scenarios we can come up with is, of course, finite. Our coverage (i.e. the amount of cases we can come up with to evaluate a design vs the amount and variety of possible scenarios the system intrinsically has) always runs behind. In plain words: we can only partially evaluate what we design. A corollary can also be: the verification system is yet another system we plug to the system-of-interest.

Question is, what happens to what is left unevaluated?

All this impacts the overall development process of space systems: a great deal of effort, resources, and time needs to be allocated for designing and building the verification infrastructure.

6.3 Simulation

I always find quite amusing the metaphor of comparing brains to computers, or computers to brains. Such comparison is preposterous: computers are, luckily for us, rather stupid. This means, they are very easy to fool. The fact they get faster but not necessarily smarter plays in our favor here. For example, imagine a computer controlling a chemical process in some industrial plant. Monitoring a chemical process for sure will need reading temperatures, pressures, flows, etc. For the process to be controlled, the computer needs to read those kinds of variables of interest from the physical environment. This requires, first, converting the real physical quantities into either voltages or currents that the computer will turn into digital. Then, the control system governing the plant will process those converted readings according to some criteria, and it will apply the actuation needed to correct potential deviations from a desired set-point. The control criteria usually require monitoring the output variable and comparing it to the set-point; the error between those two defines the actuation strategy. This is closed-loop control or feedback control; we spoke about this before. Closed-loop control heavily relies on information flowing back and forth, information that is used for decision-making. Note from Fig. 6.1 the division between the computer and the physical boundaries right at the sensor and actuator layer.

The typical building blocks of a generic cyber-physical system are:

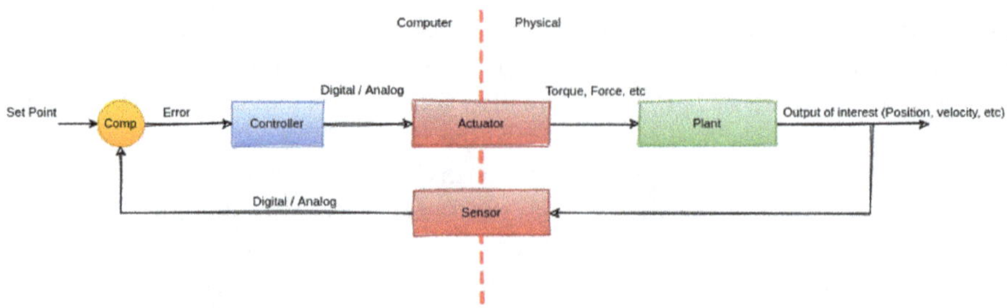

Fig. 6.1 A generic closed-loop digital control system (also called a cyber-physical system, or CPS)

- **Comparator**: It provides a signal which is proportional to the difference between the set-point and the output to be controlled.
  - Input: The desired set-point, which is the ultimate objective of the control system; i.e. the desired value the control system is expected to reach. The set-point is in the computer domain, and can be either analog or digital.
  - Output: The signal which is the difference between the set-point and the measured output. This signal is in the computer domain, and can be digital or analog.
- **Controller**: It computes the actuation needed for the output of the plant to reach the set-point, within some margin. The controller actuates not only upon the immediate value of the error signal but also on its variations and longtime evolution. The controller lives fully in the computer domain.
  - Input: signal from the comparator
  - Output: actuation criteria.
- **Actuator**: It affects the physical property according to some control signal. The actuator is a mixed-domain device: partly computer-domain and partly physical-domain.
  - Input: Signal command which quantifies the amount and type of actuation. This signal is in the computer-domain, and can be analog or digital.
  - Output: A physical quantity which will ultimately interact with the environment (heat, force, torque, etc.).
  - Actuators are never perfect, and they contain nonlinearities, dead bands, and failure modes.
- **Plant**: This is the actual physical process to control. The plant resides fully in the physical domain or physical universe.
  - Input: Physical variable to alter in order to make the plant transition from an initial state to another state.
  - Output: Physical quantity to keep under certain desired margins.

- **Sensor**: It observes a physical property and provides a signal proportional to the observation. The sensor is a mixed-domain device: partly computer domain and partly physical domain.
  - Input: The physical variable to observe; this is in the physical domain.
  - Output: A signal which is proportional to the physical variable observed. This is in the computer domain, and can be analog or digital.
  - Note that real sensors are far from perfect and they show biases, noises, nonlinearities, and multiple types of failure modes. Some sensors are complex subsystems composed with digital embedded computers and software, which means they are susceptible to bugs, bit-flips, resets, etc.

A digital control computer produces and consumes numbers; typically, it reads and writes such numbers in a memory or in processor registers. The numbers the control software needs to consume can come from actual sensor readings, or they could be numbers manually written by us: the computer will not be able to tell the difference. But what if we had a device capable of creating consistent but *bogus* values in a way we could inject them into the control computer replacing the ones coming from the real sensors? Would the computer be able to recognize they are coming from our device, i.e. artificially created, instead of coming from the real physical process? The answer is: no, it would not be able to tell the difference. And this is the core of what simulation does, in a nutshell. The space industry applies this approach for evaluating spacecraft software before being shot to space, but any organization dealing with cyber-physical systems (missiles, drones, rockets, cars, chemical plants) need to do the same. For space, the development process requires adding different bogus numbers generators to inject consistent values in some specific inputs of the spacecraft computers to make the system believe it is in space whilst it is sitting in a clean room. This does not come without challenges though, since it exerts a great deal of pressure on the synthetic simulation environments for them to be realistic and precise, which can turn them to be extremely complex and hard to maintain. Moreover, if only the problem would be replacing those numbers in memory, the issue could be perhaps simpler. The real pain is that control software quickly detaches from the "number producer/consumer" approach and it grows too aware of the underlying architecture, and this dependency becomes problematic when such underlying architecture is not always present upfront when the development starts. The number producer/consumer is an oversimplification as well. The way those numbers are refreshed in memory (time delays) can affect the way the control laws work. In short, the control software "functional coupling" to the physical environment is multilayered and highly context-dependent. The number of things in the middle between the control laws and the environment is staggering. No wonder it is very cumbersome to transition control software from pure software environments to operative environments (Fig. 6.2).

Classic space invests millions in complex and highly realistic simulation systems, whereas NewSpace tends to take a radically different approach, usually going for lightweight and ultra-low-cost approaches. Classic space simulation environments and ground support equipment (GSE) can take entire rooms in facilities,

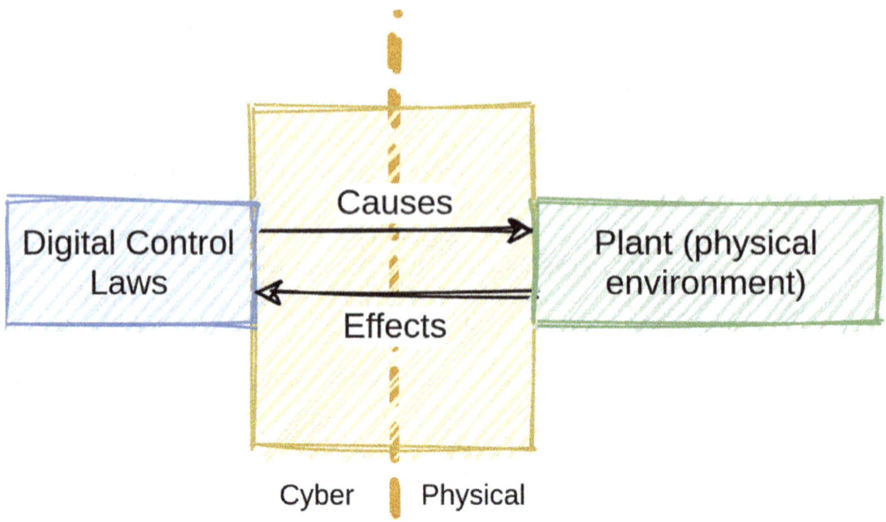

Fig. 6.2 A cyber-physical system at its highest level of abstraction. Between the digital control laws and the physical environment, there are multiple layers of things

comprising many racks, industrial computers and power supplies, imposing considerably logistic challenges during launch campaigns. In contrast, and largely due to budgetary constraints, NewSpace simulation environments run in small, cheap embedded devices, and often fit in a small suitcase (Fig. 6.3). We will discuss the pros and cons of these two extreme approaches in this chapter.

Just as computers cannot tell if a value in memory represents an actual reading from a physical variable or if it was manually generated by us, software does not really care on top of what platform it is running, provided all dependencies are met (libraries, headers, system calls, etc.). If we manage to meet all those dependencies, we can eventually make a piece of software run just anywhere. This is what is called portability, and somehow the reasoning behind virtualization, containers, etc. A piece of software responsible for controlling a spacecraft runs on very specialized computers, which are robustized for space, mission-critical applications, etc. But, as it is just software, we can manage to make it run on more regular computers, like a cheap laptop. The key here is, again dependencies. There are two levels of dependencies: compile time and runtime. One thing is to fool software to make it compile; a different story is to fool it to make it run and run consistently (i.e. with good numerical results). If only the former is accomplished, most likely it will crash. At compile time, the software is basically expecting pieces of a puzzle such as more source code, compiled libraries, etc. At runtime, the software is expecting certain values in the right places and for that it expects to use services from the underlying software layers, such as an operating system (if there is one, this is optional). If any of that is missing, the software will just not run. What is more, for software that runs intensive numerical computations, the underlying architecture can make a

Fig. 6.3 Simulation environments in NewSpace often run in small embedded computers (*credit* "Raspberry Pi" by Flickr user Gijs Peijs and licensed under CC BY 2.0)

big difference: data types, floating point unit precision, data widths, all this needs to be consistent between the final platform (the on-board computer) and the testing environment.

When the control system is out in its real environment, functional consistency is provided by the physical world itself. But if we are to run the control software in a simulated *reality*, then it means we will have to replace the physical world (or the numbers coming from that physical world) with some sort of artificial replacement. This means we will have to interface the control software to other software to provide those services and values. Let's diagrammatically see the difference. In the real environment, a cyber-physical system employs the next building blocks, as minimum (Fig. 6.4).

From Fig. 6.4, the legend "Air gap" is meant to illustrate that the control computer and the sensors and actuators are physically separate entities. If we now use blocks and terminology relevant for a spacecraft attitude control subsystem, the equivalent block diagram looks like in Fig. 6.5.

Note that in Fig. 6.5, there is also a thin layer between the drivers and the input/output hardware (IO HW) which runs on the control computer. This layer is usually called board-support package (BSP), and it is in charge, among other things, of defining the routing of pins and configuration of processor peripherals before software can safely run.

As it can be seen, a digital control system is a layered system, or a stack. The outer layers are the ones which reach closer to the physical universe, and it gets inwards

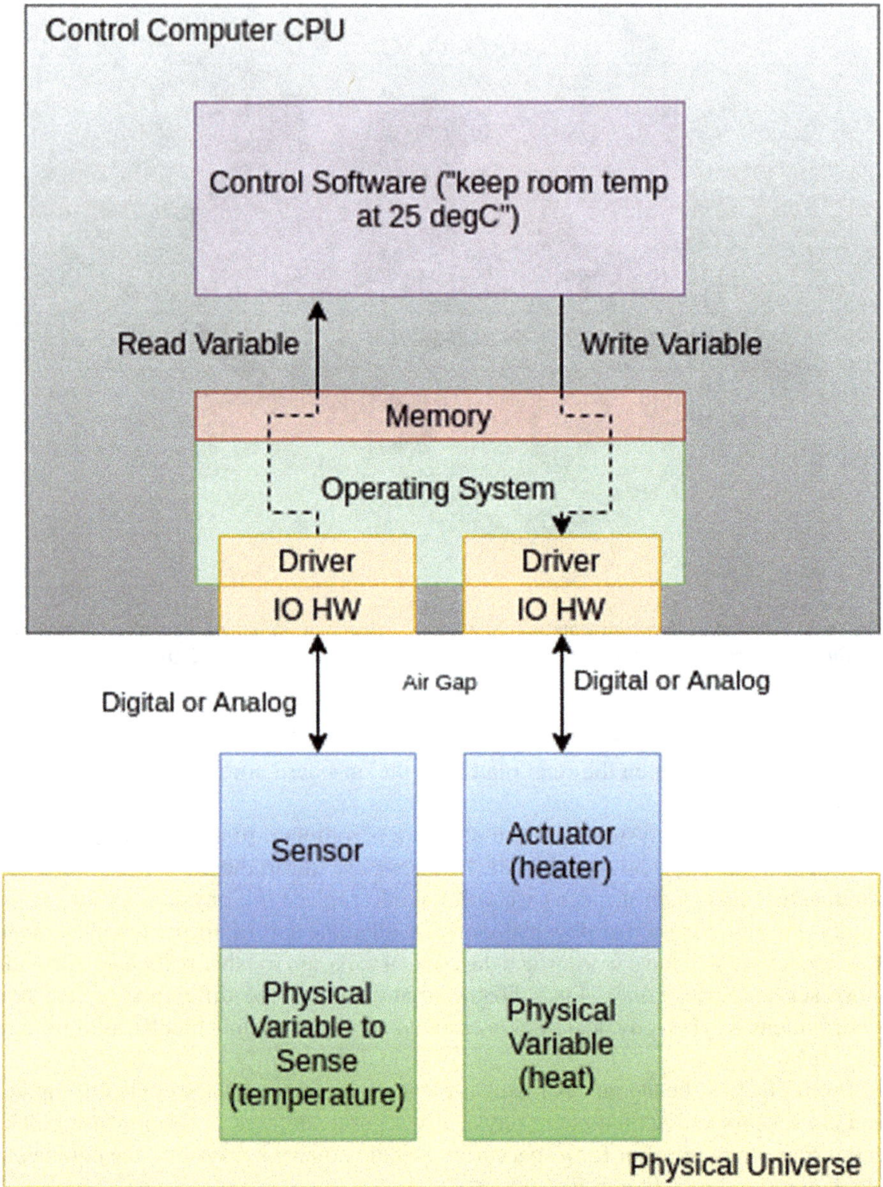

Fig. 6.4 Building blocks of an example digital thermal control system in its operational environment (Type I)

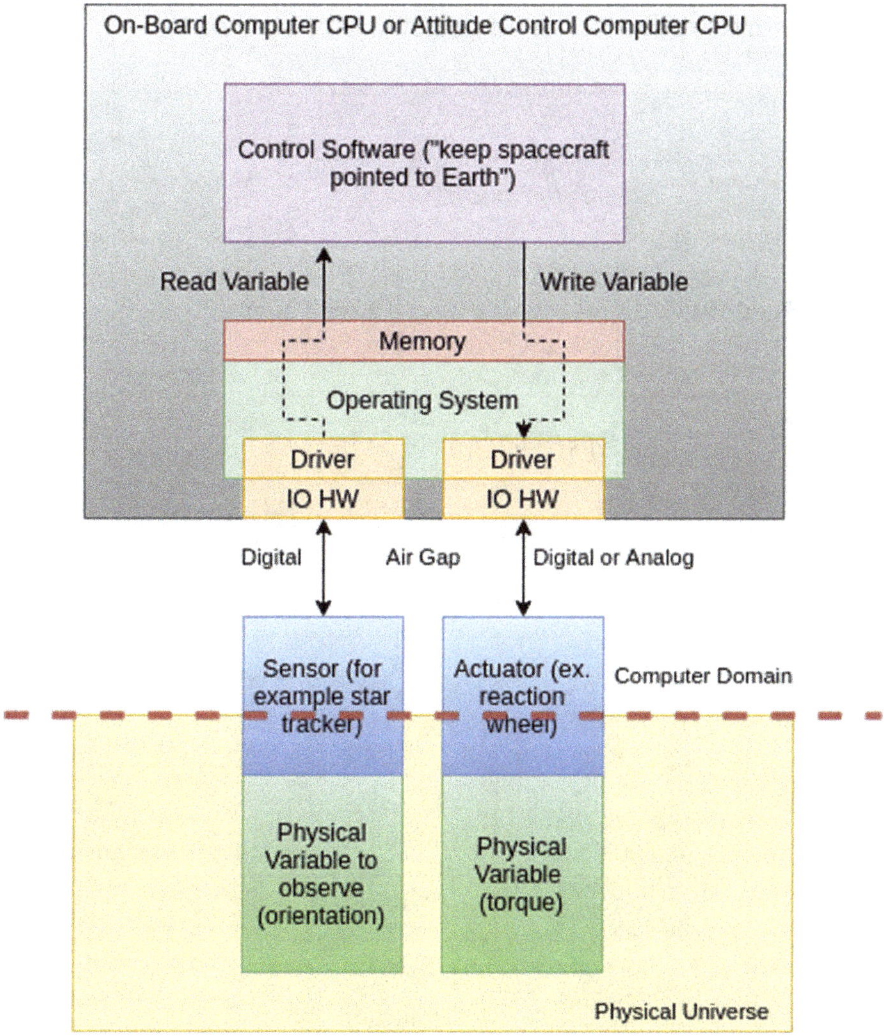

Fig. 6.5 Building blocks of a spacecraft control system in its real operational environment (Type I)

toward the computer domain. In order to fool the control computer and make it believe it is running as in its operational environment, there are different layers we can start replacing with simulators. As stated before, the complexity of the simulation infrastructure grows with the system of interest. But we will start the opposite way: by analyzing how the control system works in its operational environment at its full extent, and then from there replacing bit by bit toward a full digital simulator version. For the sake of clarity of diagrams, we will use one sensor (a star tracker) and one actuator (reaction wheel). In Fig. 6.6, we replace the "Physical Universe"

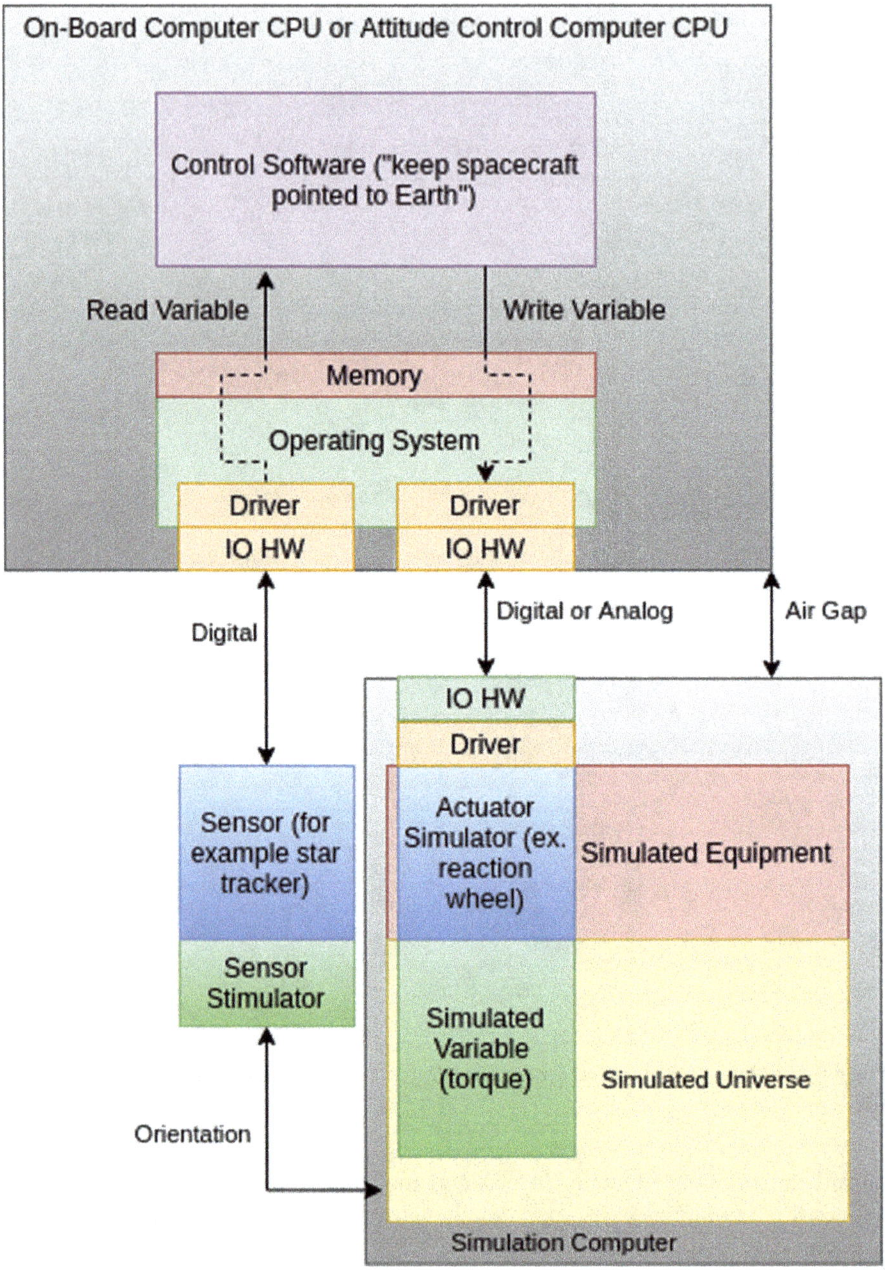

Fig. 6.6 Hardware-in-the-loop simulation, when feedback stops coming from real physical universe (Type II)

by a simulated universe. This sounds a bit too presumptuous, but, is rather simple. It is enough to simulate what the spacecraft needs to know about the universe around it. This is: a position in an orbit, and a physics engine which will re-create the same reaction to torques and forces as a satellite in orbit would experience. This is very similar to physics engines in video games. A computational simulation of physics laws is a numerical representation of causes and effects which mimics what happens in the physical world. More simply put: in reality a torque acting on a body axis will most likely make it rotate due to acceleration and speed. In a computer environment, all this happens in a spreadsheet-like environment where the torque is input as a number, and accelerations and speeds are calculated as outputs. As long as we feed these numbers to the control system consistently, it has no way of knowing if it comes from a real universe or a spreadsheet. Computer stupidity at its best.

At the same time, let's assume we do not have the real actuator (reaction wheel), so since we must continue our work, we replace it for a simulated one. This is a software routine that reads the commands from the control computer (using still the same interfaces), and it outputs a simulated torque which is fed to the simulated universe. In this example, we decide to keep the real sensor (star tracker) but we stimulate it with simulated data coming from the simulated universe in order to close the loop. This type of simulation approach is called "hardware in the loop" since it contains most of the hardware that will eventually go to orbit. It usually comes in the latest stages of development, when all the hardware and subsystems are procured. There are many different "depths" of HiL simulation, as we will see in the next step.

Now a quick mention on form factor. Note that in Figs. 6.5 and 6.6 we have not specified how these elements are laid out. Therefore, the control computer could be installed inside an actual spacecraft body, or it could be sitting on a table; for simulation infrastructure, it does not really make a difference how things are laid out, but just about how things are connected and how signals flow. In general, flight computers and subsystems start to take the final form as the development matures. Flatsats become engineering models (i.e. "Iron birds"), and eventually engineering models become flight models.

Then, let's assume we do not have a real start tracker (sensor), so we replace it with a numerical simulator (Fig. 6.7). This simulator reads the orientation from the simulated universe and it outputs toward the control software using the exact same interface as in the actual system (this can be MIL-1553, RS-422, SpaceWire, etc.). Note that the control software might need to command the star tracker as well (star trackers can have different modes, configurations, etc.), so these commands should be implemented by the sensor simulator for full fidelity simulation. This is when things between simulation and reality start to depart from each other. Note that, so far, the control software could have been running without a single clue and some of the parts had been replaced by simulated counterparts. But now, if the simulated sensor will not be able to receive all commands as the real one (for the sake of simplicity), then this could impact the way the control software works. For example, if the control software execution depends at some point of the result of a command to the star tracker, if that command is not implemented in the sensor simulator, then the control software might see its execution affected.

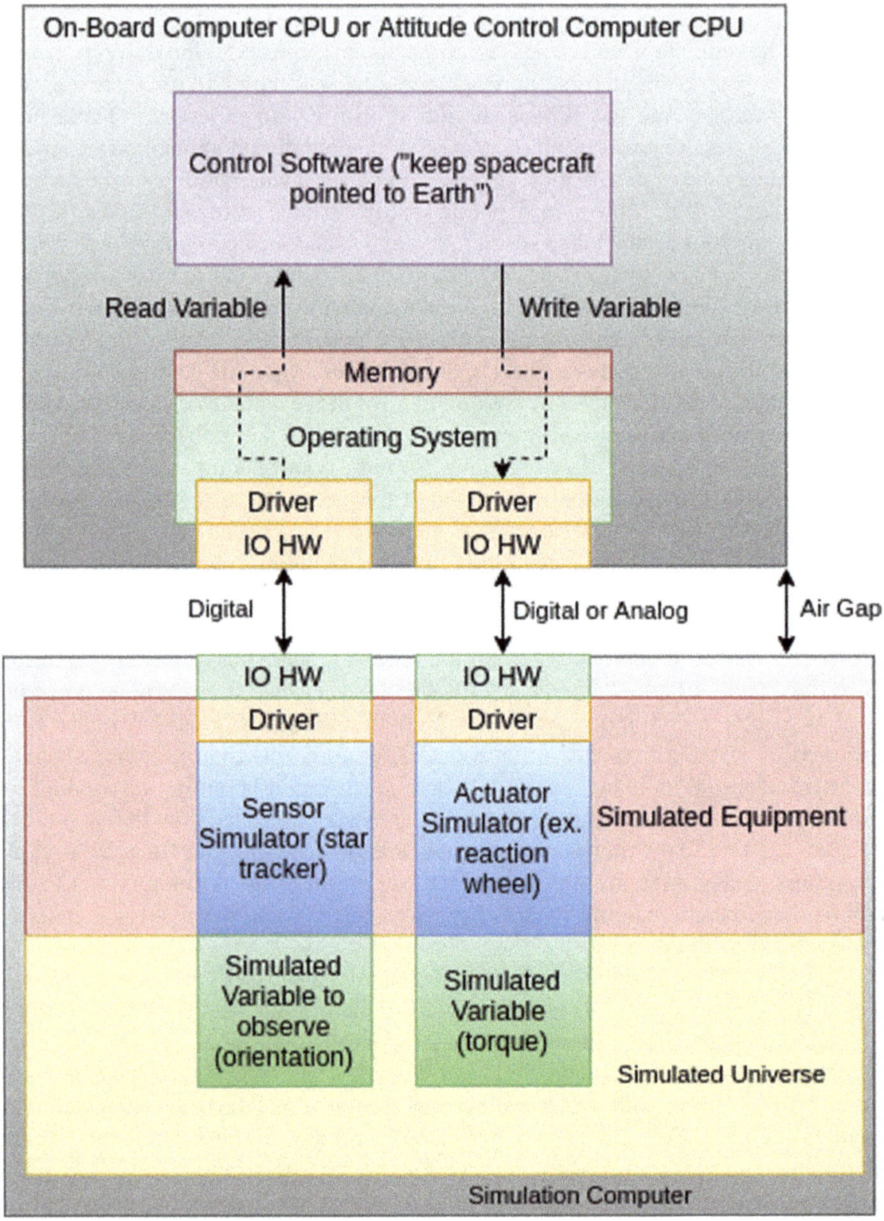

Fig. 6.7 HiL simulation with equipment fully simulated, still using real interfaces and two computers separated by an air gap (Type III)

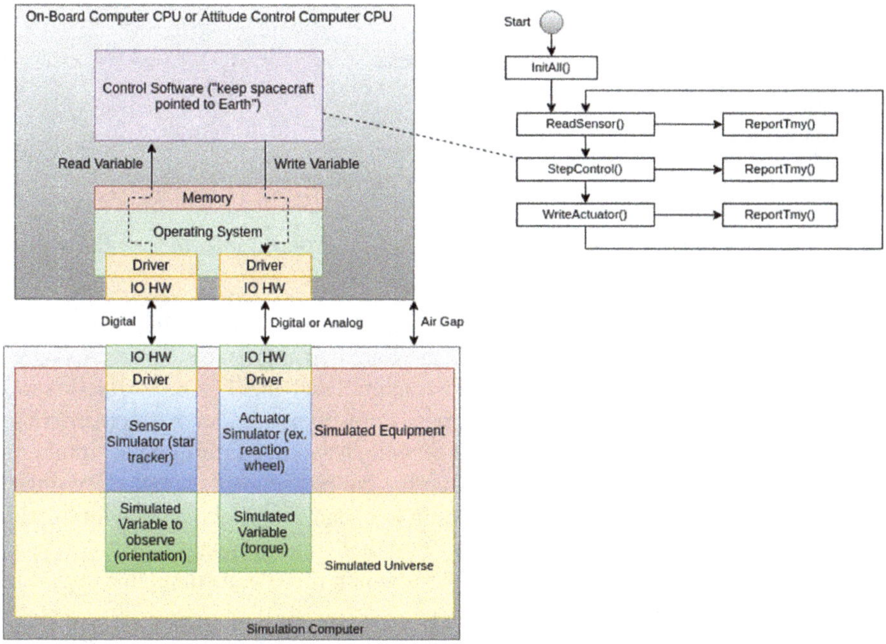

Fig. 6.8 Control loop typical functions

The least realistic the simulators will be, the higher the impact on the control software, the more extensive the changes it will have to withstand in order to make it work, hence the farthest it departs from the version which will fly. This is a trade-off which will be pervasive while developing simulation infrastructure. A simplified depiction of the functions the control software will run is (Fig. 6.8):

As depicted from the figure, the most salient functions are:

- InitAll()
  - Sets the system for work. Initializes memory and drivers.
- ReadSensor()
  - This function provides all readings from all sensors present. This includes both digital as analog.
  - It reaches the sensors through drivers.
  - Optionally includes a call to ReportTmy() in order to inform how the reading of the sensors went.
- StepControl()
  - This function computes the corrections needed in order to match a desired set-point (for spacecraft control, this means a specific orientation in three-axis in some frame of reference).

   - Optionally includes a call to ReportTmy() in order to inform how the step of the control loop went, if there were errors, etc.
- WriteActuator()
   - This function writes the torques and forces calculated from StepControl() into all actuators present. This can be either digital or analog.
   - It reaches the actuators through drivers.
   - Optionally includes a call to ReportTmy() in order to inform how the writing of actuators went.

If no real sensors nor actuators are connected in the loop, such equipment must be simulated with numerical simulators. And there are two approaches for this: using the same interfaces as in the real thing (Fig. 6.7) or simulated one (Fig. 6.9). Classic space tends to use the first approach, and NewSpace tends to use the second, for simplicity and cost reasons. The main disadvantage of using an ad hoc interface for simulation like in Fig. 6.9 is that all the driver layer on the control software is bypassed during functional verification. The advantage is simplicity and cost effectiveness. Simulating equipment using the real interfaces requires signal conditioning for analog sensors and disparate communication I/O equipment which usually accumulates in various rack-mounted cases (see again Fig. 6.3a).

One big impact for the simulation infrastructure is when the on-board computer (OBC) is still not available (due to lead time issues or similar). Since the software development cannot stop to wait for the CPU to arrive, this means that all the software stack needs to run elsewhere in order to permit the overall implementation and verification effort to carry on. There are two approaches to this:

- Run it on an emulator of the OBC CPU (if available)
- Port it to other architecture if emulator is not available.

If we happen to find an emulator for the CPU that our on-board computer uses, then this means we can install on it both the control software and the operating system. The operating system will most likely have to be modified to boot on the emulator, given that there will be for sure differences between the actual board and the emulated hardware. But more changes will also have to be made. For example, the drivers that were communicating with real interfaces in the HiL scenario will no longer run, since there is no underlying hardware to support that. If this driver was for a communication board (e.g. MIL-1553 or CAN), then the simulated drivers will have to take the data coming from the control software and pass for example to a shared memory (Fig. 6.10). This shared memory will also be accessed by the simulated equipment. If the real driver is for an analog input/output, the new driver will have to provide simulated analog-to-digital (ADC) or digital-to-analog (DAC) conversions in a way the control software will be affected the least. There will be impact for sure, because the real driver surely had low-level mechanisms as interrupts that the simulated driver will not have. If the CPU emulator is not available (some legacy space systems work with very old processors with architectures that are very different to the ones that are ubiquitous in space nowadays), then there are two options: (a)

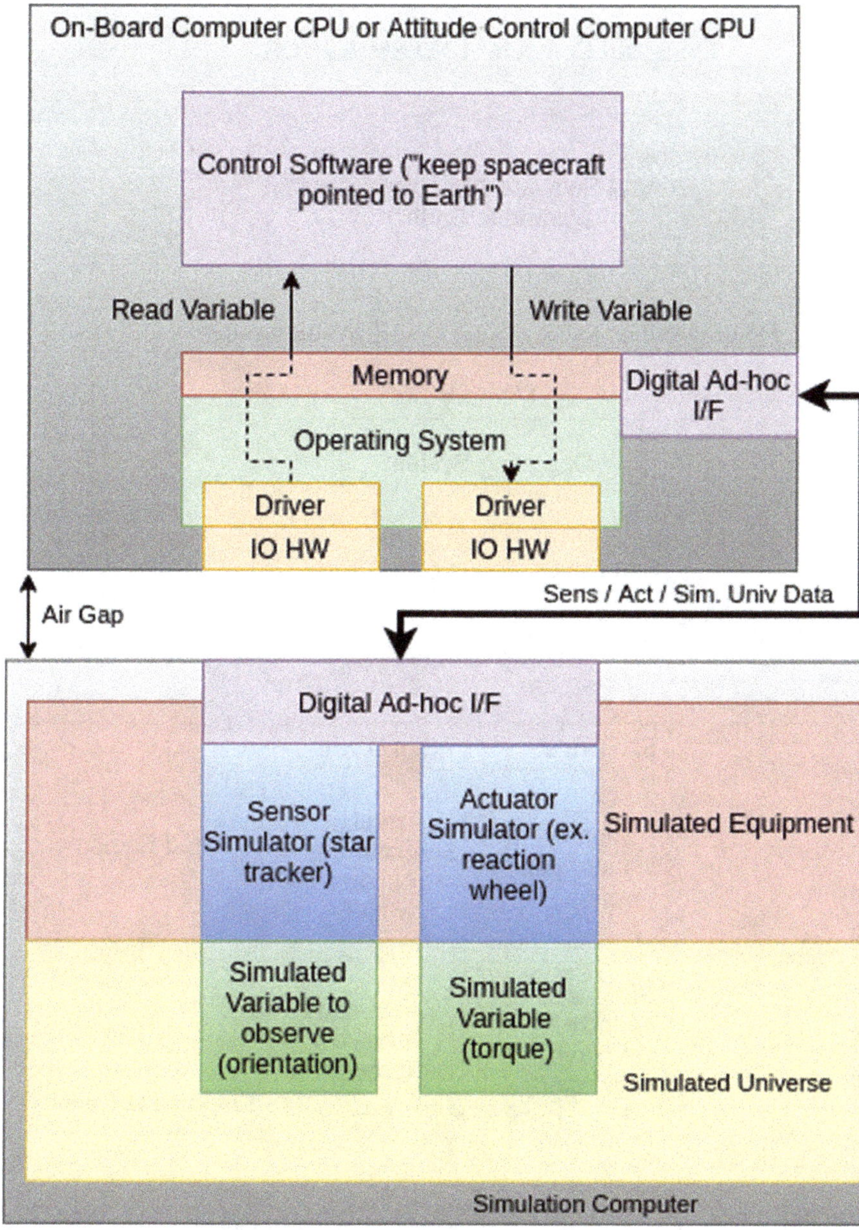

Fig. 6.9 Using an ad hoc digital interface for equipment simulation (Type IV)

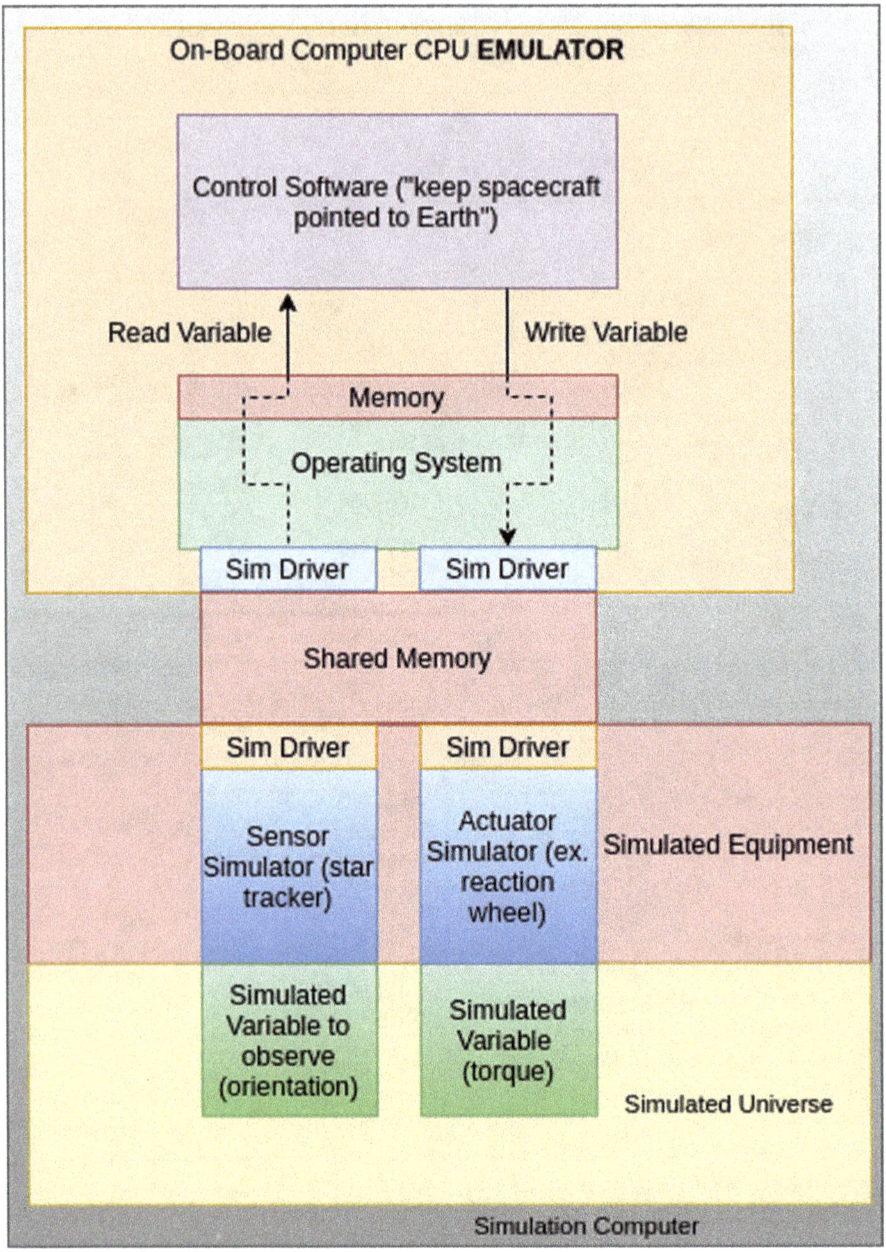

Fig. 6.10 Software-in-the-loop simulation, with control software running on a CPU emulator (Type V). Note that no more air gaps exist; all this runs on the same computing host. Also note that a shared memory is just an example. It could be any other IPC (interprocess communication) mechanism

Create an emulator from scratch (highly discouraged, this can take a long time and be very complex), or port the control software to another architecture. For the latter, it is usually needed to port it to the host architecture where the control design activity takes place, which usually is a workstation. Porting the control software to a workstation comes with a myriad of challenges, since it requires cross-compiling from the OBC original architecture to the workstation architecture. The control software now must run on the host workstation operating system, so all the calls in the original software which do not match the signature of the host operating system must be modified. Also, since now everything is software and running on top of the same CPU, many middle-layers become pointless, like the driver layer. The control software can now exchange information directly with the simulated universe through variables in a shared memory (if they run in separate processes), or just variables in memory if they run inside the same process but as different threads. This direct transfer of data between control software and the simulated universe has better performance for running accelerated time simulations (one key feature in order to simulate scenarios for the whole mission lifetime which can span several years). Note that for Type V versions, there could be an air gap between the emulated CPU and the simulated universe, for example if both are connected using TCP, UDP, or similar.

The way we have presented all these simulation scenarios was in a backward manner, as stated before. The reason for that was to show how to transition from a digital control system in its intended operating environment all the way to the designer's own simulation environment PC. In that transition, plenty of modifications must be done on the software in order to make it believe it is running in space. The changes introduced greatly impact reliability since big chunks of code must be conditionally compiled or removed in order to work in a changing context (Fig. 6.11).

It is worth noting that there could be one subtype (Type VIa) which is very similar to Type VI but the simulated universe and the control software do not run on the same host machine. Note that the hosts can be either two different physical computers connected by means of some TCP/IP connection over Ethernet or two virtual machines communicating through raw sockets or similar. Note how the Types VI and VIa were totally stripped from any driver layers and the software entities talk very straight to each other (Fig. 6.12). We can take one final step in terms of abstraction levels and define a simulation environment where the abstraction is so high that the control software and the simulation environment do not utilize sensors and actuators but just exchange physical variables back and forth; this is depicted in Fig. 6.13. This is the most minimalistic simulation environment we can come up with, where the control software and the simulated universe exchange information in simulated physical values, bypassing the sensor and actuators layer. This is usually called algorithm-in-the-loop and it is very similar to what you could get by using some MATLAB or Octave model or similar. This type of configuration is used at very early stages of the control design process, where even sensors and actuators are not chosen yet. This environment enables preliminary assessments about solar panel configurations, power budget analysis, and similar. In this very lightweight simulation environment, the fact no sensors and actuators are present means the control software "cheats" in order to read and write physical variables directly from

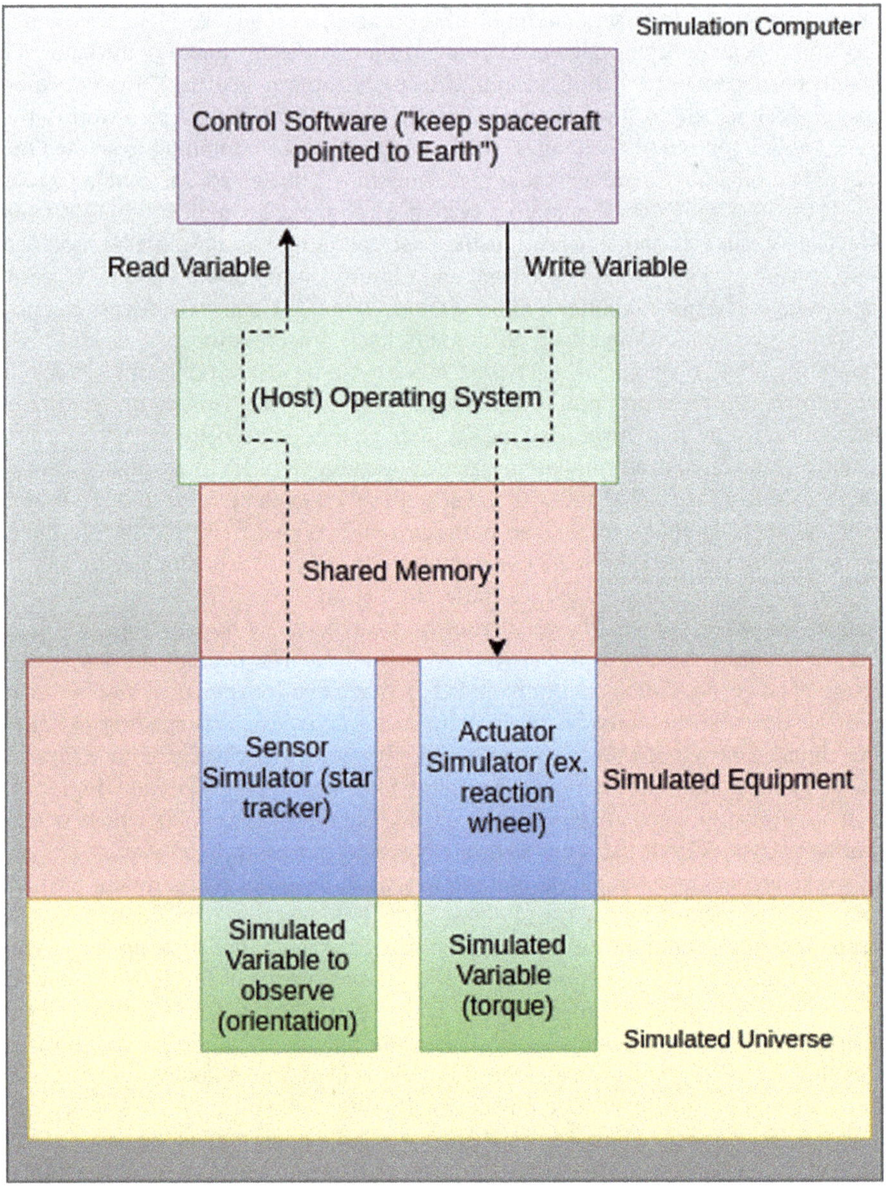

Fig. 6.11 Software-in-the-loop simulation, with control software running on the same processor CPU as simulated equipment and universe (no emulator); Type VI. Note that equipment simulation and control software can exchange data directly through shared memory or similar IPC schemes

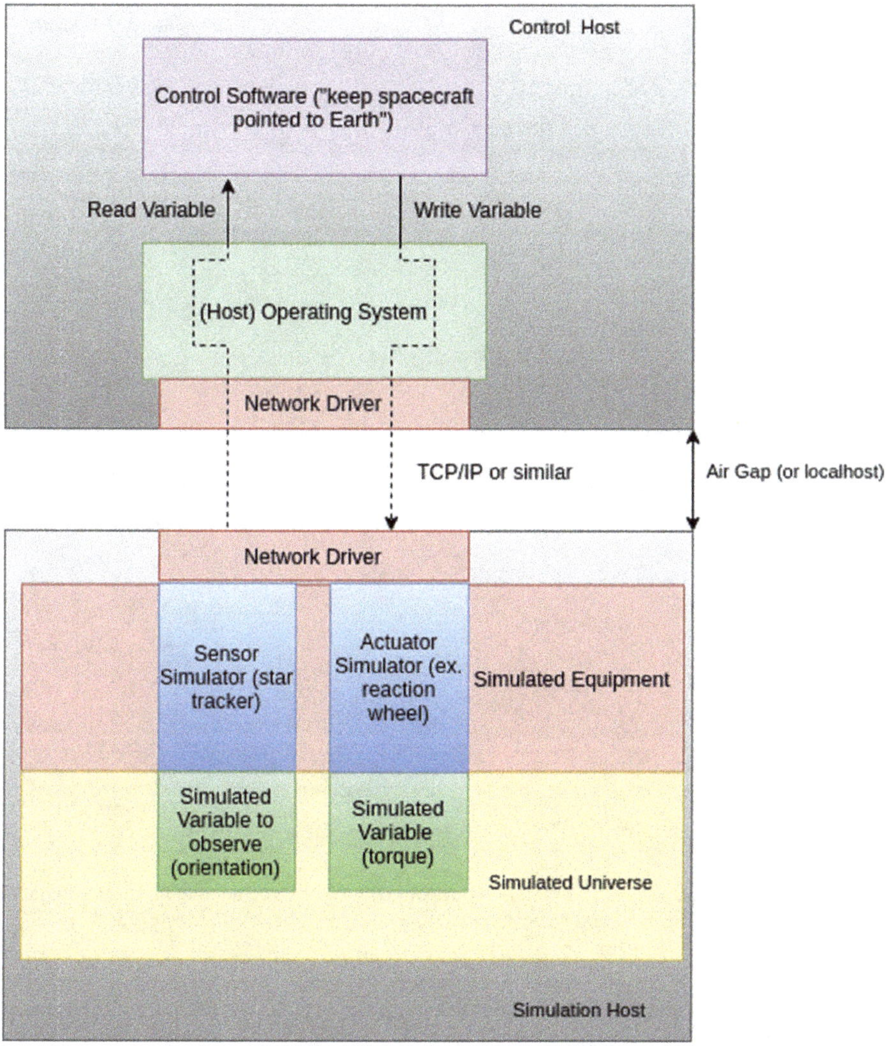

Fig. 6.12 Software-in-the-loop with control software and simulation environment running on different hosts. Note that this can also run on one single computer using localhost or virtual machines (Type VIa)

the simulated universe. This is impossible in real environments: sensors and actuators are our only means to interact with the physical universe around us.

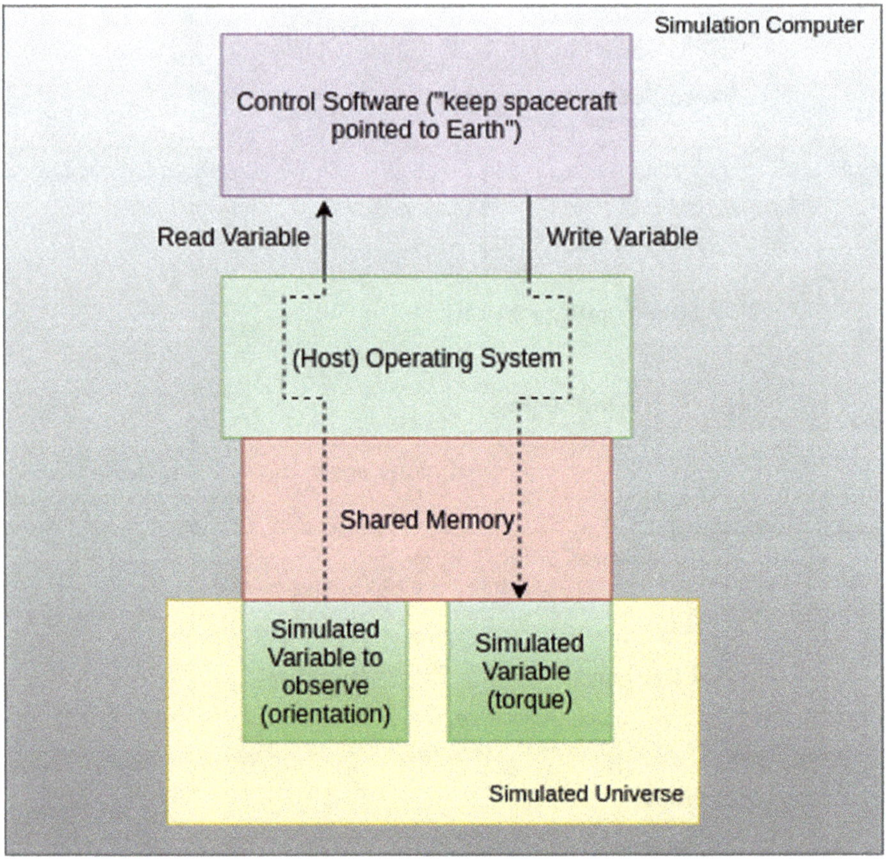

Fig. 6.13 The most minimalistic simulation environment. Note there is no equipment simulation layer. Control algorithms read and write simulated physical parameters directly (Type VII)

6.3.1 Simulation-Centric Design

As seen, developing software for digital control or cyber-physical systems involves multiple steps to transition it from purely software environments all the way to the target operative environment. The approach of adding simulation/emulation layers to the control algorithms can only be done properly if it is added as part of the software design process. This means, in order to make the control software *fooling* process more straightforward and reliable, it is key to design it in a way its architecture is aware of the multiple transitions from Type VI (fully software simulation, running on a host workstation) down to Type I (a control system running on its real environment). In other words, from a regular laptop all the way up to space. Moving from one type of simulation scenario to the next, one must be as seamless as possible, to reduce the amount of refactoring and changes introduced to the code when transitioning

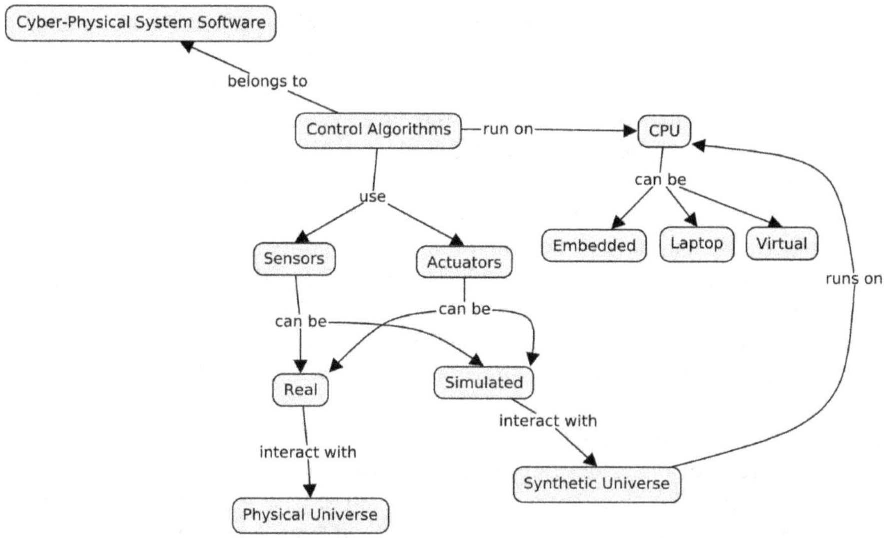

Fig. 6.14 Concept map of a digital control system and the simulation environment

between steps. Ideally, refactoring between simulation stages or types should be zero. The configuration decisions and selections to make the seamless progression of the software scenarios to happen requires a combination compile time (in most cases) and some runtime modular approaches. For example, conditional compilation is typically used to select which layers the control algorithms will be compiled with. Runtime approaches can also be used, for example using initialization files to select the way the control algorithms will communicate with the rest of the layers (Fig. 6.14).

6.3.2 Physics and Spacecraft Simulation Engines

For the control laws of a cyber-physical system to converge, the algorithms need to be fed with consistent data, regardless of whether running in the real or synthetic environment. Control algorithms could not care less. When running in a synthetic environment, this means that the simulated sensors and actuators must be fed with data matching replicas of the physics laws. Otherwise the control software will quickly detect inconsistent physical behavior as a failure and change its course of action. To simulate the physics, basically we must create a digital equivalent of a mass orbiting another mass (spacecraft orbiting Earth if we assume an Earth orbiting mission). This per se is not enough, since we must also allow the mass to rotate around its axes following torques applied. Forces acting on this mass should (even though this might not be a hard requirement) translate the mass in inertial space and allow it to describe a different orbit. We must add perturbations to the orbits described, as it happens in the physical universe. Since orbit perturbations are mostly a product of environmental factors (magnetic field, gravity contribution of other bodies, atmospheric drag, etc.) this means we will have to re-create all those environmental effects and make them affect and alter our mass's orientation in three axes but also its position in orbit. We need to keep track of the position of the Sun as well, in some frame of reference, for power and thermal analyses. At the same time, a single mass point is usually not enough for conceptual design of spacecraft. Evaluating antennae, solar panels, and payload orientation is of key importance throughout spacecraft design, hence the simulation environment must be able to support creating multibody structures which can also show both rigid and flexible behavior, in order to fine tune control laws. It can be already seen that basically a "physics engine" must be devised. It is typical that NewSpace organizations jump to grow this capability from scratch. This can take a very long time and a great deal of development effort, and it is at the same time a double-edge sword: if the simulation environment and the system of interest are both under development, this means having two immature systems coupled together. The simulation environment's first goal is to permit assessment, evaluation and testing of the system of interest. If both are being developed, the level of confidence in the simulation environment is low, hence results found during verification cannot be fully trusted unless such trust in the simulation environment is gained in due time. In short, who verifies the verification environment?

To solve this challenge of the two systems being developed at once, NewSpace organizations must rely on proven and existing "physics engines" and avoid growing them from scratch. There are great open-source options that NewSpace companies can use to jumpstart their designs, notably 42 (fortytwo[1]).

6.3.2.1 The Answer Is Always 42

Fortytwo (NASA Software Designation GSC-16720-1) is an open-source, very comprehensive yet lightweight, general-purpose simulation environment of attitude and trajectory dynamics and control of multiple spacecraft composed of multiple rigid or flexible bodies, created by Eric Stoneking. Fortytwo is a "Type VI" full software design tool which accurately re-creates spacecraft attitude, orbit dynamics, and environmental models. The environment models include ephemerides for all planets and major moons in the solar system. Fortytwo is so complete that it also supports 3D visualization through a lightweight OpenGL interface (which can be disabled if needed). The simulator is open-source and portable across computing platforms, making it customizable and extensible. It is written to support the entire design cycle of spacecraft attitude control, from rapid prototyping and design analysis, to high-fidelity flight code verification (Stoneking 2010–2020). Fortytwo can run fully standalone or be integrated to other environments as a member of a distributed simulation environment by means of its IPC (interprocess communication) facilities using a set of TCP/IP sockets. Fortytwo can be integrated to external tools for extending

[1] https://software.nasa.gov/software/GSC-16720-1.

subsystem simulation such as power subsystem (e.g. feeding it to a spice model using NGSPICE or similar), radio link subsystem, etc. This can be done as a post-processing step (ingesting 42 data after simulation is done) or co-simulation. Fortytwo can also be used as the physics engine of real-time hardware-in-the-loop (HiL) environments. **42 is released under NASA's Open Source Software Agreement.**

Features

Fortytwo is highly configurable, which allows the user to create from rather simple, rigid body single-satellite scenarios all the way to constellations of multibody spacecraft. A summary of features is listed next.

- Multiple spacecraft, anywhere in the solar system
  - Two-body, three-body orbit dynamics (with seamless transition between)
  - One sun, nine planets, 45 major moons
  - Minor bodies (comets and asteroids) added as needed
  - RQ36, Eros, Itokawa, Wirtanen, etc.
- Multibody spacecraft
  - Tree topology
  - Kane's dynamics formulation
  - Each body may be rigid or flexible
  - Flexible parameters taken (by m-file script) from Nastran output (.f06 file)
  - Joints may have any combination of rotational and translational degrees of freedom.

More features:

- Supports precision formation flying
  - Several S/C may be tied to a common reference orbit
  - Encke's method or Euler-Hill equations used to propagate relative orbit states

    Precision maintained by judicious partitioning of dynamics

      Add big things to big things, small things to small things
- Clean FSW interface facilitates FSW validation
- Open Source, available on sourceforge and github
  - Sourceforge.net/projects/fortytwospacecraftsimulation
  - Github/ericstoneking/42

Architecture

One of the outstanding features of 42 is that it is a self-contained software. It does not contain strange dependencies to obscure libraries nor abandoned projects; it contains in itself all it needs to run; building 42 is very straightforward. Figure 6.15 depicts 42 architecture.

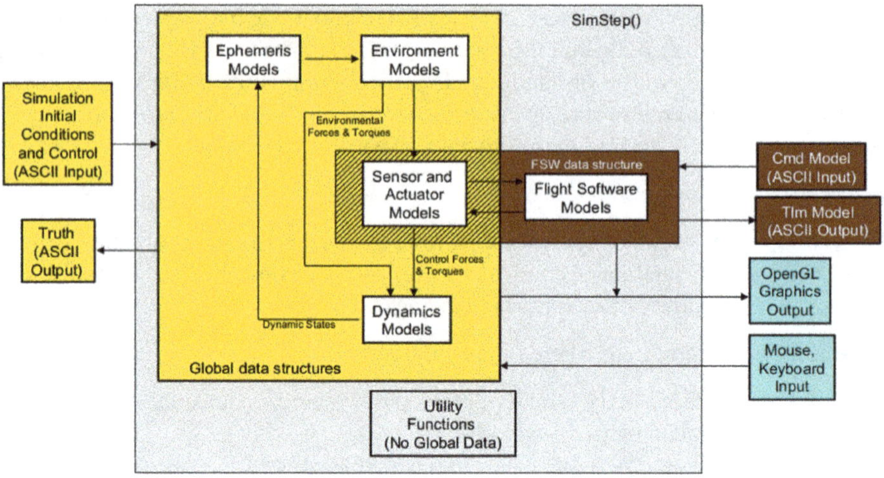

Fig. 6.15 42 Simulator architecture (copyright: Eric Stoneking, used with permission)

Data in 42 is structured in a hierarchical way. These data structures are populated according to the input files configurations and refreshed as the simulation runs. The simulation can be executed in real time or accelerated time (Fig. 6.16).

Environmental models

- Planetary ephemerides

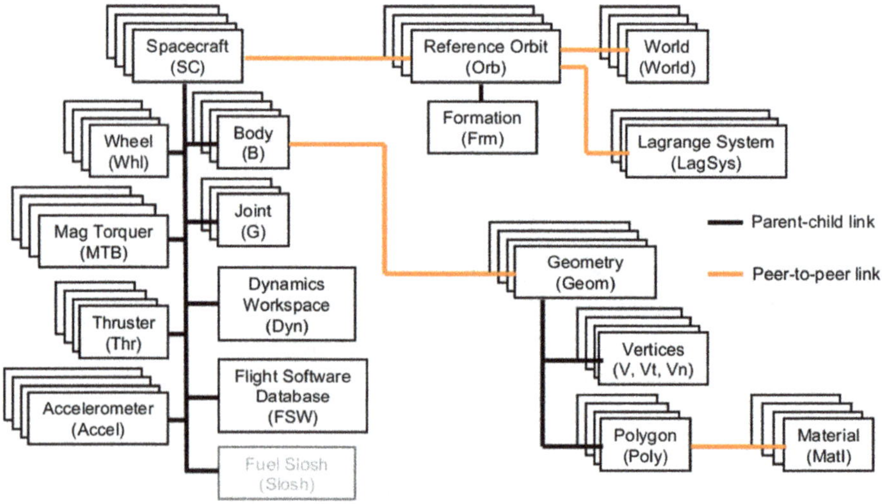

Fig. 6.16 42 Global data structure relationships and hierarchy (copyright: Eric Stoneking, used with permission)

```
<<<<<<<<<<<<<<<<<  42: The Mostly Harmless Simulator  >>>>>>>>>>>>>>>>>
************************* Simulation Control *************************
REAL                            !  Time Mode (FAST, REAL, or EXTERNAL)
6000.0   0.01                   !  Sim Duration, Step Size [sec]
1.0                             !  File Output Interval [sec]
FALSE                           !  Graphics Front End?
Inp_Cmd.txt                     !  Command Script File Name
*************************** Reference Orbits *************************
1                               !  Number of Reference Orbits
TRUE   Orb_LEO.txt              !  Input file name for Orb 0
***************************** Spacecraft ****************************
1                               !  Number of Spacecraft
TRUE  0 SC_CfsSat0.txt          !  Existence, RefOrb, Input file for SC 0
***************************** Environment ***************************
03 21 2016                      !  Date (Month, Day, Year)
12 00 00.00                     !  Greenwich Mean Time (Hr,Min,Sec)
0.0                             !  Time Offset (sec)
USER_DEFINED                    !  Model Date Interpolation for Solar Flux and AP values:
230.0                           !  If USER_DEFINED, enter desired F10.7 value
100.0                           !  If USER_DEFINED, enter desired AP value
IGRF                            !  Magfield (NONE,DIPOLE,IGRF)
8   8                           !  IGRF Degree and Order (<=10)
8   8                           !  Earth Gravity Model N and M (<=18)
2   0                           !  Mars Gravity Model N and M (<=18)
2   0                           !  Luna Gravity Model N and M (<=18)
FALSE   FALSE                   !  Aerodynamic Forces & Torques (Shadows)
FALSE                           !  Gravity Gradient Torques
FALSE   FALSE                   !  Solar Pressure Forces & Torques (Shadows)
FALSE                           !  Gravity Perturbation Forces
FALSE                           !  Passive Joint Forces & Torques
FALSE                           !  Thruster Plume Forces & Torques
FALSE                           !  RWA Imbalance Forces and Torques
FALSE                           !  Contact Forces and Torques
FALSE                           !  CFD Slosh Forces and Torques
FALSE                           !  Output Environmental Torques to Files
******************** Celestial Bodies of Interest ********************
MEAN                            !  Ephem Option (MEAN or DE430)
FALSE                           !  Mercury
FALSE                           !  Venus
TRUE                            !  Earth and Luna
FALSE                           !  Mars and its moons
FALSE                           !  Jupiter and its moons
FALSE                           !  Saturn and its moons
FALSE                           !  Uranus and its moons
FALSE                           !  Neptune and its moons
FALSE                           !  Pluto and its moons
FALSE                           !  Asteroids and Comets
***************** Lagrange Point Systems of Interest ******************
FALSE                           !  Earth-Moon
FALSE                           !  Sun-Earth
FALSE                           !  Sun-Jupiter
************************ Ground Stations ***************************
5                                           ! Number of Ground Stations
```

Fig. 6.17 Simulation descriptor file example (Inp_Sim.txt)

– VSOP87

  From Meeus, "Astronomical Algorithms"
  Expressed in heliocentric ecliptic frame
  Good enough for GNC validation, not intended for mission planning

– DE430 Chebyshev coefficients

  From JPL
  Expressed in ICRS frame (aka J2000 equatorial frame)

More accurate than VSOP87

- Gravity models have coefficients up to 18th order and degree
  - Earth: EGM96
  - Mars: GMM-2B
  - Luna: GLGM2
- Planetary magnetic field models
  - IGRF up to 10th order (Earth only)
  - Tilted offset dipole field
- Earth atmospheric density models
  - MSIS-86 (thanks to John Downing)
  - Jacchia-Roberts Atmospheric Density Model (NASA SP-8021)
- New models easily incorporated to the source tree as the state-of-the-art advances.

Dynamic Models

- Full nonlinear 6-DOF (in fact N-DOF) dynamics
- Attitude Dynamics
  - One or many bodies

    Tree topology (no kinematic loops)
- Each body may be rigid or flexible
- Joints may combine rotational and translational DOFs
  - May be gimballed or spherical
- Slosh may be modeled as a pendulum (lo-fi, quick to implement and run)
  - 42 may run concurrently with CFD software for hi-fi sloshing simulation
- Wheels embedded in body
- Torques from actuators, aerodynamic drag, gravity-gradient, solar radiation pressure, joint torques.
- Orbit dynamics
  - Two- or three-body orbits
  - Encke or Euler-Hill (Clohessy-Wiltshire) for relative orbit motion (good for formation flying, prox ops)
- Forces from actuators, aerodynamic drag, non-spherical gravity, third-body gravity, solar radiation pressure.

Frames of Reference

42 defines several fundamental reference frames, and notational conventions to keep quaternions and direction cosines sorted out.

- Heliocentric Ecliptic (H)
  - Planet positions expressed in this frame
- Each world has an inertial (N) and rotating (W) frame
  - For Earth, N = ECI (True of date), W = ECEF
  - N is the bedrock for orbits, S/C attitude dynamics
  - True-of-Date < - > J2000 conversions are provided
    - Star vectors provided in J2000 (from Skymap), converted to H
    - Planet ephemerides are assumed given in true-of-date H
    - Transformation from N to W is simple rotation, implying N is True-of-Date
    - TOD ↔ J2000 conversions
- Each reference orbit has a reference point R
  - For two-body orbit, R moves on Keplerian orbit
  - For three-body orbit, R propagates under influence of both attracting centers (as point masses)
  - S/C orbit perturbations integrated with respect to R
- Associated with each R is a LVLH frame (L) and a formation frame (F)
  - F is useful for formation-flying scenarios
  - F may be offset from R, may be fixed in N or L
- Each spacecraft has one or more Body (B) frames and one LVLH frame (L)
  - L(3) points to nadir, L(2) points to negative orbit normal
  - SC.L is distinct from Orb.L, since SC may be offset from R.

Configuring 42: Input Files

To get 42 to run, one must get acquainted with its initialization files. These are plain text files which contain all the values needed for configuring the spacecraft mass properties, physical configuration. The hierarchy of initialization files is shown in Fig. 6.18. From the figure, it can be seen that the Inp_Sim.txt is the highest in the hierarchy. This file defines simulation general configurations such as duration, step size, graphic front-end, and a reference to the Inp_Cmd.txt file which contains time-tagged commands. Then it goes to orbit information (it contains also a reference to the Orb_*.txt descriptor file for further orbit details), spacecraft configuration (number and a reference to the SC_*.txt spacecraft descriptor file(s)), environmental models (magnetic field model, gravity, perturbations, etc.), celestial bodies to be added, and finally ground stations. An extract of a typical Inp_Sim.txt is shown in (Fig. 6.17).

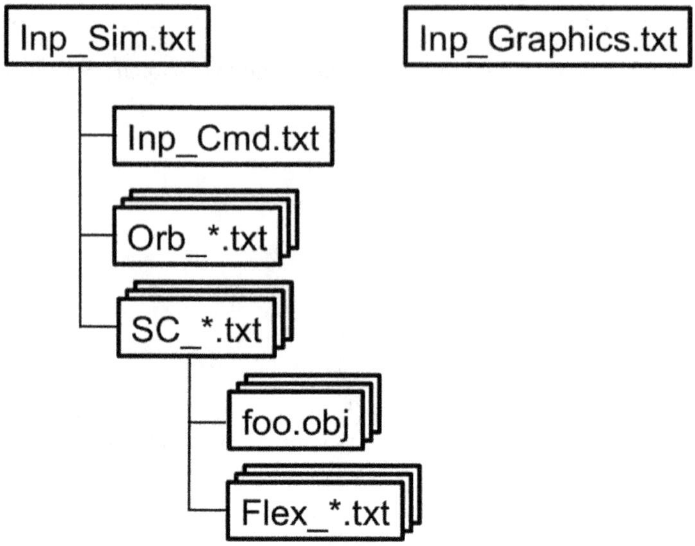

Fig. 6.18 42's Input File Hierarchy (copyright: Eric Stoneking, used with permission)

The spacecraft descriptor file (SC_*.txt) describes the geometry and the locations and number of sensors and actuators, as well as simulated noises, random walks and drifts of sensors. The SC descriptor file also contains references to 3D model files to be used in the graphical front-end, as well as a descriptor for flexible mechanics. For a detailed description of the spacecraft configuration file, refer to 42 documentation or browse example descriptor files in the repository.[2] Same for orbit and commands descriptor files.

Simulation Loop

At every simulation step, 42 executes the following activities:

1. Initialize
   a. Read user inputs
   b. Set things up
2. Call Ephemeris: Where is everything?
   a. Sun, Earth, Moon, etc.
   b. Orbits
   c. Spacecraft
3. Call Environment Models: What forces and torques exerted by the environment?

[2]https://github.com/ericstoneking/42.

4. Call Sensor Models
   a. Input truth
   b. Output measurements
5. Call Flight Software Models
   a. Input Measurements
   b. Process Control Laws, etc.
   c. Output Actuator Commands
6. Call Actuator Models
   a. Input Commands
   b. Output Forces and Torques
7. Integrated Dynamics: How does S/C respond to forces and torques?
   a. Integrate dynamic equations of motion over a timestep
8. Goto 2.

Simulation Outputs

Fortytwo creates a set of output files (*.42) with results from simulations. These files are plain text files which can be easily imported to parsers for post-processing and plotting.

6.3.2.2 Basilisk

Just to add more evidence on how unreasonable it is to develop spacecraft simulation from scratch, another simulation environment will be listed here, called Basilisk.[3] The Basilisk astrodynamics framework is a spacecraft simulation tool developed with an aim of strict modular separation and decoupling of modeling concerns regarding coupled spacecraft dynamics, environment interactions, and flight software algorithms. Modules, tasks, and task groups are the three core components that enable Basilisk's modular architecture. The Basilisk message-passing system is a critical communications layer that facilitates the routing of input and output data between modules. One of the most salient characteristics of Basilisk is that it is wrapped in Python by means of Swig. This way, Basilisk users can run simulations using an interpreted language like Python and abstract themselves from the underlying complexity. The software is developed jointly by the University of Colorado AVS Lab and the Laboratory for Atmospheric and Space Physics (LASP), and released under ISC Open Source License. The resulting framework is targeted for both astrodynamics research modeling the orbit and attitude of complex spacecraft systems, as well as sophisticated mission-specific vehicle simulations that include hardware-in-the-loop scenarios (Fig. 6.19).

[3]https://bitbucket.org/avslab/basilisk.

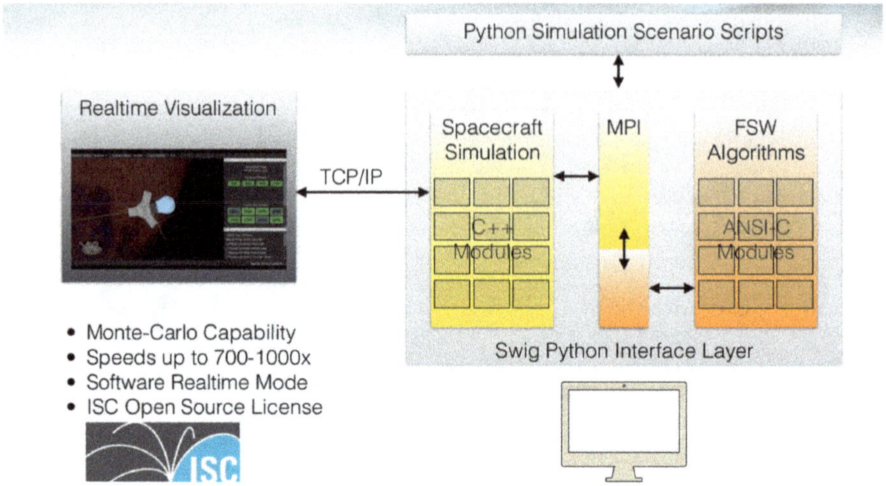

Fig. 6.19 Basilisk architecture (*credit* Dr. Hanspeter Schaub, used with permission)

Basilisk also comes with a comprehensive visualization engine called Vizard.[4] Vizard is based in Unity[5] and it can display in a three-dimensional view the Basilisk simulation data.

Software Stack and Build

The core Basilisk architectural components and physics simulation modules are written in C ++11 to allow for object-oriented development and fast execution speed. However, Basilisk modules can also be developed using Python for easy and rapid prototyping and C to allow flight software modules to be easily ported directly to flight targets. There exists a range of space environment models, such as the various planetary global reference atmospheric models, which are written in FORTRAN. Simulation developers can implement a simple C-code wrapper to translate the execution control and data flow to and from a module, thereby transparently integrating their FORTRAN code base as a Basilisk module. Whereas Basilisk modules are developed in several programming languages, Basilisk users interact with and script simulation scenarios using the Python programming language (Fig. 6.20).

Main building blocks of Basilisk

Basilisk is built on top of three main architectural blocks (Kenneally et al. 2020):

- Components:

A Basilisk simulation is built up from modules, tasks, and task groups. A Basilisk module is standalone code that typically implements a specific model (e.g. an actuator, sensor, and dynamics model) or self-contained logic (e.g. translating a control

---

[4]http://hanspeterschaub.info/basilisk/Vizard/Vizard.html.

[5]https://unity.com/.

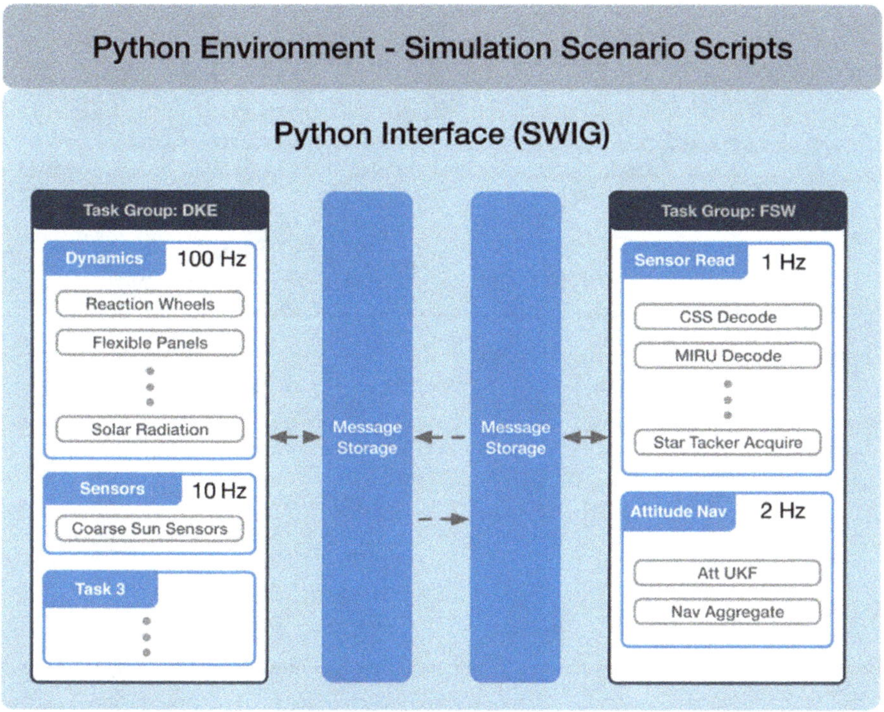

Fig. 6.20 An example layout of a complete Basilisk simulation where each element of the system has SWIG-generated Python interfaces available in the Python environment (*credit* Dr. Hanspeter Schaub, used with permission)

torque to a reaction wheel (RW) command voltage). A module receives input data as message instances by subscribing to desired message types available from the messaging system. Similarly, a module publishes output data as message instances to the messaging system. Tasks are groupings of modules. Each task has an individually set integration rate. The task integration rate directs the update rate of all modules assigned to that task. As a result, a simulation may group modules with different integration rates according to desired time-step fidelity. Furthermore, the configured update/integration rate of each task can be adjusted during a simulation to capture increased resolution for a particular phase of the simulation. For example, a user may increase the integration rate for the task containing a set of spacecraft dynamics modules, such as flexing solar panels and thrusters, in order to capture the high-frequency flexing dynamics and thruster firings during Mars orbit insertion. Otherwise, the integration time step may be kept to a longer duration during the less dynamically active mission phases such as interplanetary cruise. The execution of a task, and therefore the modules within that task, is controlled by either enabling or disabling the task (Fig. 6.21).

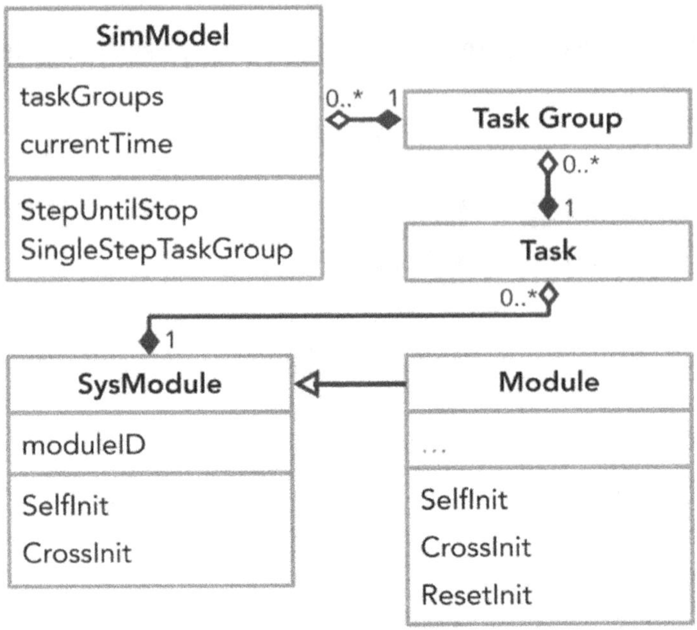

Fig. 6.21 Class diagram of Basilisk task group, tasks, and modules (*credit* Dr. Hanspeter Schaub, used with permission)

- Messaging System:

The Basilisk messaging system facilitates the input and output of data between simulation modules. The messaging system decouples the data flow between modules and task groups and removes explicit intermodule dependency, resulting in no runtime module dependencies. A Basilisk module reads input message instances and writes output message instances to the Basilisk messaging system. The message system acts as a message broker for a Basilisk simulation. The messaging system employs a publisher–subscriber message-passing nomenclature. A single module may read and write any number of message types. A module that writes output data, registers the "publication" of that message type by creating a new message-type entry within the message system. Similarly, a module that requires data output by another module subscribes to the message type published by the other module. The messaging system then maintains the message types read and written by all modules and the network of publishing and subscribing modules.

- Dynamics Manager:

The third and final piece of Basilisk's modular design is the implementation of the dynamics manager. The spacecraft dynamics are modeled as fully coupled multibody dynamics with the generalized equations of motion being applicable to a wide range of spacecraft configurations. The implementation, as detailed in (Schaub et al. 2018), uses a back-substitution method to modularize the equations of motion and leverages the resulting structure of the modularized equations to allow the arbitrary addition of both coupled and uncoupled forces and torques to a central spacecraft hub. This results in a highly decoupled dynamics management which does not need to know what type of *effectors* or the simulated spacecraft contains. An effector is a module that impacts the translational or rotational dynamics. Effectors are classified as either a state effector or a dynamic effector. State effectors are those modules that have dynamic states to be integrated, and therefore contribute to the coupled dynamics of the spacecraft. Examples of state effectors are reaction wheels, flexible solar arrays, variable-speed control moment gyroscopes, and fuel slosh. In contrast, dynamic effectors are modules that implement dynamics phenomena that result in external forces or torques being applied to the spacecraft. Examples of dynamic effectors include gravity, thrusters, solar radiation pressure, and drag.

6.3.2.3 Conclusion

Despite differences in their architectures and conceptions, both 42 and Basilisk are tools space systems engineers cannot afford *not* to have in their toolbox: they are open-source and they are continuously improved and maintained. These tools can be used for early-stage analysis, preliminary physical configuration, design automation, design optimization, resource planning, collection planning, maneuvering planning, actuator dimensioning, and a very long list of etcetera. NewSpace must use simulation tools that are proven and actively kept, and throw away the idea of burning thousands of man hours on reinventing the wheel, because chances are also there for our homegrown simulation experiments to underperform compared to these.

Both 42 and Basilisk are open-source projects which take large amounts of effort from their creators and developers, so they **must be supported and properly credited**.

6.4 A Collaborative Networked Simulation Environment for Digital Twins

Every organization designing and building cyber-physical systems for aerospace applications (spacecraft, rockets, missiles, drones, UAVs) needs to procure synthetic simulation environments throughout the whole life cycle: from high-level analysis at early stages all the way to hardware-in-the-loop (HiL environments). Literally every single organization designing cyber-physical systems rely on simulation infrastructure. Surprisingly, many of them grow this infrastructure from scratch, which takes time and it can prove to be quite costly and complex. Moreover, two organizations creating cyber-physical systems which are meant to interoperate will most likely grow separate simulation environments which will not be able to connect each other.

Then, perhaps the next step in cyber-physical systems engineering is to create collaborative, networked simulated space which would make it possible to field different "digital twins" and be able to interoperate them in pure computer domain.

If this environment existed, it would allow different organizations to deploy their systems and interact with other systems in a physics-accurate virtual way. This would require an accurate simulation of the inner workings of the system (and its subsystems) as well as simulating the interaction (causes) to the surrounding virtual physical environment and re-create its feedback (effects), as well as the interaction with other systems present in the environment.

For example, an organization offering a relay comms constellation in geostationary orbit could virtually test their constellation with customer spacecraft; this would allow all actors involved in this shared simulation to refine their concept of operations and verify their designs, also by including virtual ground stations to assess coverage and service level agreements. Or, a launcher company could run virtual launches; such virtual launches and orbital payload deployments could help space orgs to rehearse and train their workforce for early orbits operations, but also test ops automations, processes and tools, way before they deploy to real orbit.

Moreover, virtual ground stations would also take part of the virtual simulation environment by creating virtual links to the simulated flying satellites and systems. This way, operators could verify their systems practically end-to-end (including ground station APIs) but in a purely synthetic manner.

This collaborative environment would offer private simulation instances (where actions would only affect one single organization), or shared ones where actions could impact others.

Taking the concept even further, this synthetic environment could also encompass life cycle stages such as assembly, integration, and verification (AIV), where organizations could use virtual reality to integrate virtual subsystems provided by suppliers, which would help identify integration problems early in the process.

References

INCOSE. (2015). *INCOSE Handbook* (4th ed.). Wiley. ISBN:978-1-118-99940-0

ISO. (2015). *ISO/IEC/IEEE 15288:2015 Systems and software engineering—System life cycle processes*. International Standards Organization.

Kenneally, P., Piggott, S., & Schaub, H. (2020, September). Basilisk: A flexible, scalable and modular astrodynamics simulation framework. *Journal of Aerospace Information Systems, 17*(9). https://doi.org/10.2514/1.I010762

Klesh, A. T., Cutler, J. W., & Atkins, E. M. (2012). Cyber-physical challenges for space systems. In *IEEE/ACM Third International Conference on Cyber-Physical Systems* (pp. 45–52). https://doi.org/10.1109/iccps.2012.13.

Kopetz, H. (2018). *Simplicity is complex: Foundations of cyber-physical system design. Springer*. https://doi.org/10.1007/978-3-030-20411-2.

Project Management International. (2017). *A guide to the Project Management Body of Knowledge (PMBOK guide)*. PMI.

Schaub, H., Allard, C., Diaz Ramos, M., Kenneally, P., & Piggott, S. (2018, December). Modular software architecture for fully coupled spacecraft simulations. *Journal of Aerospace Information Systems*, *15*(12). http://dx.doi.org/10.2514/1.I010653.

Stoneking, E. (2010–2020). 42: A general-purpose spacecraft simulation. https://github.com/ericstoneking/42

Chapter 7
NewSpace Bullshit

"Just put something that sounds appealing but not so appealing in a way they will start asking questions"
—Ex-boss to a baffled me, many years ago

Abstract In 2016, the word *post-truth* was selected "word of the year" by the Oxford Dictionary. Post-truth is a concept that denotes the vanishing of the shared standards for objective evidence and the intentional blur between facts, alternative facts, knowledge, opinion, belief, and truth. In NewSpace in particular, due to its complexities and sci-fi *halo*, facts often give way to over-decorated fictions, which can stretch long distances unchecked. These fictions, shaped under the pressure of funding needs and the urge of growing like there is no tomorrow, create some particular conditions where fads, fashions, imitation, and humbug are the norms.

Keywords Humbug · Post-truth · Bullshit · Agile · Meetings · Waterfall

Bullshit is all around us. It is part of our lives, and an instrumental means for socializing. Friends around a barbecue will bullshit each other about their fishing accomplishments, or their romantic adventures; kids will try to bullshit each other about skating skills or vandalistic achievements. In these contexts, we are allowed to say things we do not really believe in, as a sort of a game (Spicer 2018).

Although there is nothing inherently wrong with bullshitting a friend to have a laugh, things took a turn when at some point bullshit found its way into the workplace. In the informal happenings mentioned before, bullshit is never taken seriously, nor the bullshitters expect their bullshit to be taken seriously. In the workplace, in contrast, bullshit is often produced with the expectation of being seriously consumed. This created the need to build bullshit defenses to filter it out from actual valuable information. Bullshit became a frequent noise on the line, and since our filters can never be perfect (but also bullshit got smarter), it started to sneak in, affecting our decision-making. Like an Ivy plant you did not prune in your summer cottage, it spread and encroached (Fig. 7.1).

So why is there bullshit everywhere? Part of the answer is that everyone is trying to sell you something. Another part is that humans possess the cognitive tools to

I. Chechile, *NewSpace Systems Engineering*,
https://doi.org/10.1007/978-3-030-66898-3_7

Fig. 7.1 Bullshit is like Ivy, if you do not prune it in time will take over (credit: "Ivy house" by Alizarin Krimson, licensed under CC BY-SA 2.0)

figure out what kinds of bullshit will be effective. A third part is that our complex language allows us to produce an infinite variety of bullshit (Bergstrom and West 2020).

Like it was not challenging enough having to develop complex technology such as space systems on top of razor-thin budgets, on crazy short timelines and no revenues, NewSpace organizations need to deal with industrial loads of bullshit; both as producers and consumers. The startup world (hey, the whole world, including this book) is bullshit rich, and NewSpace is no exception. The fact most people ignore how space stuff works and tend to be amazed *by default* because of rockets and stuff makes the conditions for bullshit spreading close to ideal. Thing is, as much as bullshitting can be a practice that startups are somehow "forced" to perform due to social/business/context pressure, they can easily get so hooked into it until the point bullshit takes the helm, and that's a point of difficult return.

NewSpace did not only change the way space systems are made but also changed the way things about space systems are communicated. In a type of industry where engineering excellence has been the norm for decades, NewSpace has in turn chosen the shallow approach of sugar-coating and over-decorating reality. Press releases suddenly preclude engineering progress and milestones.

In NewSpace, often technical facts are relative, evidence has become seldom relevant and blatant overpromising has become the rule, in a very strange choreography where founders, investors, and prospective customers dance a sort of bullshit *minuet* together. But investors are no fools: they just *choose* to play along and to willingly munch the bullshit, for a while. They may even invest considerable amounts of capital on basically slides. But sooner than later they will want to see what is behind those slides.

7.1 Slides Engineering (Bullshit as a Service)

Oh, the pitch decks and websites. Top-notch templates. Superb graphic design and very sophisticated CSS[1], all at the service of bullshit. Nothing like conveying a tornado of buzzwords, with a nice design. Slides and slides showcasing shiny spacecraft that only exist in the domains of Blender. Smooth websites playing high-definition videos in the background, showcasing happy people smiling and sitting around a table with distant and intentionally defocused whiteboards crowded with colorful sticky notes (sorry, *information radiators*), as we are invited to scroll down endlessly, struggling to understand what they exactly do, while being beamed with bombastic words such as "revolutionary", "groundbreaking", "next-generation", "digital" and of course "democratizing". Or, the urge to declare everything *as a service* (aaS). We are either witnessing a golden era of the service industry sector, or a dark era of rigurosity in using terminology. It is high time startup world returns the word *service* back to where it belongs: the service industry. Let services be the ones offered *as a service*. If you are designing and building spacecraft, you are not providing a service, you are providing a tangible product, like shoes are. This should not stop you from offering services on top of your product (that is fine, just like you buy a post-sales support service for your car). But it is about time to stop saying "space as a service"; we all die inside every time we read it.

Slides are bullshit missiles; the autobahns of crap. Slides are the powerhouse of nice-looking, logo-driven vacuous non-sense. I must admit I have seen incredibly beautiful slides in my career, with gorgeous fonts I had never seen before. Branding can be gorgeous at times. Once I took a photo of a slide, which was so nicely designed that I wanted to keep a memory of it. I cannot remember at all what content was in the slide, unfortunately, but the color choices and fonts were superb.

Many space startups are just a slide deck, and nothing else. Most of the constellations NewSpace has been promising are just slide constellations. They have no actual traction on any real and tangible technology they can showcase whatsoever, but they do have slides saying how they are coming to disruptively democratize us from the tyranny our eyes are too blind to see. They promise wonderlands that usually are 2–3 years from now. And the buzzwords, of course. They flow like in a fountain: "synergize", "maximize", "aggregate", "lean", "actionable", "orchestrate", "proven", "agile", "cutting-edge"; the list can go on and on. There is no inherent

[1]Cascading Style Sheets (CSS) is a style sheet language used for describing the presentation of a document written in a markup language such as HTML. CSS is an ubiquitous technology of the World Wide Web, alongside HTML and, sigh, JavaScript.

problem with passionately explaining your idea aided with visually appealing material. But smoke and mirrors cannot become the norm. Smoke and mirrors can be an initial platform from where to start walking, but never the place to stay. Investors might knowingly swallow the seed-stage of your ridiculously abstract humbug for the sake of giving leeway to an idea that *may* look interesting and turn into something, but they will never stop longing to see real tangible technical progress. NewSpace may have changed many things, but it has not changed the fact you need competence to realize your crazy dreams in orbit. You cannot get into space riding a pitch deck.

There is a time for promises, but then it is time to show progress. As Linus Torvalds[2] famously said: "Talking is cheap, show me the code".

7.2 Democratizing Bullshit

Since a few years ago, there has been a trend (yet another one) where every single NewSpace decided to put themselves as the ones coming to democratize something. Do a quick Google search with the keyword "democratizing space" (including the quotes) and you will get an impression on how many have been claiming to come to democratize it, in a silly fad where the main victim is the word democracy. In any case, the only thing NewSpace startups have truly *democratized* is the lack of rigurosity in using some terms. If we grab a dictionary and we look for the verb "to democratize", we get a simple: "to introduce a democratic system or democratic principles into something". What has NewSpace democratized after all? NewSpace as an industry has lowered some barriers for doing space, making it more accessible for companies to enter the game in ways 10 or 15 years ago would have been unthinkable. But, strictly speaking, that is not NewSpace industry's merit, but a merit of the classic space industry that changed its mindset, saw the value and the business in having lots of small spacecraft as customers, and opened the door with rideshares launchers and lower launch costs. Classic space was the true democratizer of space in that sense, not NewSpace. NewSpace does not even democratize things it could democratize, for example by increasing knowledge sharing, collaborating on standardizations, and fostering cross-organizational collaboration. On the contrary, NDAs are rapidly flying both ways at the second email: "confidential", "do not disclose"; "trade secret". Fine, they may be just protecting their secret spice, ok. Has NewSpace then democratized access to their own products even? For example, Morrison (2020) states about the experience of purchasing satellite imagery data:

> It's closer to a hostage negotiation than a sales transaction. There will be email introductions, phone calls, lists of demands, let-me-get-back-to-you-on-that-s, appeals to lawyers, requests for paperwork, and ultimately…a ransom. Often, the captive is never released.

[2]Linus Torvalds is a Finnish American software engineer who is the creator and, historically, the principal developer of the Linux kernel, which is the kernel for GNU/Linux operating systems and other operating systems such as Android and Chrome OS.

It does not really scream "democratizing", does it? Should it be *bureaucratizing* instead? What has NewSpace truly democratized remains a largely unanswered question. NewSpace democratizing *grandeur* remains, as of today, just bullshit.

7.3 The Agile Versus Waterfall Bullshit

It is easy to blame Agile. Agile has sparked lots of opinions; both pro and against. This section is not about calling Agile bullshit, but to call bullshit one of the typical arguments Agile people use when it comes to defend it: the everlasting antagonism with waterfall.

Agile is a religion on its own, but again you cannot blame religion itself, you can only blame the dilettante fanatics who misinterpret it and go to defend it like there is no tomorrow. It is not Agile's fault that many have used it as a distraction to divert the attention away from their own incompetence about how to develop complex technology. It is not Agile's fault that it has been turned into a lucrative industry on its own and as such it needs to justify itself to exist, because entities will always try to preserve the problem to which they are the solution for.[3] Agile coaches, certifications, scaled-frameworks, thousands of books, courses. Agile snowballed. And the main responsibility falls on tech managers incapable of calling out Agile when it was not bringing any tangible benefit but on the contrary altering for the worse things that used to work well. This section is mostly for them: to those waving the Agile flag just to avoid being called "old fashioned" or "waterfall-ey", those afraid of saying that the Emperor is naked.[4]

I have seen Agile working reasonably well, and I have seen Agile working badly. In the more than 18 years being around technology development in many different areas of industrial automation, space and telecommunications, I have never seen Agile or Scrum working *extremely* well. In the situations I saw it working *ok*, those were rather simple projects with rather small architectures and rather atomic tasks with simple dependencies, with very few disciplines involved, and an overall short duration. Some studies describe successful application of Agile methods in complex aerospace projects, such as fighter aircrafts. A quick look at the published papers indicates an extensive tailoring and refactoring of the Agile methodology overall, for

[3]This is known as the Shirky Principle.

[4]Emperor's clothes tale: Two swindlers arrive at the capital city of an emperor who spends lavishly on clothing at the expense of state matters. Posing as weavers, they offer to supply him with magnificent clothes that are invisible to those who are stupid or incompetent. The emperor hires them, and they set up looms and go to work. A succession of officials, and then the emperor himself, visit them to check their progress. Each sees that the looms are empty but pretends otherwise to avoid being thought a fool. Finally, the weavers report that the emperor's suit is finished. They mime dressing him and he sets off in a procession before the whole city. The townsfolk uncomfortably go along with the pretense, not wanting to appear inept or stupid, until a child blurts out that the emperor is wearing nothing at all. The people then realize that everyone has been fooled. Although startled, the emperor continues the procession, walking prouder than ever.

example introducing additional iterations and stages ("increments", "development steps", etc.), and a scheme of synchronization between teams. Often, Agile ends up being a *skin* toward the outside whereas to the inside it shows large tailoring to cope with the inherent challenges of multidisciplinary systems: where does the threshold lay where a methodology stops being truthful to its origins to become something else remains unspecified. Methodology *mutation* is a largely under researched topic.

Agile and Scrum, as any other process, may work or not. For some reason, other processes and methodologies would be trashed easier than Agile if it proved ineffective; but not Agile. It seems to be a sort of shame for managers to un-Agile a project, even if it really explodes in their faces it is making things worse. If Agile does not work, the problem is always elsewhere, the problem is never Agile. You just probably did not get it right. Agile always works.

Agile has made a living out of antagonizing with the so-called waterfall model. Waterfall is Agile's archenemy, the evil of all evils, and a strong foundation of its narrative in general. Waterfall is what Agile came to save us from. Ok, but what if I told you that waterfall is an imaginary monster in the closet that never existed? What if literally no project in the entire world has ever been or could ever be waterfall even if it wanted to? In short:

1. There's no such thing as the *Waterfall* approach, and there never was.

Which means a large amount of all slides decks on Agile are wrong, since they all start with the "Agile vs Waterfall" pseudo-dilemma. The whole confusion still points to a largely misinterpreted paper by Dr Winston Royce titled "Managing the Development of Large Software Systems" (Royce 1970). A 50-year-old paper in which the word "waterfall" was never used, and a paper that actually acknowledges the iterative nature of software engineering.

Royce begins with a very minimalistic conception of how things need to be done (Fig. 7.2). He uses two stages: analysis and coding. Oversimplifying? Yes. Inaccurate? No. We have mentioned before we always need to analyze before implementing. Royce's approach is aligned to that. What software engineer in the world starts coding

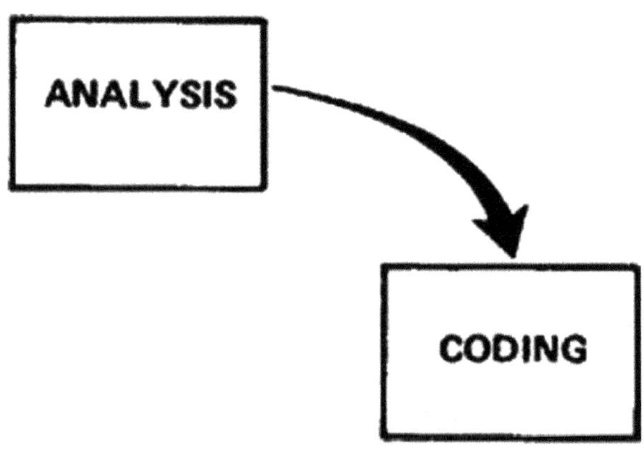

Fig. 7.2 Minimalistic process from Royce's paper. Nothing that does not represent how things work in reality! (Royce 1970)

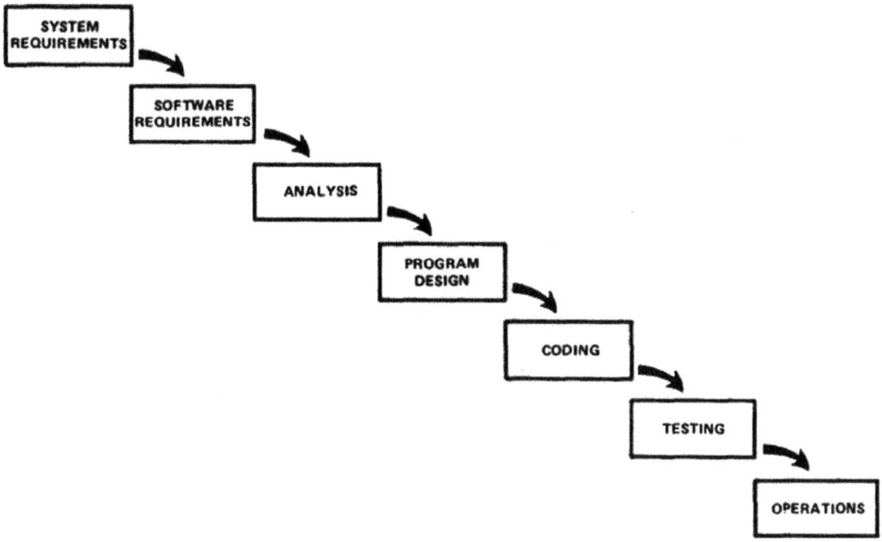

Fig. 7.3 The cascade of steps Royce's defined in his paper (Royce 1970)

before analyzing what needs to be done? What software engineer in the world does not define her data structures before starting to code any behavior?

Then, Royce commits the sin of extending a bit of his thinking and comes up with the diagram shown in Fig. 7.3.

The sequence of steps shown in Fig. 7.3 has been reproduced *ad nauseam* by Agile evangelists: "Behold to the one-way evil which makes things never come back". Like a one-way staircase to hell. Many of them surely did not read the paper or they just stopped reading the paper right here. But the paper goes on. The stages Royce described were (and are) sound in terms of the actual activities that need to take place in any software or engineering endeavor. But then Royce states:

> The testing phase which occurs at the end of the development cycle is the first event for which timing, storage, input/output transfers, etc., are experienced as distinguished from analyzed.

He is right, when you reach testing is when you have the whole thing together and you can really gauge performance, and continues:

> The required design changes are likely to be so disruptive that the software requirements upon which the design is based and which provides the rationale for everything are violated. Either the requirements must be modified, or a substantial change in the design is required. In effect the development process has returned to the origin and one can expect up to a 100-percent overrun in schedule and/or costs.

He acknowledges that late problem findings are troublesome. That was true in 1970 as it is today in 2020. He correctly points out that a finding at a very late stage can cause overrun in time and cost. So, Royce is the first critic ever of the so-called waterfall approach. He quickly acknowledges that engineering is a highly iterative

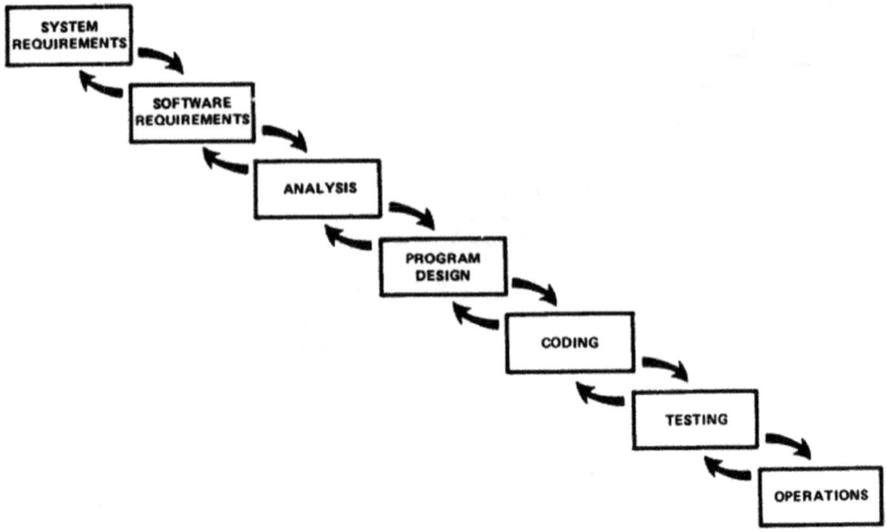

Fig. 7.4 Iterations between stages in Royce's model

activity, so there must be some path back to previous stages. So, he adds feedback loops to the graph (Fig. 7.4).

Note that this one with the iterations is never shown in any Agile slides, in a strange case of collective selective ignorance. Royce's main flaw, perhaps shaped by the computing context at the time, is to expect (or *hope* as he phrases it) that iterations will be "confined between two subsequent stages". He tends to underestimate the complexity of the life cycle of software and overlook the fact that feedback loops might be from everywhere to everywhere, and recursively across the architecture.

Besides Royce's paper, in reality, there is no rigorous definition of the waterfall approach, besides what everyone parrots. Although we cannot find a rigorous definition, the key attribute of the so-called "waterfall approach" seems to be extreme inflexibility. In particular:

- Once you have completed a phase, its results are *frozen*. You cannot go back and revise anything based on changing needs or fresh insights.
- Until the results of the current phase are complete and approved, you may not start on any work that properly belongs to the next phase or any later phase.

Presumably, the waterfall metaphor was suggested by the inability of water to flow uphill. Once you have passed a given point, there is no going back. Who would ever design a system in such a single-directional way? Who would interpret and accept such rigidity in the development life cycle? The waterfall approach exists only in the minds of those who condemn it (Weisert 2003).

We commented in a chapter before, that subdividing the life cycle in stages is done to manage things better, just as we do with our own lives. Separating life cycles in stages or chunks is part of any problem-solving activity.

Reality is, we never "freeze" the users' requirements or a system specification or any other phase deliverables. Of course, there is a cost associated with modifying work that is already been completed and approved. A well-conceived life cycle methodology gives the organization a rational choice based on the costs and expected benefits of making or not making any non-trivial change to previously accepted results. Furthermore, nothing stops the project manager from initiating advance work on aspects of phase $n+1$ before phase n is completed, except for the obvious risk that such work may turn out to be invalid if phase n is still too immature. Again, the organization should be given a rational choice whether to take that risk in the interest of schedule compression.

Finally, all life-cycle methodologies (including Agile ones) encourage or at least permit partitioning a large project into multiple smaller projects, which can then proceed independently through the remaining phases. Phase disciplines, when practiced with sensible judgment and flexibility, are a *good* thing with no offsetting downside. They are essential to the success of large projects. Bogus attacks on the non-existent "waterfall approach" can obscure a new methodology's failure to support long-established sensible practice.

7.4 Meetings

Engineering is a collaborative adventure but that does not mean engineers must do everything together. A good deal of the creative work related to engineering design is done individually. I have preached enough about how social interactions define and shape success in multidisciplinary technical projects, but that does not invalidate the fact engineers need time alone to figure things out. As needed as meeting up with teammates can be, in order to build a house someone will eventually have to start piling up bricks. A thick part of the design activity requires an engineer sitting in front of a screen thinking and designing/coding/laying things out, and this requires focus and a special kind of mental state. During these hours working alone, people are ideally in a state that psychologists call flow. Flow is a condition of deep, nearly meditative involvement. In this state, there is a gentle sense of euphoria, and one is largely unaware of the passage of time. For anyone involved in engineering design tasks, flow is a must. Unfortunately, you can not turn on flow like a switch. It takes a slow descent into the subject, requiring 15 min or more of concentration before the state is locked in. During this immersion period, you are particularly sensitive to noise and interruption. A disruptive environment can make it difficult or impossible to attain flow (DeMarco and Lister 1999, 61).

When it comes to brainstorming, interface agreement, reviews, roadmapping, and strategic alignment, collective collaboration is critical, which requires gathering people in some forum, either a meeting room, a video call, or a mixture of both. These types of events are naturally noisy, chatty, and rich in opinions coming and going; these forums often foster good ideas and critical thinking. One can quickly see that meetings are both a blessing and a curse for engineers. Lately, more a latter than the former.

I find it quite interesting that meetings have gained such a bad reputation, representing the opposite of usefulness. Most of them are plainly considered a waste of time. Just google "meetings are a", and Google autocompletes with "waste of time". "Too many meetings" is a popular complaint in the workplace, as well as "too many unproductive meetings". Instead of being spaces to foster collaboration and streamline decision making, meetings have apparently become the medium by which a great deal of bullshit spreads, and a perfect source of flow disruption. Meetings have turned into boring, endless slide-showing, self-aggrandizing, rank-pulling sessions where most people bring their laptops to do something else and management-speak is all you can hear. Nine out of ten people daydream in meetings, while 73% claim they do other work (Pidgeon 2014). Calendars quickly get filled up with fixed overhead in the form of bi-weekly or weekly "sync ups", "catch ups", "tag ups", "stand ups", "updates", "huddles", and the like. Is that what pushes projects forward? When does the team get time to sit down and design?

Two factors to analyze meetings and their lack of usefulness are: attention and spontaneity. While we do engineering, we perform our tasks in a project, which involves many others, and surely our work will show dependencies to those others. Sooner than later we will need something from someone, or someone will need something from us. We cannot really anticipate what we are going to need from a colleague or a team, or what they will need, until those needs become a reality. Such needs, whatever they are, become roadblocks in the project progress and as such they have to be cleared away, otherwise it means the work is stuck. Hence, we request the attention of the person or team who knows about that topic or issue, to provide us with the information needed for us to proceed; asking someone for their time means asking for one of their most valuable resources. But attention is not only time; attention means also requesting their mental engagement to properly think about questions that we are asking for. Ultimately, you can have someone's time, but having someone's attention is a slightly different thing. But to get someone's attention for helping you get rid of what is blocking your path ahead, you technically do not have to drag them in a room. On the contrary, technical exchanges usually are more difficult in meeting rooms for engineers. A visit to your colleague's office or desk (if collocated) is way more straightforward than sitting in a meeting room, where browsing through code or walking through schematics is more straightforward than on a projector. For a colleague not collocated, a call should do, when you get to share your screen. In these cases, a simple calendar invite can be used to give "heads up" to your colleague you need their attention; this approach ensures you are not randomly breaking their hard-to-gain flow. This is somewhat similar to the red/green tiny lights that supermarket cashiers turn on when they need change or a cancellation from the supervisor. Supervisor is surely busy but remains reasonably attentive, for as soon as she frees herself up she will assist whoever needs her, without having to interrupt what she is doing right away.

After the supervisor assists the cashier in need, light goes off, the cashier carries on, things move ahead, customers are happy. This follows a very simple pattern:

it is need-driven. This is a spontaneous request for attention and resources. This way of requesting resources is what computing calls "interrupt based". Interrupt is a mechanism in which a device notifies the CPU that it requires its attention. Interrupt can take place at any time. So, when the CPU gets an interrupt signal through the indication interrupt-request line, the CPU stops the current process and responds to the interrupt by passing the control to the interrupt handler which services the device. An important factor for interrupts-based computing is what is called "context switching". When the CPU gets interrupted, it saves its current context in a set of registers and jumps to the interrupt routine where a different context will be used. Once the interrupt is serviced, the CPU goes back to where it was, and retrieves the context as it was before the interrupt happened. What we have described is exactly what happened between us and our colleagues. This is the most efficient way of requesting resources, both for humans and computers: you only disrupt the flow of others and make them switch context when you have a clear need which is impeding you to advance your work, otherwise you just let them be.

The other way of asking for resources is by polling. In a polling mechanism, one periodically checks and interacts with the target in order to get something for that resource: whether it is just a status or actual information. This is the typical bi-weekly "sync-up" approach. These types of meetings are basically polling schemes equivalent to a worker periodically asking someone else "do you need something?". The polling method can be more illustratively explained using traffic lights. Think when you stop with your car at a traffic light: you continuously look at the lights to check when it will turn green for you to go ahead. You *poll* the traffic light. Would not be great if you could do something else in the meantime and get notified when it turns green? Certainly, even though we know it would not be safe according to traffic rules; there is probably a very good reason why traffic lights have to be polled, there is probably nothing more important you should be doing at that time other than looking at those colored lights. Polling makes sense in some scenarios where attention must be continuously directed somewhere. But there is no real scenario why our interactions while we do engineering need to work in such a way. Attention (and its associated context switching) are a resource that must be carefully protected. A person sitting in a "polling" meeting is not being able to apply her brain power somewhere else. Polling meetings usually create the right environment for bullshit to ferment. When nothing really useful can be said, the opportunity for half-baked ideas and partial truths arises. Some organizations, as they grow, suffer a creep of *polling* sessions on a very frequent basis, causing pointless context switching and incurring on costs of people meeting for nothing specific. Unless you need something concrete from me: why having a "weekly" then? To double check? I am capable of letting you know! In the supermarket example, the polling method would mean the supervisor asking the cashier every 10 minutes: do you need something? It boils down to synchronicity and granularity/frequency. Synchronicity is needed in collaborative enterprises, due to the fact outcomes are the product of the interaction and agreements of many actors. Hence, syncing all those contributions in some way definitely adds value. If it becomes too frequent, it will start affecting the outcomes because a. most of the time will be spent just in synchronizing and b. work does not progress enough to make the sync any worth. If syncing is not frequent enough, outcomes are affected since work from multiple actors and disciplines would not integrate well if they go long distances without agreements. Unfortunately, this indicates a trade-off is needed. Different disciplines have different time constants. Software can have shorter cycles, which means sync frequency can be higher. Space systems engineering can have slower cycles which means sync frequency needs to be scaled down.

One could also argue that also meetings are not the place where innovation precisely blooms. Meetings are often very structured, and they tend to turn into rigid forums where detouring from the defined "agenda" for spontaneous brainstorming are usually seen as a bad sign: "let's discuss this in a separate meeting" the organizer will say in order to look he or she is in control. Discussions that foster disruptive and inventive ideas often take place in informal, unstructured environments. Informal social networks are important for innovation because they expose individuals to new ideas rather than simply facilitating collaboration (Andrews 2019).

In conclusion, for early stage orgs, both good and bad meetings have implications. Bad meetings are just money wasted, and money wasted means a shortening runway. Could we estimate how much it cost a bad meeting? The key is how to define a bad meeting in a reasonably accurate manner, since it can also be *cool* to hate meetings; in private surveys employees offered accounts of effectiveness in meetings that were quite favorable. When asked about meetings in general from a productivity perspective, a significant majority responded positively (Rogelberg et al. 2007). In terms of cost, according to Keith (2015) bad meetings cost the US economy each year somewhere between $70 and $283 billion. Good meetings, as needed as they can be at times, can be a source of *flow* disruption and act as innovation scarecrows (Fig. 7.5).

7.5 Grow Like There Is No Tomorrow

As systems grow in size and complexity, they tend to lose basic functions (Gall 1975). As most remarks from John Gall's *Systemantics*, it is in a somewhat humorous way. Growing is *ok*. Early stage organizations need to grow as a baby needs to grow: healthy and steady. When it comes to defining what a startup is, growth always comes up; it is very coupled to the startup term itself. The whole startup concept seems to be very deeply, intimately connected to it. Such growth-oriented school of thought can be summarized in few bullets (readers are invited to research by themselves as well):

- Startups are meant to grow fast.
- The most essential thing about a startup is growth: all the rest comes after.
- Growth is founders' main guidance, and a compass to make almost every decision.

From this, one can quickly get a notion the whole startup concept is somewhat *explosive* in itself. Grow big, grow fast; like a sort of blast. Organizations can be explosive in their strategies and wishes, but can technology grow at such a similar

Fig. 7.5 Meetings scare innovation away

pace? Space technology can progress reasonably fast, but not as fast. Learning takes time, quality takes time, reliability takes time, process traceability and repeatability takes (a long) time.

Is it that NewSpace and startups are not as compatible concepts as we may have initially thought?

The type of talent NewSpace requires is not easy to find. Attitude control engineers, digital control systems engineers, flight dynamics engineers, integration engineers, simulation engineers, systems engineers. These are not profiles universities produce in mass, like software engineers, or lawyers. Due to this difficulty, NewSpace organizations often settle to hire non-space people and train them from scratch to become space savvy, and that training takes time and effort. Growth, often used as a key performance indicator in startups, cannot be easily used in that sense in NewSpace. On-orbit performance is the actual NewSpace KPI; you can grow the org at the speed of light but if you do not perform in orbit you're done. On-orbit success cannot be easily correlated as a function of growth *per se*, but as a function of knowledge, processes and more fundamentally as a function of lessons learned. Learning lessons takes time, and lessons learned can only come from failures and mistakes made. A vertiginous growth path combined with mistakes and failure sounds a bad combo. As much as we want to make space startups run at the *SaaS* or *App* startup pace, space technology remains highly dependent on processes and organizational strategies and cultures, which are not fast as a lightning. This is perhaps the heritage and legacy of the classic space achievements of the last 60 years or so. Space is hard.

Organizational growth never comes for free. Growth brings overhead. Anyone who has ever managed a team of dozens or hundreds would know what every single percentage of growth means in terms of management, communication, and human factors operations needed. Linear growth means more coordination, more meetings, less flow; less flow means less engineering. With less engineering is it difficult to think about success in NewSpace. It is easy to ask or suggest growth when you are not supposed to be the one handling that growth on a daily basis. Those who have witnessed an organization go from 20 to 200 people would know the massive surge of management effort such a leap brings. Linear growth increases costs accordingly, culture diffuses, and bullshit jobs flourish. You would think that the last thing a profit-seeking firm is going to do is shell out money to workers they do not really need to employ. Still, somehow, it happens (Graeber 2018). By Graeber's estimation, these jobs are essentially meaningless—as in, if they all disappeared tomorrow, hardly anyone would take notice. The worst part? The workers performing these jobs know all this, and they are essentially trapped in a prison of pointless labor. Bullshit jobs are usually a sign things are going corporate. As investment rounds go by, the company suddenly has more cash than they know what to do with. The logical answer? Go on a hiring spree. When the "go big" approach to business tickles into hiring, a company may end up with role confusion or several workers performing different variants of the same job (Prupas 2019).

Continuous linear growth can introduce statistical mirages as well, which managers can rapidly pick up for their own benefit since those mirages show things they like to see, creating strange feedback loops. For example, electronic surveys for the workplace. These surveying companies offer (and not for a small price) a nice-looking, sophisticated questionnaire to gauge employee's engagement and motivation levels. They produce fancy reports which usually compare trends from current to previous periods, to help their customer companies understand how things are evolving over time. For companies that are rapidly growing, year-to-year comparisons are very misleading since they may have doubled the size in that time, or even more. As newcomers join, they fill their surveys while being in the *honeymoon phase*. This could lead managers to believe things are great. Truth is, things are perhaps great for the newcomers, but might not be so great for the old timers, who are the ones carrying a great deal of the weight growth brings, by training the new personnel and witnessing the mutation of the culture. Those old timers are the ones who possess the highest amount of knowledge; an old timer leaving can inflict considerable *brain damage* to the organization. Good thing is that organizations can recover from such damages in a reasonably short time. How? With a knowledge management approach.

Linear growth as a metric is a terribly bad metric. Goodhart's law states that as soon as a metric is picked as an indicator, we start to make decisions just for the sake of keeping the metric nice and safe. For example, headcount growth plans are typically done in such a way that some manager estimates a year in advance the headcount growth for that whole period, which then a higher up reviews and approves. If the year goes by and the headcount growth does not match the predictions, a manager may very well decide to "loosen up" the hiring criteria (i.e. lowering the bar) just to make sure he or she complies with the estimations given a year ago. Faced upon the risk of being flagged as an underachiever or a bad planner, the manager adjusts the selection criteria to meet the metric, by allowing under qualified resources to enter the organization.

If organizations, as we discussed, are an intricate structured system composed of people and technology, how does growth spread across the board? Does the whole organization grow at once? If your organization is 100 people, and you have 10 departments with 10 people each, does a 50% growth mean every single department grows the same amount? Growing an organization is no different than growing a building. You cannot just add more and more floors on top of each other and expect the whole thing will remain stable and safe. Every new floor you add you need to make sure the foundations are ready to take the extra load. Adding load without robustizing the foundations eventually leads to structural collapse. This is what architects and structural engineers check in buildings. Who architects the organization for growth?

The correlation between organizational growth and revenues is always debatable. Provided the org has reached some "critical mass" in terms of structure (good-enough facilities, the most essential departments, competences, capabilities, etc.), and provided there is market fit for the product, the question the organization needs to ask itself is: how to increase revenues, *all other things equal*? This is what the word *scaling* is all about. Often, *growing* and *scaling* are used interchangeably, but they refer to different things. Scaling means getting the most of what you already have, without having to bloat your structure like there is no tomorrow. For very classic businesses, for example a restaurant, physical size growth, and headcount growth *may* directly correlate to revenues. For technology and knowledge-intensive enterprises, if revenues only grow with physical and headcount growth, something is odd.

The big question is: how to scale? How to make the most of what already exists? There are myriads of books on that, and all the recipes around will not work without tailoring to specific contexts, but what can help is: a. better processes (highly manual processes can make it too difficult to handle the most basic activities, and the illusion of bringing more people to solve it might make it worse); b. better tools, c. better training, and d. good information structuring and knowledge management to increase collective understanding.

Finally, growth *at all costs* overlooks the fact that all organizations are part of an environment, which not only contains finite resources that need a sustainable use but also contains other organizations trying to do their own thing. Goes to say, organizations cannot grow unaware of the environment they are embedded in, for their growth also depends on many external factors they cannot control nor directly influence. While the main startup mantra goes:

> If you get growth, everything else falls into place

A different approach could also be:

> If everything else falls into place, you get growth.

7.6 Politics and Drama

Last but not least, a word on politics and drama in startups. There is a naive negative perception of politics in the workplace. It is seen as a *bad thing*. But, in anything, if there are a bunch of people gathered for x or y reason, there is politics. Thinking that organizations do not grow a political scene is plain naive, or stupid. There is no such thing as *no politics* in the workspace, as well as there is no such thing as *no drama* in the workplace. We are political by design. Politics, understood as the power of influencing others for your own benefit, is at the very core of our social interactions, as well as conflict. In the workplace, like in any other aspect in our lives, there are people we get along with and there are some others we do not. There are adversaries or opponents in our way of getting to our goals (a promotion, a project, etc.), and circumstantial alliances we engage in for the sake of building power or our sphere of influence. While working with others we are set to compete for resources (budget, stock options, manpower). It is commonly said that engineers tend to stay away from politics and focus more on technical things, which is not true. Politics is embedded in projects and in the design process as well. We as engineers will try (consciously or unconsciously) to influence decision-makers in order to get our ideas through vs others'. That is a political strategy as well.

Underlying the political game, lie our own psyches. We bring to work our heavy baggage of life with us. A blend of our own insecurities, fears, and past experiences. We bring our personal tragedies and losses, our addictions, our depressions, and jealousies. All our reactions can be tracked down to something that has happened to us in our past and shaped us to react in the ways we react. Being fired, being demoted, being neglected, being left; all this is wired in our brains and it will dictate how we will behave in our current and future work endeavors. We must understand this when we see a startling reaction from someone which we cannot easily explain from our own mind schemas and worldviews.

In a young organization, the leaders must not only be technical leaders but also full-time *shrinks*, and this is of paramount importance; sometimes people just need a valve to vent out and managers must leave everything they may be doing, listen, and provide support. But this can turn out to be heavy for the leaders as well. Coaching, listening, and counseling the team can take up most of the time for respected leaders in growing orgs, and quickly suck up their stamina. Leaders must be psychologically contained and heard as well by upper management. Psychological safety in the workplace must assure clear paths upwards but also sideways.

This section is a bit of an outlier in this chapter: politics and drama are not bullshit, even though they attract it if their existence is not acknowledged or handled. They are real and need proper attention. Politics and conflict must not be ignored. The only office without political conflict is an empty office. Power, leadership, performance, relationships, motivations, scorn. They exist and ignoring them is ignoring human norms.

References

Andrews, M. (2019). *Bar talk: Informal social interactions, alcohol prohibition, and invention.* SSRN. https://ssrn.com/abstract=3489466.

Bergstrom, C. T., & West, J. D. (2020). *Calling bullshit - The art of skepticism in a data-driven world*. New York: Random House.

DeMarco, T., & Lister, T. R. (1999). *Peopleware: Productive projects and teams* (2nd ed.). Dorset House Publishing Company.

Gall, J. (1975). *Systemantics: How systems work and especially how they fail.* Quadrangle/The New York Times Book Co.

Graeber, D. (2018). *Bullshit jobs: A theory*. Penguin House.

Keith, E. (2015, 12 4). *55 million: A fresh look at the number, effectiveness, and cost of meetings in the U.S. [Blog post]*. Lucid Meetings Blog. https://blog.lucidmeetings.com/blog/fresh-look-number-effectiveness-cost-meetings-in-us.

Morrison, J. (2020). *The commercial satellite imagery business model is broken.* Medium. https://medium.com/@joemorrison/the-commercial-satellite-imagery-business-model-is-broken-6f0e437ec29d.

Pidgeon, E. (2014, November 17). *The economic impact of bad meetings.* https://ideas.ted.com/. https://ideas.ted.com/the-economic-impact-of-bad-meetings/.

Prupas, J. (2019, May 28). *Why startups breed 'Bullshit Jobs,' and how to fix it.* Underpinned. https://underpinned.co/magazine/2019/05/why-startups-breed-bullshit-jobs-and-how-to-fix-it/.

Rogelberg, S. G., Scott, C., & Kello, J. (2007). The Science and Fiction of Meetings. *MIT Sloan Management Review, 48*(2), 18–21.

Royce, W. (1970). *Managing the development of large software systems.* http://www-scf.usc.edu/~csci201/lectures/Lecture11/royce1970.pdf.

Spicer, A. (2018). *Business bullshit*. Routledge. ISBN: 978-1-138-91166-6

Weisert, C. (2003). *There's no such thing as the Waterfall approach!* Information Disciplines. http://www.idinews.com/waterfall.html.

GPSR Compliance
The European Union's (EU) General Product Safety Regulation (GPSR) is a set of rules that requires consumer products to be safe and our obligations to ensure this.

If you have any concerns about our products, you can contact us on

ProductSafety@springernature.com

In case Publisher is established outside the EU, the EU authorized representative is:

Springer Nature Customer Service Center GmbH
Europaplatz 3
69115 Heidelberg, Germany

www.ingramcontent.com/pod-product-compliance
Lightning Source LLC
Chambersburg PA
CBHW060514181125
35560CB00003B/110
* 9 7 8 3 0 3 0 6 6 8 9 7 6 *